Advanced Fossil Energy Utilization

Topical Conference at the 2010 AIChE Spring Meeting and 6th Global Congress on Process Safety

San Antonio, Texas, USA
21 – 25 March 2010

ISBN: 978-1-61738-437-0

Printed from e-media with permission by:

Curran Associates, Inc.
57 Morehouse Lane
Red Hook, NY 12571

Some format issues inherent in the e-media version may also appear in this print version.

Printed by Curran Associates, Inc. (2010)

For permission requests, please contact AIChE
at the address below.

AIChE
3 Park Avenue
New York, NY 10016-5991

Phone: (203) 702-7660
Fax: (203) 775-5177

www.aiche.org

Additional copies of this publication are available from:

Curran Associates, Inc.
57 Morehouse Lane
Red Hook, NY 12571 USA
Phone: 845-758-0400
Fax: 845-758-2634
Email: curran@proceedings.com
Web: www.proceedings.com

TABLE OF CONTENTS

BOILER BROMIDE ADDITION - A SURVEY ON BROMINE BASED MERCURY ABATEMENT FROM FLUE GASES AT COAL COMBUSTION AND WASTE INCINERATION PLANTS

Prof. Dr. Bernhard W. Vosteen, Vosteen-Consulting GmbH, Cologne (Germany)

Extended Abstract:

This keynote lecture will report on bromine based mercury abatement from coal combustion and waste incineration flue gases, especially in its form of Pre-combustion Bromide Addition or Boiler Bromide Addition. The technology was invented by the author in 2000 - still in his time with Currenta GmbH & Co OHG, Leverkusen (Germany), and patented by her in many countries. Bromine-containing compounds as e.g. aqueous NaBr, $CaBr_2$ or HBr solutions respectively are added to the coal or waste feed or are injected elsewhere into the high temperature combsution gases to enhance mercury oxidation by gasphase bromination during boiler passage and e.g. at the high-dust SCR-DeNOx-catalysts further downstream.

Pre-combustion/ Boiler Bromide Addition is commercially applied in Germany, mainly at waste-to-energy plants (since 2001), and in the States as well, mainly at coal fired power plants (since 2005), and has been proven and established as a simple and inexpensive retrofit method for effective co-benefit mercury abatement from flue gases in existing dry and/or wet emission control systems. Often, only 25 – 100 ppmw bromine addition per feed already allows for 100 % mercury oxidation and thus high dry and/ or wet mercury removal > 90 % (feed to stack).

Alstom Power Inc., Knoxville TN, has been granted an exclusive marketing rights license from Vosteen Consulting GmbH for joint exploitation and commercialization of this technology in the United States and Canada. Southern Company Services, Birmingham AL, and US EPRI, Palo Alto CF, got R&D licenses.

Results from more than 20 large scale tests and selected permanent applications in Germany and USA, as well as results from accompagnying bench scale laboratory tests conducted at the Martin-Luther-University Halle-Wittenberg (Institute Prof. Dr.-Ing. habil. Heinz Köser) are demonstrating the effectivity of this technology; it is important that the first contact of added bromine compounds with the flue gases containing mercury vapor and sulfur dioxide should happen at high temperatures.

In case of co-benefit mercury capture by absorption at wet flue gas desulphurization systems (WFGD), the dissolved oxidized mercury Hg^{2+} might be reduced and then partially reemitted as elemental mercury Hg° - back into the scrubbed gas. There are several means to avoid mercury reemission from WFGD - as mercury complexation by halogenides, mercury precipitation by sulfur organic or inorganic precipitation agents as TMT™15, PRAVO™100 (Precipitation Agent Vosteen), or mercury chemisoption by suspended pulverized activated carbon (PAC). Evonik Energy Services, Essen (Germany), and Vosteen Consulting have jointly developed a process, combining bromide addition to the coal feed and suspended PAC addition directly to the WFGD scrubber slurry, that ensures both > 90 % mercury removal (coal to stack), low-mercury gypsum and – if wanted – a low-mercury FGD-fines sludge fraction, as well. Full scale testing of the new process was performed at an industrial wet bottom boiler during several weeks. The waste water treatment downstream is working with selective precipitation and separation of the almost mercury-free gypsum, the low-mercury sludge fraction and the residual high-mercury sludge fraction – i.e. a small amount of mainly mercury-rich PAC, that either can be thermally treated to recover the mercury and recycled afterwards to the fire-box, or properly disposed off, see attached Figure.

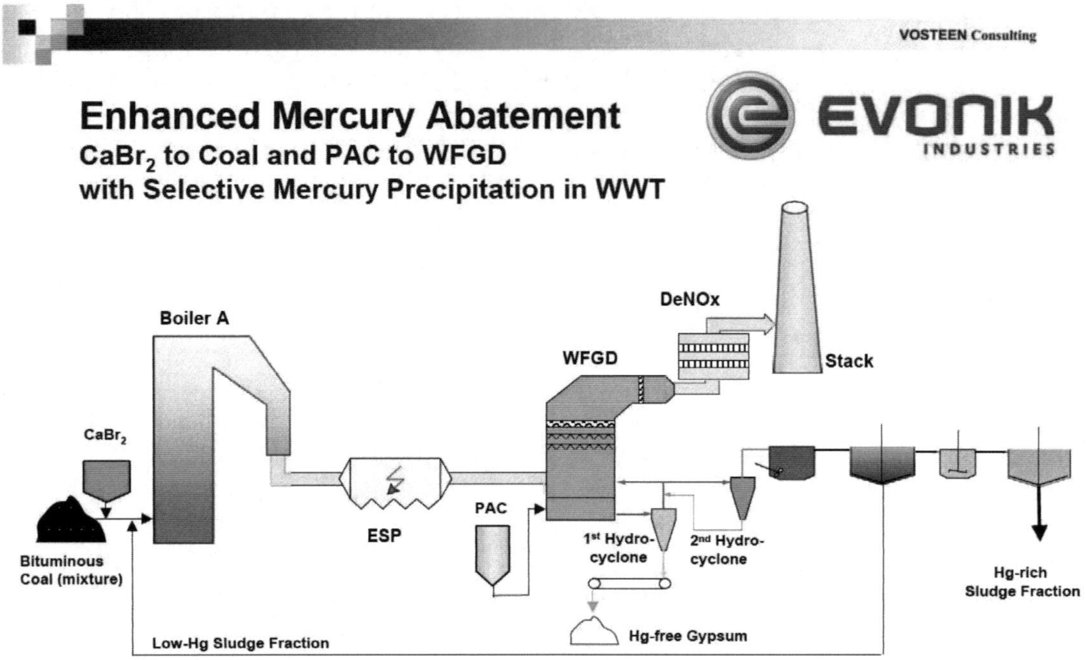

Modeling Mercury Chemistry in Coal-Fired Air Preheaters: Effects of Sulfur Trioxide and Surface Area

Brydger Van Otten, Andrew Fry, Constance Senior
Reaction Engineering International, 77 W. 200 S., Suite 210, Salt Lake City, UT 84101

Abstract

Mercury emissions from coal-fired power plants are a concern to both the state and federal governments in the U.S. US EPA had proposed to regulate mercury emissions from coal-fired power plants using the Clean Air Mercury Rule (CAMR). This two-phase program allowed trading of mercury emissions, similar to the SO_2 trading program already in place. As of now, all aspects of CAMR have been vacated by a court ruling, and the EPA has committed to proposing draft mercury emissions regulations for coal-fired power plants by March, 2011. As of October 2009, nineteen states had passed their own mercury emissions regulations.

Mercury exists as the elemental form (Hg^0) in the high-temperature regions of coal-fired boilers. As the flue gas is cooled, a series of complex reactions begin to convert the Hg^0 to gaseous oxidized forms (Hg^{2+}) and particulate-bound mercury (Hg_p). The extent of conversion of Hg^0 to Hg^{2+} and Hg_p depends on the flue gas composition, the amount and properties of fly ash and the flue gas temperature and quench rate. The speciation of mercury in coal combustion flue gas affects the performance of activated carbon (a dedicated mercury control technology) and the removal of mercury by wet FGD scrubbers (a "co-benefit" approach to mercury control).

In order to prepare for the existing and impending mercury emission regulations, utilities must have useful tools for compliance planning. REI's MerSim™ mercury simulation tool includes homogeneous and heterogeneous oxidation kinetics, adsorption on fly ash, oxidation across SCRs, and removal and re-emission across wet FGD scrubbers. This model can be used by utilities, given inputs that are generally available to them. On the surface, this requirement might seem obvious and inconsequential; however, matching models for the complex chemistry of mercury in practical combustion systems with the information that is generally available to plant engineers is a difficult undertaking.

We have endeavored to create fundamentally based submodels for mercury behavior, but these models must be based on input data that a plant engineer can reasonably supply. An integrated model requiring inputs that are not available or difficult to obtain will not be a useful compliance tool. Therefore, a balance must be struck between the complexity of the submodels and the complexity of the input parameters, in order to provide utilities with a useful and accurate tool.

Detailed homogeneous and heterogeneous kinetic pathways for mercury oxidation are included in the MerSim integrated power plant model. Overall results of the model have been previously reported.[1] However, details of some of the key submodels have not always been reported in detail. Innovation and improvement continue in the integrated model. Recently, a more detailed submodel has been implemented to predict both mercury oxidation and adsorption across regenerative air preheaters.

Using full-scale mercury speciation data assembled from previous DOE- and EPRI-funded programs,[1] one can see that regenerative air preheaters promote formation of oxidized and, to a lesser extent, particulate-bound mercury. Figure 1 illustrates these points with observed values of oxidized and particulate-bound mercury at air preheater outlets.

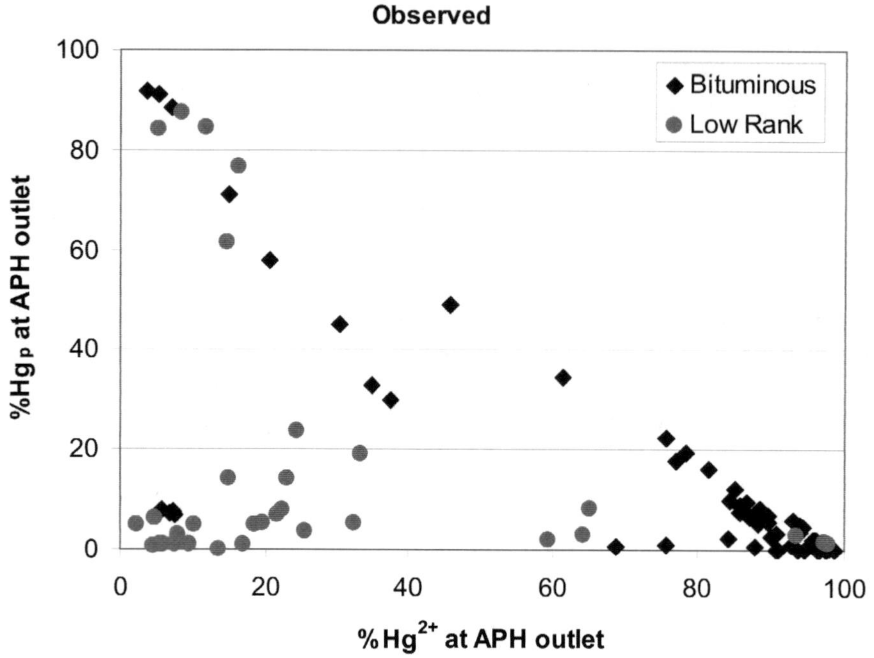

Figure 1. Observed percentage of oxidized and particulate-bound mercury at air preheater exits from full-scale coal-fired power plants.

Our approach was to use the heterogeneous and homogeneous rates for mercury oxidation and the rate of mercury sorption, which are currently used to predict in-flight mercury transformations in the duct and ESP submodels, to account for deposition of fly ash on the air preheater surfaces. Furthermore, detailed SO_2 oxidation kinetics (the Leeds mechanism[2]) were added to the gas-phase kinetic model, to accurately predict the concentration of SO_3 at the air preheater inlet.

Heterogeneous reactions of mercury are assumed to take place on the surface of unburned carbon particles. Unburned carbon particles from coal combustion are larger than inorganic fly ash particles and have a lower density.[2] A mean diameter of 50 microns and a density of 0.4 g/cm^3 would be reasonable for unburned carbon. Using these mean values for unburned carbon and ash loading, the external surface area of unburned carbon in the duct (in m^2/m^3) can be calculated. The air preheater surface area is much greater than the external surface area of the unburned carbon that is suspended in the flue gas. Thus, unburned carbon that deposits on the air preheater surface area represents more available surface area. The available surface area for heterogeneous mercury reaction is thus assumed to be the sum of the in-flight ash and the ash deposited on the air preheater surfaces.

If we can predict accurately the concentration of SO_3 in the air preheater, and the loss of SO_3 in the air preheater, then we can begin to outline a way to predict the effect of SO_3 on oxidation and adsorption of mercury on unburned carbon in the air preheater. Note that H_2SO_4 and SO_3 will be used interchangeably here. SO_3 is the oxidized sulfur species created in the flame or across the SCR, but it is converted to H_2SO_4 at lower temperatures. Since concentrations of SO_3/H_2SO_4

will be expressed in ppmv, the concentrations in the flue gas will be equivalent, no matter which form is thermodynamically stable.

Most power plants do not operate such that the air preheater exit temperature is below the dew point of H_2SO_4. However, much of the surface within the air preheater is below the dew point, because of contact with the relatively cold air stream. Thus, there is loss of H_2SO_4 within an air preheater, even when the exit temperature in the flue gas is above the dew point. H_2SO_4 is condensing on the surfaces inside the air preheater. It must, therefore, condense on unburned carbon on those surfaces. The condensation of SO_3 should remove sites for mercury oxidation and adsorption. Therefore, the reactive surface area in the air preheater available for reaction with mercury was reduced in the model, based on the predicted condensation of SO_3 in the air preheater.

The results of the calculation of mercury oxidation across the air preheaters in full-scale power plants are shown in Figure 2, in which calculated oxidation is compared against observed oxidation at the air preheater outlet. Further details of the model and of predictions of mercury emissions from coal-fired power plants will be presented.

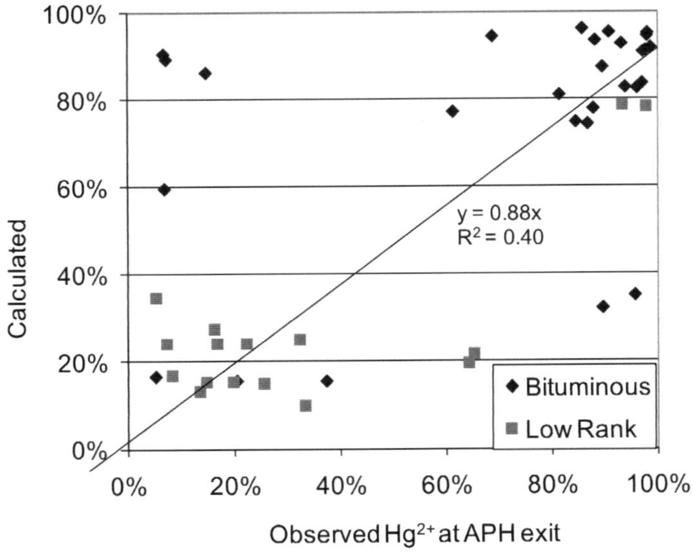

Figure 2. Observed and predicted oxidized mercury at air preheater outlet.

References

1. Senior, C., Fry, A., Montgomery, C., Sarofim, A., Wendt, J. Modeling Tool for Evaluation of Utility Mercury Control Strategies." Presented at the DOE-EPRI-U.S. EPA -A&WMA Combined Power Plant Air Pollutant Control Symposium - The Mega Symposium, Baltimore, MD, August 28-31, 2006.
2. http://www.chem.leeds.ac.uk/Combustion/Combustion.html.
3. Belbla, V.H.; Martin, C.E.; Baldrey, K.; Lindsey, C.; Altman, R. The Effects of Unburned Carbon Properties on Electrostatic Precipitator Performance Modeling. Presented at the DOE-EPRI-U.S. EPA -A&WMA Combined Power Plant Air Pollutant Control Symposium - The Mega Symposium, Baltimore, MD, August 28-31, 2006.

Speciation Control of Mercury for Coal-Fired Power Plants

Jinghui V. Lu[a], John Meier[b], Jitendra Shah[a] and Bruce A. Keiser[a]

[a] *Nalco Company*
1601 West Diehl Road
Naperville, Il 60563

[b] *Nalco Mobotec, Inc.*
2540 Camino Diablo Blvd, Suite 200
Walnut Creek, CA 94597

INTRODUCTION

Mercury species in flue gas from coal-fired power plants consist of three types, ionic (Hg^{2+}), elemental (Hg^0), and particulate (Hg^p). Of these elemental mercury is the most difficult to capture with available technologies. Ionic mercury, however, can be removed by AQCD (Air Quality Control Devices) such as cold side ESP (Electrostatic Precipitator), Fabric Filters (FF) or wet Flue Gas Desulfurization (FGD) scrubbers. Mercury speciation in flue gas correlates with the amount of chlorine content in coals.[1] Bituminous coals typically contain higher chlorine content (i.e. >200 ppm) and thereby yield a higher ionic mercury fraction in the flue gas. In contrast, sub-bituminous and lignite coals typically contain lower amounts of chlorine (i.e. <50 ppm) and as such generate flue gas containing a higher fraction of elemental mercury.[2] Therefore, a strategy to reduce mercury air emissions at coal-fired power plants is to cause a shift in mercury speciation in the flue gas toward higher ionic mercury content. Such a shift should increase mercury capture by existing AQCDs. Such an approach is particularly appealing to those plants utilizing low-chlorine fuels.

Nalco Mobotec, Inc. a patented process, which involves_MerControl® technology, (i.e., MerControl® 7895) for improving mercury capture efficiency across existing AQCDs. MerControl7895 is an aqueous solution that is easily applied and effective. A one-week demonstration of the benefits of MerControl 7895 was completed recently at a coal-fired power plant. The purpose of the demonstration was to increase mercury capture across the existing cold-side ESP and wet FGD to a mercury air emission of 0.008 lb/GWh or less. To reach this limit, mercury capture needed to improve from the baseline value of ~60% to at least 90%.

EXPERIMENTAL

Demonstration Unit

The test boiler was a 195 MW pulverized coal (PC) unit equipped with an SCR for NOx control, an ESP for particulate control, and a wet FGD scrubber for SOx with forced oxidation. Figure 1 below illustrates the site layout as well as the locations of mercury measurement, sample collection, and addition points.

Figure 1. Configuration of the test unit (Coal to Stack)

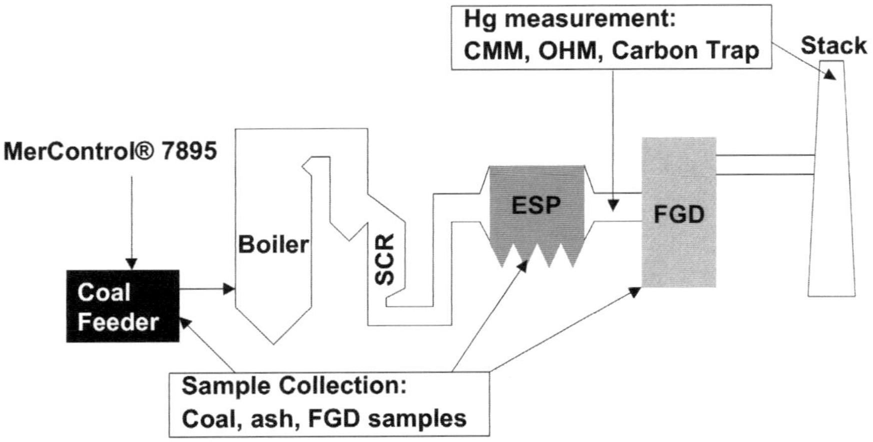

Mercury Measurement Procedure

Three different methods were implemented to measure mercury in the flue gases: Continuous Mercury Monitors (CMMs), Ontario Hydro Measurement (OHM), and Method 30B Carbon Trap Sampling. All gas phase mercury measurements were corrected to 3% O_2 on a standard basis, using O2 measurements taken at the sample locations.

Coal and Fly Ash Analysis

Pulverized coal and fly ash samples were also collected. All these coal and ash samples were sent to an independent laboratory for analysis. Proximate and mercury analysis were performed on all coal samples while ultimate, bromine and chlorine analysis were only performed on selected coal samples. Fly ash samples were measured for mercury and carbon content.

RESULTS AND DISCUSSION

Demonstration of MerControl® 7895 for Improved Mercury Capture

Flue gas mercury was measured at the stack using three methods: CMM, OHM, and Method 30B Carbon Trap. The trends of the respective methods are shown in Figure 2 below. All three methods showed the same trend and correlation to variation in feed of MerControl 7895 over time. The OHM results were systematically lower than the Method 30B Carbon Trap and CMM measurements. Mercury emissions at stack measured by Carbon Trap were about 44.4% higher compared to OHM while the averaged CMM data corresponding to each OHM run was 18.2% more than OHM data. The cause of the observed bias is unclear.

Figure 2 also shows that mercury emission at the stack decreased as MerControl 7895 feed increased. When MerControl 7895 feed was set lower (e.g. at 11/14/09 14:50), the stack mercury emission increased. Throughout the demonstration, mercury air emissions at the stack inversely correlated with 7895 dose.

Figure 2. Stack Mercury Emission Measured by Three Methods (Corrected to 3% O_2). Feed rate of MerControl 7895 shown by the red bar is reported in ppm product in coal (weight basis), as shown in the y-axis on the right.

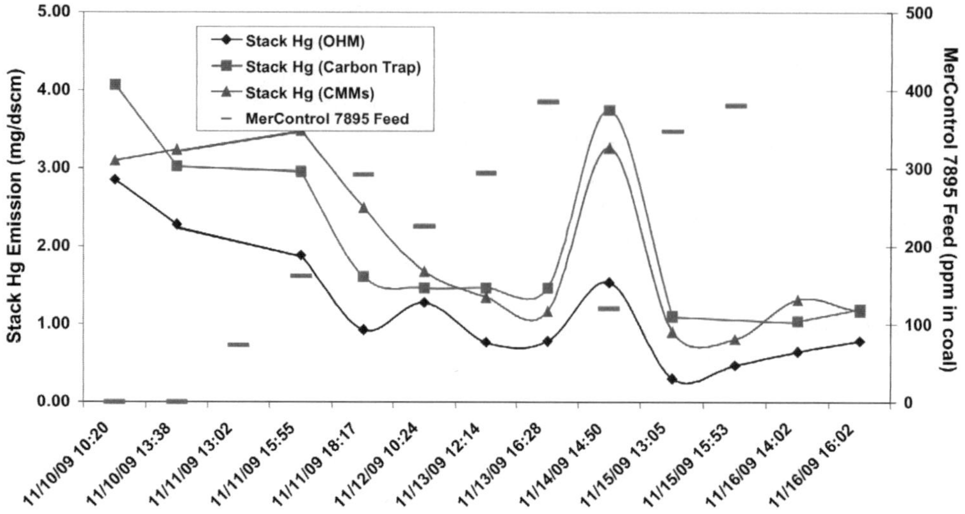

The goal of the demonstration was for the stack gas mercury content to be at or below 0.008 lb/GWh. The stack gas mercury content in lb/GWh is compared to the feed rate of MerControl 7895 in Figure 3 . The OHM method during the test was most reliable and was used in the following analysis. There is a definite inverse correlation between mercury stack emissions and 7895 feed rate. The OHM data correlation factor to 7895 feed has an R-squared value of 0.91. Above a feed rate of 265 ppm MerControl 7895 on coal, the stack mercury emission was below the target 0.008 lb/GWh.

Figure 3. Impact of MerControl® 7895 on Stack Mercury Capture

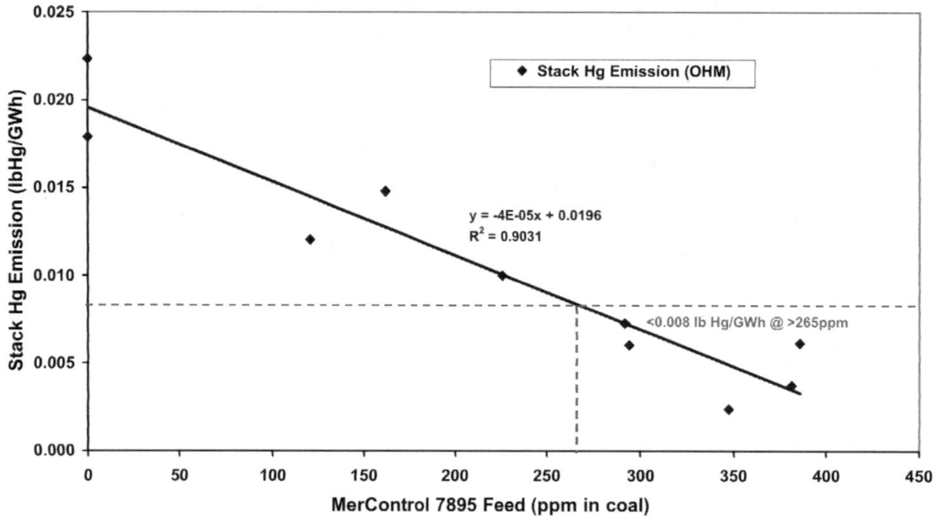

Percent mercury captured from coal to stack was also calculated using the stack mercury concentration and wet FGD inlet mercury concentrations and is shown in Figure 4. At baseline, the flue gas contained about 80% ionic mercury resulting in mercury capture of 63% and 60.4% by OHM and Method 30B Carbon Trap respectively. With addition of 7895, both percent of ionic mercury and percent mercury captured increased, as reported by the three measurement methods. OHM and Carbon Trap results were used in Figure 4. As the dose of 7895 varied from 50 ppm to 400 ppm product in coal the percent of ionic mercury increased from 80% to 94% (OHM). Simultaneously, percent mercury capture increased to 94.3% (OHM) and 90.2% (Carbon Trap). Mercury capture exceeding 90% consistently was achieved at 340 ppm 7895 in coal based on the OHM methods.

Figure 4. Impact of MerControl® 7895 on Mercury Oxidation and Capture. Percent mercury oxidized in the flue gas at the wet FGD inlet is shown by the red lines and reported on the primary y-axis on the left; percent mercury captured is shown by the blue triangle (Method 30B Carbon Trap) and the green square (OHM) and reported on the secondary y-axis on the right.

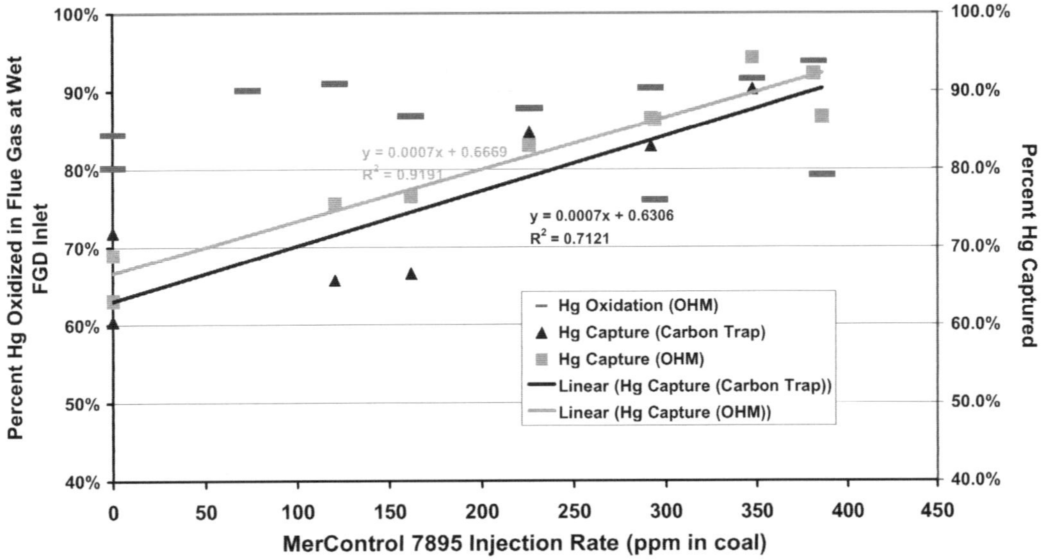

CONCLUSION

The demonstration of MerControl 7895 at a 195 MW coal-fired boiler successfully achieved the stated goal of reducing mercury air emissions to or below 0.008 lb/GWh (i.e. 90% or higher capture). This was accomplished by the application of 7895 at or above 265 ppm product in coal (by weight). Good agreement was observed among three measurement methods utilized during the demonstration yielding a high correlation between 7895 dose and reduction of mercury air emission.

---

1. S.A.Benson, M.J. Holmes, D.P. McCollor, J.M. Mackenzie, C.R.Crocker, L. Kong, K.C. Galbreath, K. Dombrowski, C. Richardson "Large-Scale Mercury Control Technology Testing for Lignite-Fired Utilities – Oxidation Systems for Wet FGD" *EERC and URS corporation report (September 26, 2003 – March 31, 2007) for the U.S. Department of Energy* (Cooperative Agreement No: DE-FC26-03NT41991).

Mercury Control with Regenerative Activated Coke Technology

Monday, March 22, 2010: 9:35 AM

Travis C/D (Grand Hyatt San Antonio)

H. James Peters, Hamon Research-Cottrell, Inc., Somerville, NJ

Abstract

Many issues face coal fired utilities with respect to environmental demands, some based on regulations and others based on local pressures. These include increasingly stringent control of criteria pollutants, anticipated federal and state requirements for mercury control, issues revolving around water use and water disposal, plume visibility, overall plant thermal efficiency and the future considerations for climate change issues. For utilities that burn PRB and other low sulfur coals, an exemplary reference project is the re-construction of the Isogo Power Station by J-Power. This project replaced two vintage coal fired units, and within the same site limits, more than doubled generation capacity to 2x600MW with ultrasupercritical boilers fitted with SCR, ESPs, and advanced generation ReACT™ technology. The controls provide low emissions from the high efficiency coal fired boiler plant and allow the plant to operate as the cleanest coal fired power plant in the world in terms of emissions intensity, at emissions levels that are equivalent to natural gas fired power plants. Isogo Unit #1 has been in operation since 2002 and Unit #2 started operation in mid 2009. J-Power EnTech has recently licensed the Regenerative Activated Coke Technology (ReACT™) to Hamon Research-Cottrell.

Background

The multi-pollutant control ReACT™ technology is a completely dry scrubbing system based on adsorption of SO2, SO3, and Hg and reduction of NO_x to N_2 on activated coke in a moving bed, with regeneration of the coke for return to the adsorber, and production of saleable byproduct, such as sulfuric acid, from desorbed sulfur rich gases.

The basic process concept originated with Bergbau Forschung in Germany the 1950s, saw additional development by Foster Wheeler in the 1970s in conjunction with their Resox process, and commercialization with activated coke by Mitsui Mining in the 1980s, and subsequent full commercialization by EPDC (now J-Power) in the 1990s as an advanced generation multipollutant control technology for coal fired boilers. J-Power, acquired the technology from Mitsui in 2005, now operates three full scale units at Takehara (1995), Isogo #1(2002) and Isogo #2 (2009), and has provided units for refinery, incineration, and sinter applications through its subsidiary J-Power EnTech.

J-Power's Isogo #1 and Isogo#2 burn low-sulfur coal and incorporate high efficiency ultrasupercritical boilers, combustion NO_x control, primary SCR, ESPs and ReACT™. Isogo has been recognized for emissions intensity levels for SO_x and NO_x that are the lowest worldwide for coal fired power. Isogo typically operates in the single digit ppm concentration range for SO_2 and NO_x (against permit levels of 10ppm and 13 ppm at Isogo #2), particulate at less than 5 mg/Nm3 (0.005 lb/mmBTU) and with well over 90% control of both elemental and oxidized forms of mercury.

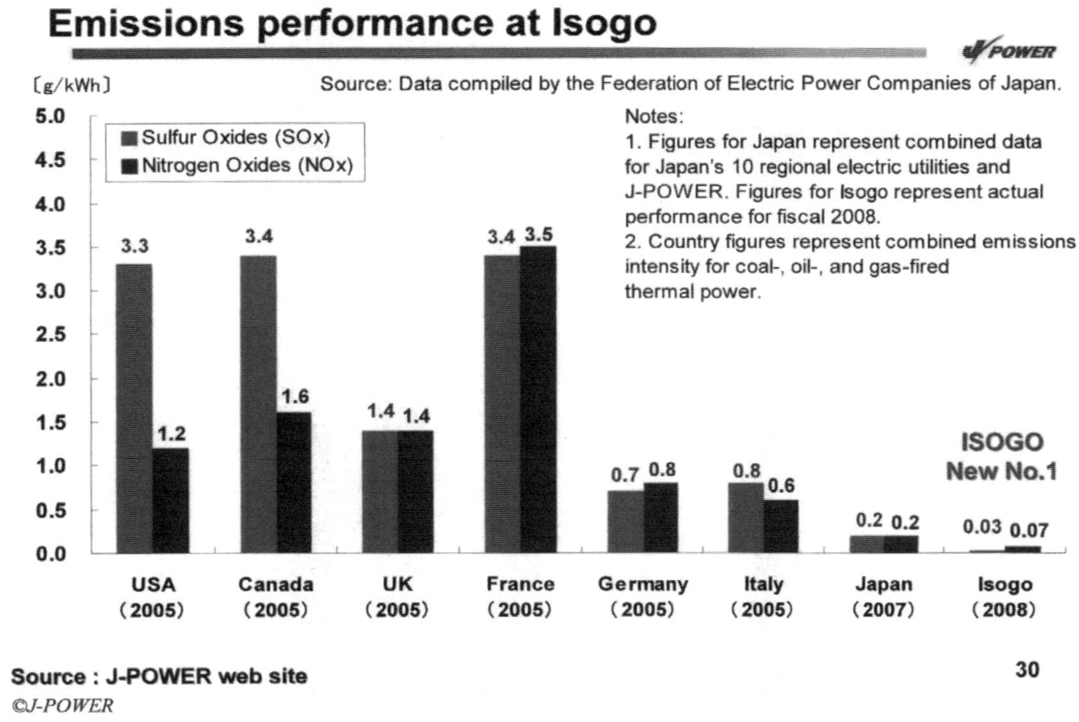

The ReACT™ process has also been already been successfully demonstrated for U.S. coal applications as an EPRI project, hosted by Sierra Pacific Power at its North Valmy Station, where the expected high levels of SO_2, NO_x, and Hg performance were readily demonstrated, consistent with commercial results at the full scale units in Japan.

This paper focuses on aspects of the activated coke technology as related to mercury control.

General process description

ReACT™ involves a three stage process. Flue gas is contacted with a slowly moving bed of activated coke. Fresh activated coke is fed to the top of an adsorption column and flows by gravity. The activated coke removes SO_2, SO_3, NO_x, Hg and other species through adsorption, chemisorption and catalytic reactions which are enhanced in the presence of ammonia. Particulate is also reduced across the moving bed by impaction on the coke pellets.

Pollutant laden activated coke is transferred to a thermal regeneration stage, where reactions are completed and pollutants are desorbed as a sulfur rich gas stream. The regenerated activated coke stream is cooled and fines are separated before return to the adsorber. Make-up coke in pellet form is added to replace the separated fines. The third stage receives the sulfur rich gas stream for saleable byproduct production.

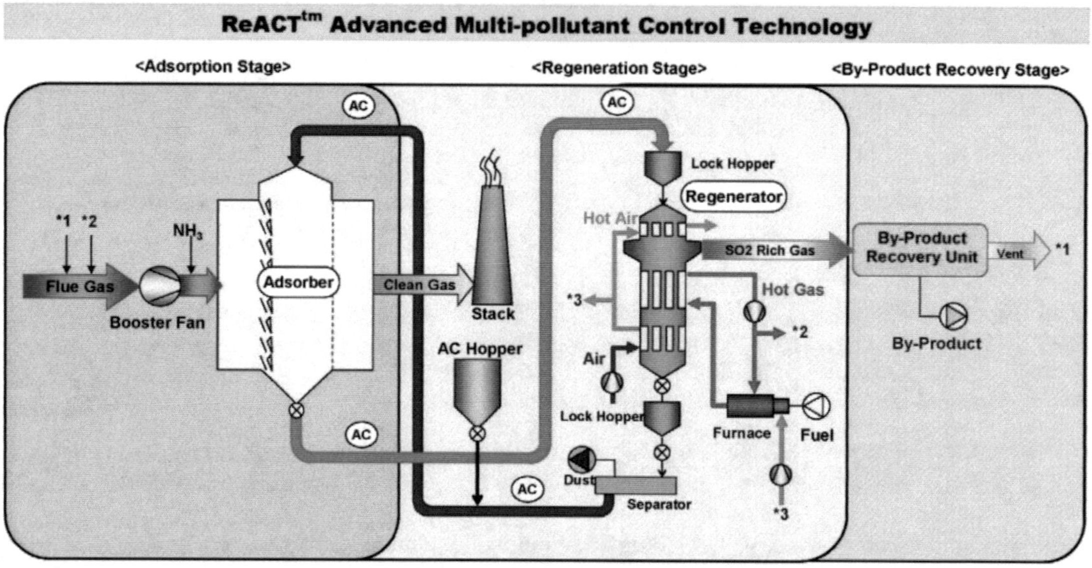

This process offers advantages for utilities, as requirements for mercury control are being added to increasingly stringent SO_2 and NO_x regulation, and where issues related to water use, downstream plume visibility, and solid waste disposal have become major factors. These advantages derive from:

- Adsorption does not require any water use
- Adsorption of SO_3 along with no humidification resolves downstream plume visibility issues
- Regeneration of sorbent greatly reduces site logistics for reagent make-up, processing and waste handling compared with other FGD processes
- Marketable byproduct is produced.
- Primary particulate is upstream (existing), preserving beneficial use of ash.
- Mercury control is obtained at >90% levels for both elemental and oxidized forms.
- Mercury laden disposal streams are minimized.

Mercury control by activated carbon

The present practice for mercury control at utility power plants typically involves the injection of powdered activated carbon into flue gas.

Activated carbon is injected into the flue gas stream upstream of particulate collection equipment as a once-through adsorbent. For target mercury removal of 90% activated carbon injection rates range from 2 to 10 lb per million acfm with some sensitivity to coal, combustion, SO_3 levels and on the type of particulate control in place. Injection rates are generally lower with fabric filters.

Contacting conditions for activated carbon in a fabric filter include short in-flight time (1-2 seconds) followed by fixed bed contact at the fabric surface of less than 0.2 seconds with average solids residence time on the fabric surface of about 4 hours. The high specific surface area of powdered activated carbon allows control of mercury at 90% levels

Mercury reduction by adsorption on activated carbon occurs co-currently with particulate removal by the ESP or FF. The result is that the mercury laden carbon is mixed with the flyash and impairs the beneficial use of flyash such as for use in cement.

Activated coke vs. activated carbon

The ReACT™ process uses activated coke in pellet form suitable for use in the moving bed adsorber. This material generally compares with powdered activated carbon as follows:

	Activated Coke	**Activated Carbon**
Shape/Diameter	Tablet/Almond ~10 mm	Powdered
Specific Surface Area	150-300 m^2/g	>1000 m^2/g
SO_2 adsorption capacity	60-120 mg/g	
Pore size	10-1000 A	10-100 A
Pore volume	0.05-0.1 cc/g	0.5-0.6 cc/g
Roga-Index (abrasion resistance)	95%	70-85% (for tablet forms)
Raw Material	Coal	Coal
Application	SO_x, NO_x, Hg removal	Hg removal

Mercury control by activated coke

The ReACT™ adsorber design is determined based on SO_2 control requirements and gas space velocity of 400 hr^{-1} and activated coke residence time of 80 hr are typical. Contacting conditions in the adsorber are approximately 10 seconds for the flue gas to transit the moving activated coke bed. Coke pellets are exposed to flue gas for 80 hours.

Regenerated (used) activated coke also has found to have superior performance compared to fresh make-up material. During the regeneration process, new carbon pores are exposed during SO_x desorption by reactions which consume some carbon, and some sorption sites become functionally enhanced by the formation of oxygen species or by trace sulfur compounds. The following figure excerpted from Reference 2 shows results of bench tests at Mitsui where near 100% control of elemental Hg persisted for used AC at longer solid contact times whereas breakthrough occurred earlier for fresh material. Similar improved characteristics for SO_x and NO_x control are seen with used AC.

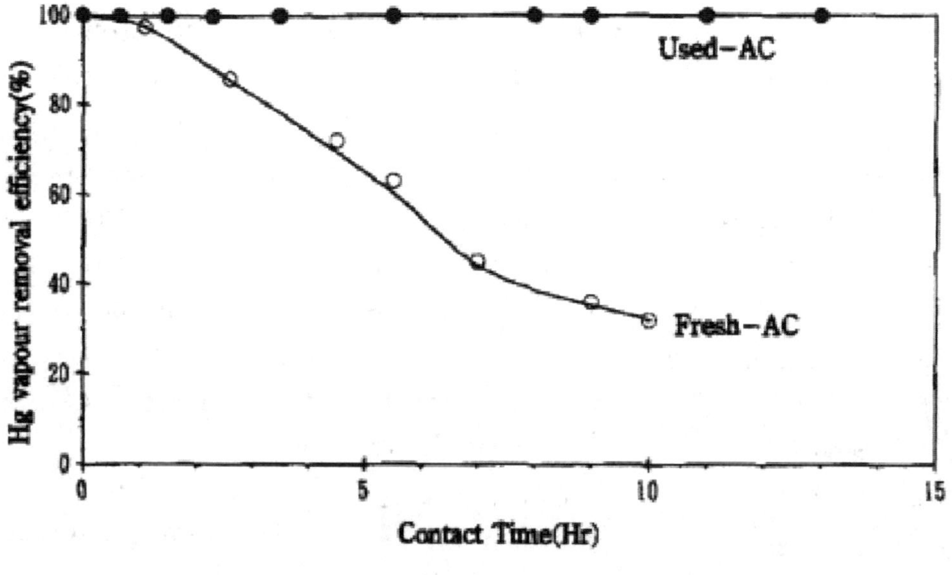

Hg(0) vapour removal test with AC (fresh versus used)

While fresh activated coke has lower specific surface area the contact time in the adsorber bed is significantly longer and the regenerated activated coke is present in considerable excess relative to mercury. This configuration results in highly efficient mercury reduction and Hg control for both elemental and oxidized forms in excess of 90% is readily achieved.

ReACT™ Activated Coke Regeneration

Following is a schematic of a typical ReACT™ regenerator which thermally desorbs sulfur rich gases.

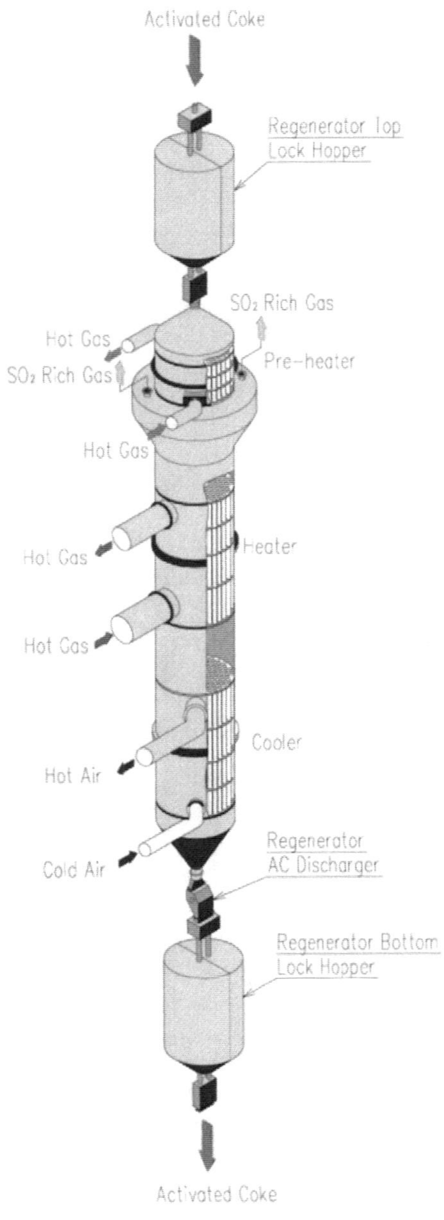

Activated coke at flue gas temperatures carrying adsorbed SO_2, NO_x and Hg is fed through a lock hopper to the top of the regenerator Activated coke moves downward by gravity flow through three indirect heat exchanger sections for preheating, heating, and cooling the regenerated coke which is then discharged through a lock hopper to fines separation and then returned to the adsorber.

A sulfur rich gas zone is located between the preheat and heating sections of the regenerator tower. The sulfur rich gas released (SO_2, N_2, CO_2, and H_2O) in this zone then flows to an acid plant for production of marketable sulfuric acid.

Prior to cooling the activated coke, ammonia gas can be sparged into the regenerator to pre-condition the activated coke for optimal SO_2 and NO_x sorption and the activated coke is cooled back to the flue gas temperature or slightly below.

The design of the heat exchanger control of the shell side heating and cooling air temperatures and flows provide the required temperature and residence time profile to complete the regeneration.

The nonisothermal kinetics related to thermal regeneration were illustrated for the Bergbau Forschung process in the following figure excerpted from Reference 1. The kinetics show that the thermal desorption can be driven to completion by heating the activated coke to 450C.

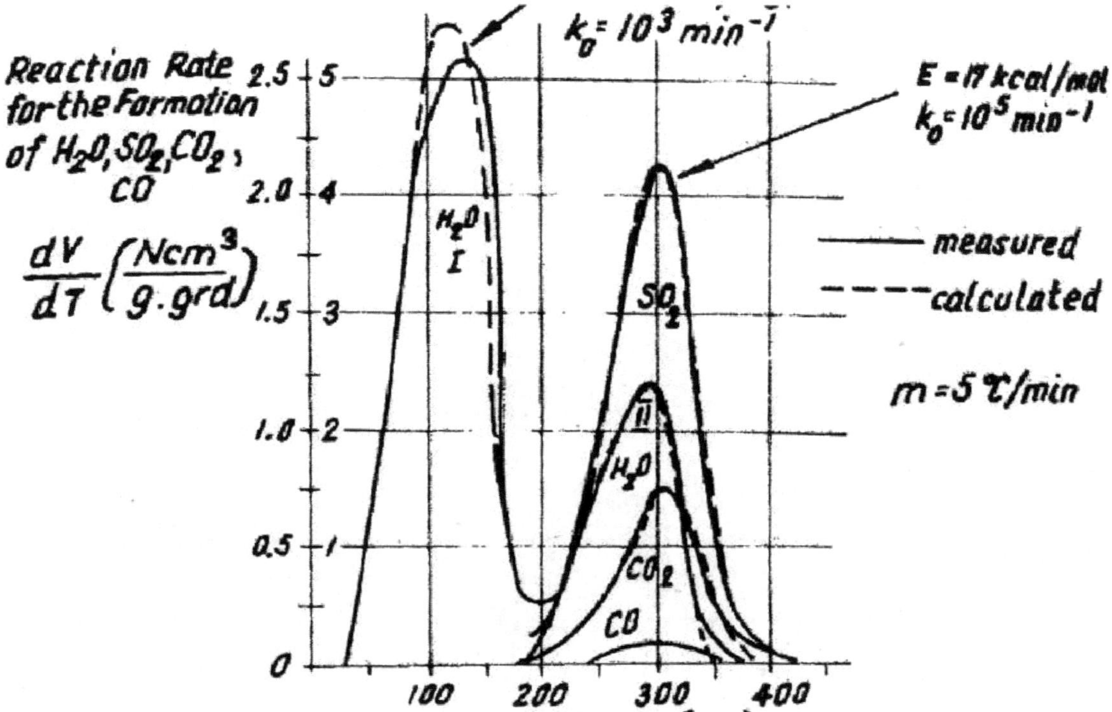

Figure 2. Nonisothermal kinetics of thermal regeneration

Mercury retention in the regenerator

In the ReACT™ regenerator temperature profiles and retention time at 450C allows the regeneration process to approach completion in accordance with the kinetic requirements.

Mercury vapors are also desorbed but mercury is held in temperature regime determined by the adsorption and thermal desorption properties of the activated coke. The longer term retention of mercury allows for improved management of mercury laden materials as only periodic disposal of activated coke from the mercury confined zone is needed.

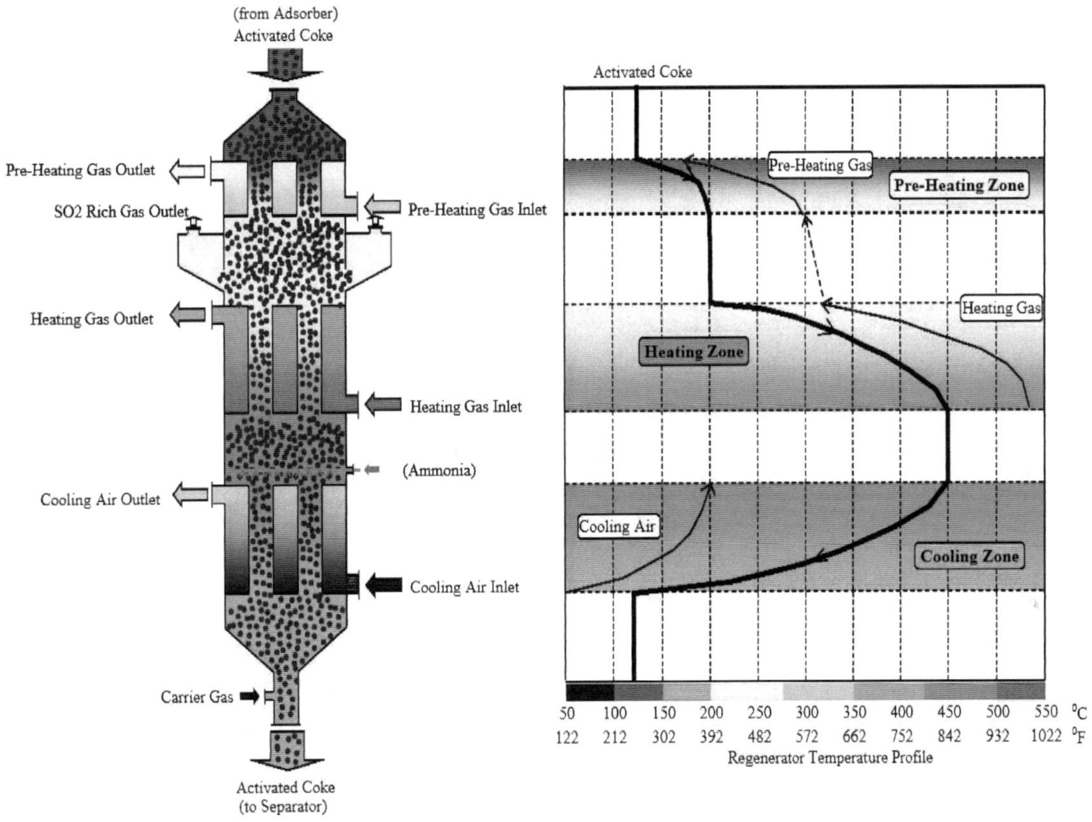

Mercury is thermally desorbed at >400C to the vapor phase. The temperature profile of the regenerator shows that this temperature is reached within the heating zone. Mercury desorption is complete at the bottom of the heating zone at 450C.

Mercury and other desorbed gases flow upward and counter-current to the activated coke flow and contact lower temperature coke surfaces. At the top of the heating zone, the activated coke at 200C has effectively re-adsorbed all the mercury vapors.

The mercury that had been originally removed from the flue gas in the adsorber vessel is effectively confined in this zone of the regenerator.

Logistics for carbon make up and disposal

Over time the mercury accumulation in this zone approaches a saturation capacity. Depending on the level of mercury in the coal this period is typically in the two to four year range. At Isogo, three years is used as a planning interval and at each of the 600 MW plants approximately 100 tons of mercury laden activated coke are disposed every three years – or approximately 0.06 tons/MW-year.

Disposal is managed during a scheduled regenerator downtime (all activated coke has been cooled) by diverting the portion of activated coke that had been in the heating zone and lower SRG zone.

A comparison of the carbon make-up and disposal requirements for a typical 200 MW plant using once-through activated carbon compared to regenerative activated coke illustrates a significant difference is the issues related to Hg-laden material disposal.

Basis:	200 MW PC boiler burning low-sulfur PRB, 75% capacity factor
Firing rate:	2200 MMBTU/hr (at 11,000 BTU/kWh)
Flue gas flow:	~800,000 acfm @ 300F
Flyash load:	2 to 3 gr/acf (45,000 tons/year @ 2 gr/acf)
Hg load:	5 to 10 lb/10^{12} TBTU

For the ReACT™ process, fresh make up activated coke replaces losses due to fines separation after regeneration and carbon consumed in regenerator reactions. Make-up for a 200MW plant is approximately 1000 tons/year (4 to 6 tons/year per MW). This mercury-free material is either disposed of, used as fuel, or may have beneficial re-use as a carbon sorbent at other sites.

For a regenerator serving a typical 200 MW coal fired boiler, this would be about 20 to 25 tons of activated coke, and if the disposal interval is two years, the disposal rate would be roughly 12 tons/year or (0.06 tons/MW-year).

For typical activated carbon systems, 90% mercury control requires injection of powdered activated carbon (PAC) into the flue gas at rates between 2 lb/MMacfm and 10 lb/MMacfm. The higher rates are generally associated with injection is upstream of an electrostatic precipitator, and lower rates with injection upstream of a fabric filter. Injection rates have some sensitivity to fuel type, boiler type and effects of increased SO_3 from upstream SCR.

When the powdered activated carbon is injected upstream of the primary particulate control, whether it be an ESP or a fabric filter, the flyash is then contaminated with mercury laden carbon which may render otherwise saleable flyash for cement or road material into landfill waste. If activated carbon is injected into a secondary fabric filter placed downstream of an ESP (which may be powered back to control the PAC/ash mixture at the FF to about 10% carbon, activated carbon usage can be minimized and flyash beneficial use maintained but with additional equipment and energy requirements.

Basis	200 MW, 800,000 acfm @ 300F, 75% capacity factor, >90% Hg control			
	Powdered Activated Carbon (PAC)			Activated Coke ReACT™
	ESP	Primary FF	Secondary FF added downstream of ESP	Removal on two year cycle
Hg removed assuming coal @ 10 lb Hg/10^{12} BTU	130 lb/year	130 lb/year	130 lb/year	>130 lb/year
PAC injection rate	10 lb/10^6acfm	5 lb/10^6acfm	3 lb/10^6acfm	---
Make-up carbon	1600 ton/year activated carbon	800 ton/year activated carbon	475 ton/year activated carbon	1,000 ton/year activated coke
Disposal – not Hg laden	0	0	41,000 tons/year as ash sales	1,000 ton/year 80% coke fines, 20% flyash 41,000 ton/year as ash sales
Hg-laden disposal	46,600 tons/year	45,800 tons/year	5,475 tons/year	25 tons every two years
Hg content ppm on carbon ppm in admixture	40 ppm on PAC 1.4 ppm in admixture	80 ppm on PAC 1.4 ppm in admixture	140 ppm on PAC 12 ppm in admixture	5400 ppm

The comparison table above shows a dramatic reduction in the volume of Hg contaminated waste for the regenerated activated coke process and into a form where off-site recovery of metallic mercury from the activated coke is also possible, eliminating re-introduction of mercury to the environment.

J-Power Isogo

The Isogo repowering project which replaced 2x265 MW conventional PC boilers and pollution control (ESP, WFGD) with 2x600MW ultrasupercritical boilers and advanced pollution control represents a world class solution for clean coal. Sited in Yokohama harbor, the plant design also considered aesthetics in terms of plant appearance, including visibility at the stack. The emissions performance cited is on a par with natural gas fired facilities and the project should be of interest for U.S. energy planning at many levels.

References:

1. Steiner, P., Juntgen, H. and Knoblauch, K., "Removal and Reduction of Sulfur Dioxides from Polluted Gas Streams", Sulfur Removal and Recovery, Chapter 15, pp 180–191 *Advances in Chemistry*, Vol. 139, American Chemical Society, April 1975

2. Tsuji, K., and Shiraishi, I., "Combined desulfurization, denitrification and reduction of air toxics using activated coke", pp 549-553, *Fuel*, Vol. 76, No. 6, 1997

3. Staudt, J., Jozewicz, W. and Srivastava R., "Modeling Mercury Control with Powdered Activated Carbon", Presented at DOE/EPI/EPA/A&WMA MEGA Symposium, May 2003

4. Hirabayashi, Y., Miya, M., and Kojima, M., "Regenerative Activated Coke Technology (ReACT): A commercial multi-pollutant technology being adapted for U.S. applications"

5. Dene, C., Gilbert, J., Jackson. K., Miyagawa, S., "ReACT Process Demonstration at Valmy Generating Station", Presented at DOE/EPI/EPA/A&WMA MEGA Symposium, May 2008

6. Derrene, S., Bustard, J., McMillan, M, Chang, R., et al, "Toxecon™ Clean Coal Demonstration for Mercury and Multi-Pollutant Control at the Presque Isle Power Plant", Fuel Processing Technology, Vol. 90, No. 11, November 2009

(24a) Energy Security and Affordability in the 21st Century

John Hofmeister, Citizens for Affordable Energy, Houston, TX

Hofmeister, former President of Shell Oil Company, and founder of the non-profit association, Citizens for Affordable Energy, tackles two of the most critical concerns facing industry today – energy availability and affordability. While green energy is helpful, it's inadequate for the demands of the future. Technology and innovation are important elements, but do they offer realistic solutions right now? Add in an aging infrastructure as a potential disabler to economic competitiveness and the picture is bleak. But it's not all doom and gloom, and it seems the answers are well within our reach. According to Hofmeister, investment in the traditional energy base is the path to a competitive affordable energy supply. And as an added bonus, necessary environmental protections can be value adding to the economy if managed wisely.

See more of this Session: Energy Leadership Keynote Address - John Hofmeister, Founder, Citizens for Affordable Energy - Sponsored by the AIChE Fuels and Petrochemicals Division
See more of this Group/Topical: Liaison Functions

Real Time Detection of Gaseous Metal Contaminants in Coal Flue and Fuel Gas

Erik C. Rupp, ORISE Postdoctoral Researcher, National Energy Technology Laboratory, Pittsburgh, PA
Evan J. Granite, National Energy Technology Laboratory, Pittsburgh, PA
Dennis Stanko, National Energy Technology Laboratory, Pittsburgh, PA

The Integrated Gasification Combined Cycle (IGCC) is a promising technology for the future use of coal to provide energy. During this process, many of the trace elements present in coal are volatilized, and remain in the gas phase through the gas turbine, only to be released in the flue gas. Trace metals are of particular environmental importance, and thermodynamic studies have indicated that Hg (as elemental Hg), Se (as hydrogen selenide, H_2Se) and As (as arsine, AsH_3), as well as P (as phosphine, PH_3) will be present as gaseous metal species. While mercury has long been of environmental concern, there is increasing evidence that the other trace metals will have a significant environmental impact if coal continues to be a primary source of energy as expected. It will become valuable to develop a method for real-time detection of the gaseous trace metals to determine the actual concentration in the fuel gas, and to determine the effectiveness of sorbents for the removal of the trace metals.

A commercial gas chromatography/ion trap mass spectrometry system has been used to detect AsH_3, H_2Se, PH_3 and elemental mercury. The gas chromatograph is a Varian 450-GC equipped with a Varian CP-Porabond Q Fused Silica column (25 m × 0.25 mm) and a VICI gas sampling valve, used for AsH_3, H_2Se and PH_3. A Hamilton 250 µL gas tight syringe was used for sampling Hg by manual injection. The mass spectrometer is a Varian 240-MS IT Mass Spectrometer equipped with internal electronic ionization. UHP helium was used as the carrier gas, as obtained from Butler Gas Products, Inc. (McKees Rock, PA). Trace contaminants were used as obtained using a VICI Metronics Dynacal permeation device (Hg, 302 ng/min at 70 °C), VICI Metronics G-Cal permeation devices (AsH_3, 487 ng/min at 25 °C and PH_3, 55 ng/min at 25 °C) and Matheson Tri-Gas cylinder (103 ppm H_2Se in N_2).

Mercury was the primary concern and proved to be the least successful and most problematic of the four targeted contaminants. Mercury was unobservable when using the gas sampling valve, as it appeared to be trapped in the tubing leading to the sampling valve, or in the sampling valve itself. Upon using the gas tight syringe, Hg was readily detectable, although the detection was qualitative only. Figure 1 indicates that mercury is detectable, although a linear response with increasing concentration is not currently possible, due to issues that remain unresolved. This may be due to difficulties in the electronic ionization of the Hg (in its elemental form), contamination of the sampling syringe or losses within the column.

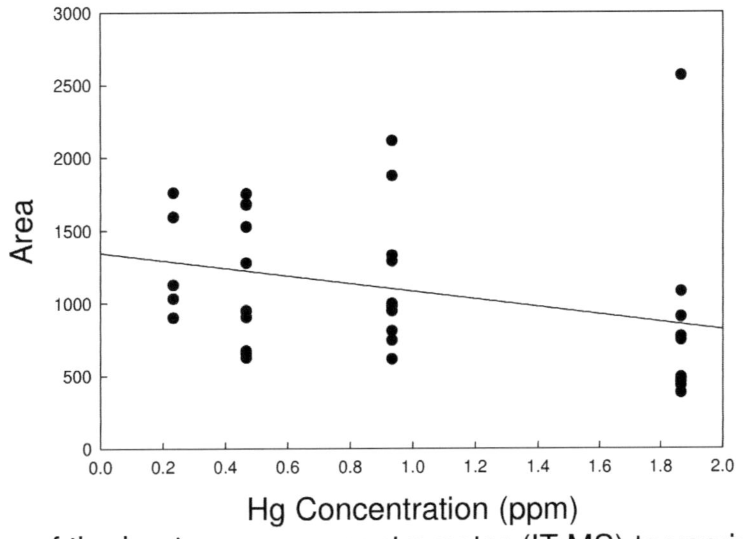

Figure 1: Response of the ion trap mass spectrometer (IT MS) to varying inlet concentrations of Hg. The line is a linear fit of the data with an R^2 value of 0.92.

Detection of the other three contaminants (AsH_3, PH_3 and H_2Se) using the IT MS system was highly successful. The detection limits for the contaminants can be seen in Table 1.

	Observed Limit (ppb)	Instrument Limit (ppb)
Hg	~200	N/A
PH_3	330	30
AsH_3	40	30
H_2Se	80	30

Table 1: Observed and instrumental detection limits on the Varion 240-MS IT in helium.

The observed detection limit indicates known concentrations that have been repeatedly observed in the laboratory. All four contaminants are in the ppb range, although Hg is a qualitative observation, and not a hard limit, based on the difficulties in quantifying the response. The instrument limit is based on an arbitrary area count of 100 for the reference ion, with the three quantifiable contaminants having a limit of approzimately 30 ppb. The arbitrary count is based on operator experience and observations of the peak shape and peak-to-noise ratio. These limits were obtained in He flow, with N_2 present in the H_2Se experiments. AsH_3, PH_3 and H_2Se all demonstrated a linear response and repeatability under the same conditions. Figure 2a-c is the calibration curves for these three contaminants.

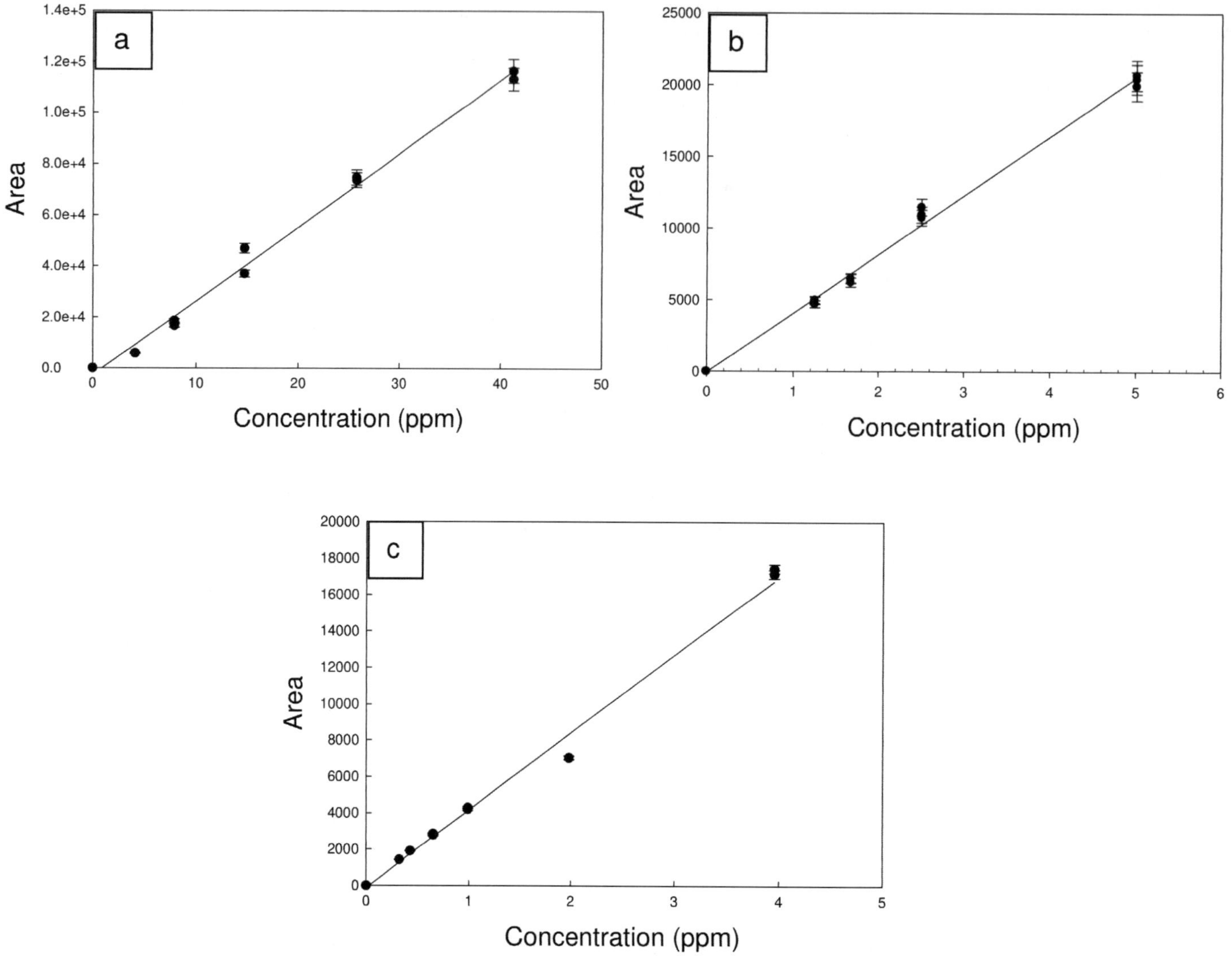

Figure 2: Linear response calibration. a) H_2Se; R^2 = 0993 b) AsH_3; R^2 = 0.995 c) PH_3; R^2 = 0.990

Each contaminant had an R^2 value above 0.99, indicating an acceptable linear response over a wide range of concentrations. As the concentration range is decreased, the R^2 value increases, as can be observed in the PH_3 concentrations below 1 ppm. The repeatability of the response was also high, with a relative standard deviation (assuming a Gaussian distribution of the response) for AsH_3, PH_3 and H_2Se of 5.1%, 1.4% and 3.9% respectively.

Although the majority of the work was done with single contaminants in a He gas, research has been performed with all three contaminants present in He, as well as individual contaminants in a simulated fuel gas containing CO_2, CO, H_2 and H_2S, and all 3 contaminants in the simulated fuel gas. Mercury was not included in this work. The presence of the other contaminants and fuel gas components did not affect the response of the MS, and it was possible to detect all three contaminants at the same time. Figure 3 is total ion count chromatogram for a He only run performed with the three contaminants, as well as the quantitation ion for each compounds (80 m/z H_2Se, 76 m/z AsH_3, 32 m/z PH_3). Each contaminant is identifiable but residence time and mass spectrums (not shown). This work has been extended to simulated fuel gas with similar results.

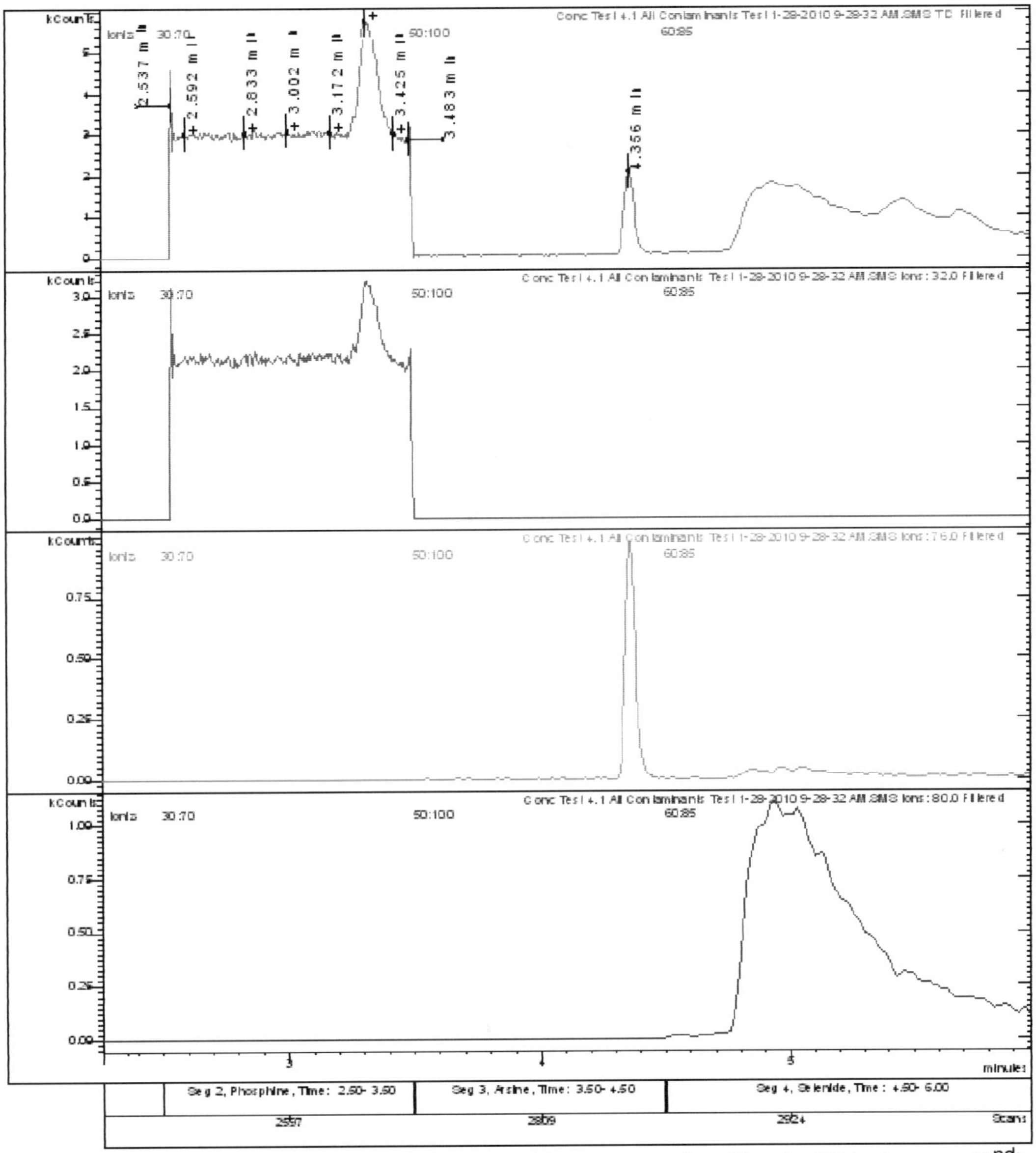

Figure 3: GC/MS chromatogram of the total ion count (red/top), PH_3 (green/2nd), AsH_3 (yellow/3rd) and H_2Se (blue/bottom).

Arsine, phosphine and hydrogen selenide were detected with a repeatable, linear response using a Varian GC/IT MS. The detection was performed in multiple gaseous environments and with multiple contaminants present. Fuel gas components did not interfere with detection. Mercury was detectable, but only qualitatively. These results indicate that this system could be used for real time detection of gaseous metal contaminants in fuel gas at a quantifiable level. There is a possibility that these results could be extended to mercury with further experimental refinements.

(36b) Recent Advances in the High Temperature Capture of Elemental Mercury by Inorganic Sorbents: Packed Bed and Flow Reactor Studies

Sung Jun Lee, Chemical Engineering, ICSE, University of Utah, Salt Lake City, UT

Yang Li, Chemical Engineering, ICSE, University of Utah, Salt Lake City, UT

Jost O.L. Wendt, Chemical Engineering, ICSE, University of Utah, Salt Lake City, UT

This paper reports on recent results relating to high temperature sorption of metallic mercury on MinPlus sorbent, an inorganic substrate derived from paper recycling residues. Results from a packed bed reactor as well as in-flow dispersed phase reactor are reported. A new type of packed bed was developed, in which 0.3 g sorbent was supported on alumina wool in an alumina lined reactor. Previous results had shown that MinPlus reacts with quartz at high temperatures, thus rendering suspect previous packed bed data that utilized quartz beads and quartz wool for support. The new data show an apparent maximum sorption of Hg at around 900C, at which temperature the bed material shrinks and by-passing occurs. Maximum sorption also occurs immediately after the sorbent has reached its maximum temperature. Thereafter a time dependent process de-activates it. The disperse phase reactor data support the hypothesis that sorption occurs only after the sorbent has been allowed to deposit on the walls for times exceeding the gas mean residence time. Data from both reactors suggest that the free SiO_2 content of the sorbent is very important, and that the performance of a poorer sorbent (resulting from changes in the paper composition from which the residue was produced) can be improved through addition of finely divided quartz powder. The data are consistent with a mechanism involving a high temperature solid reaction between CaO and (free) SiO_2 to produce a species which scavenges Hg. Spent sorbent samples are currently undergoing analysis to help close the mass balance in the packed bed, and to suggest the nature of the reaction product with mercury. These results are not available now, but will be reported at the conference.

See more of this Session: Mercury and Other Trace Elements in Fuel: Emissions, Control, and Measurement II

See more of this Group/Topical: Topical 2: Advanced Fossil Energy Utilization

On-Line Studies of Pd Based Sorbents for the Removal of Gaseous Metal Contaminants: Capture and Regeneration

Erik C. Rupp, ORISE Postdoctoral Researcher, National Energy Technology Laboratory, Pittsburgh, PA
Evan J. Granite, Dennis Stanko, John P. Baltrus, Henry W. Pennline, National Energy Technology Laboratory, Pittsburgh, PA
Hugh Hamilton, Stephen Poulston, Liz Rowsell, Andrew Smith, Wilson Chu, Johnson Matthey, London

The Integrated Gasification Combined Cycle (IGCC) is a promising technology for the future use of coal to provide clean energy in a more efficient manner. During this process, many of the trace elements present in coal are volatilized, and remain in the gas phase through the gas turbine, only to be released in the resulting flue gas. Trace metals are of particular environmental concern, and thermodynamic studies have indicated that Hg (as elemental Hg), Se (as hydrogen selenide, H_2Se) and As (as arsine, AsH_3), as well as P (as phosphine, PH_3) will be present as gaseous metal species. While mercury has long been of concern, there is increasing evidence that the other trace metals will have a significant environmental impact if coal continues to be a primary source of energy as expected. It is desirable to have regenerable sorbents for the removal of trace contaminants from high temperature syngas, in order to preserve the high thermal efficiency of the IGCC Process. While a number of sorbents, catalysts, scrubber solutions, and irradiation techniques have been developed for the capture of mercury, including activated carbons, Thief Process carbons, wet scrubbers, precious metal catalysts, and ultraviolet light at 253.7-nm, these sorbents and techniques will likely show limited capability for removing other trace metal contaminants at elevated temperatures. In this study, Pd-Al_2O_3 sorbents have been studied to determine that capability under a variety of relevant conditions.

A commercial gas chromatography/ion trap mass spectrometry system has been used to detect AsH_3, H_2Se, PH_3 and elemental mercury. The gas chromatograph is a Varian 450-GC equipped with a Varian CP-Porabond Q Fused Silica column (25 m × 0.25 mm) and a VICI gas sampling valve, used for AsH_3, H_2Se and PH_3. A Hamilton 250 µL gas tight syringe was used for sampling Hg by manual injection. The mass spectrometer is a Varian 240-MS IT Mass Spectrometer equipped with internal electronic ionization. Ultra-High Purity (UHP) helium was used as the carrier gas. Trace contaminants were introduced using a VICI Metronics Dynacal permeation device (Hg, 302 ng/min at 70 °C), VICI Metronics G-Cal permeation devices (AsH_3, 487 ng/min at 25 °C and PH_3, 55 ng/min at 25 °C) and a Matheson Tri-Gas cylinder (103 ppm H_2Se in N_2). Cylinders with a 54.00% CO, 40.01% H_2 and 5.99% CO_2 mixture and 1% H_2S with balance N_2 were used in order to create a simulated fuel gas. Pd/Al_2O_3 sorbents were used as obtained from Johnson Matthey. Palladium loadings of 2% and 20% by weight were used for the experiments, which consisted of a 10 mg packed bed supported on glass wool in a quartz tube.

Mercury, arsine, phosphine and hydrogen selenide are all detectable using the mass spectrometer. The mass spectra for the four compounds can be seen in Figure 1. Mercury can only be detected qualitatively while the other three contaminants have a repeatable, linear response on the mass spectrometer. Breakthrough curves are possible due to this linear response, and the total amount of captured contaminant can be estimated based on the difference between the inlet concentration and the outlet concentration.

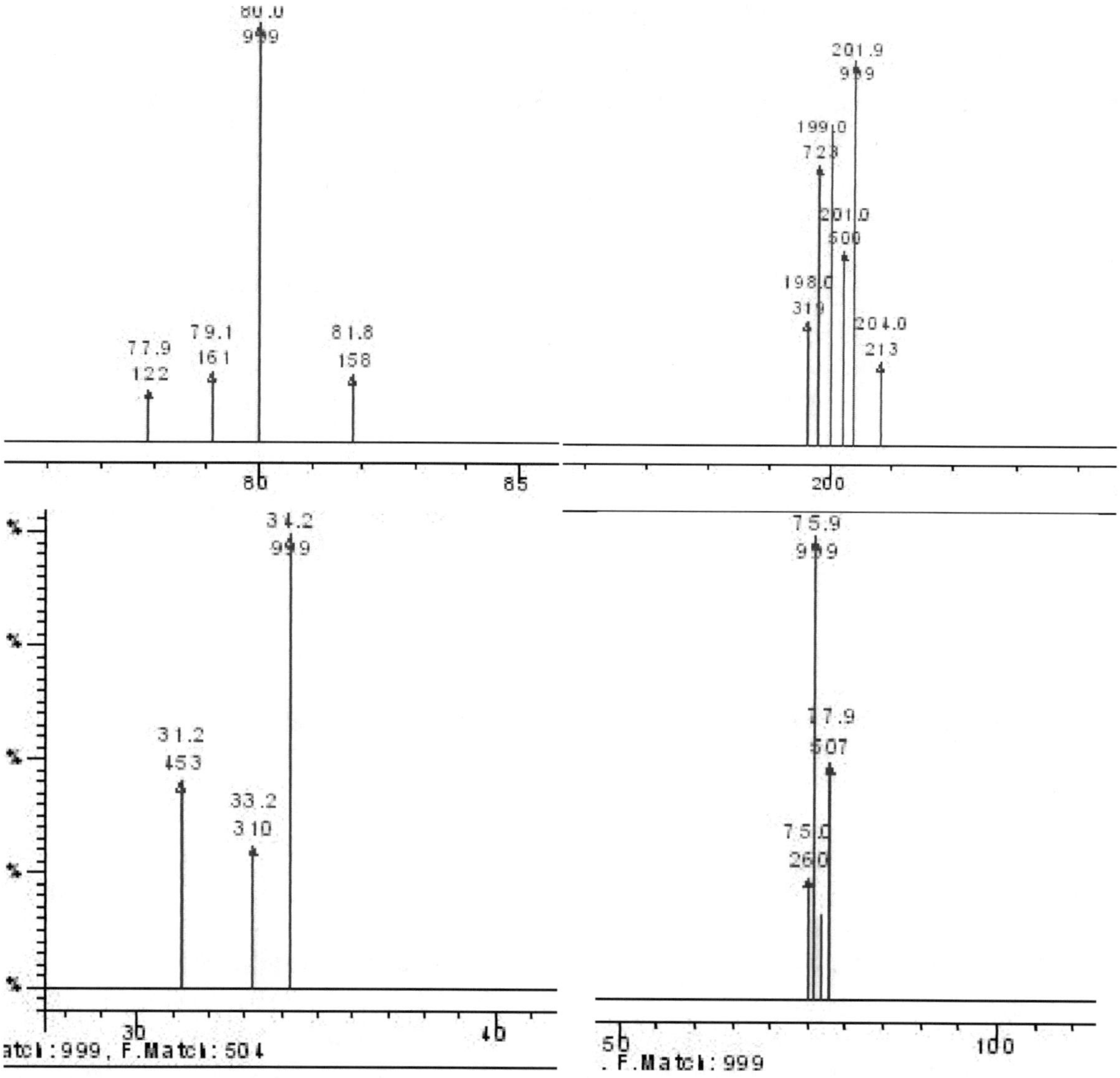

Figure 1: Mass spectra. A) Hg. B) H_2Se. C) PH_3. D) AsH_3

The initial breakthrough curves were performed on 20% Pd/Al_2O_3 at ambient temperature in a single contaminant, helium environment. Figure 2 is the AsH_3 breakthrough curve at 2.5 ppm AsH_3 over 2700 minutes using 10 mg of the sorbent. The results are indicative of several characteristics of the sorbent: 1) Even at low temperatures, the sorbent has a high capacity for the contaminants studied; 2) There is a modest level of physical adsorption, indicated by the break in the data at 1700 minutes, when the He/AsH_3 stream was removed from the bed for a period of 15 minutes; 3) Experiments must be run in a single time period, as removal of the stream from the bed can change results, including total contaminant capacity. Approximately 1.2 mg of AsH_3 was adsorbed during this experiment, although this estimate may be high, due to unrecorded desorption at 1700 minutes.

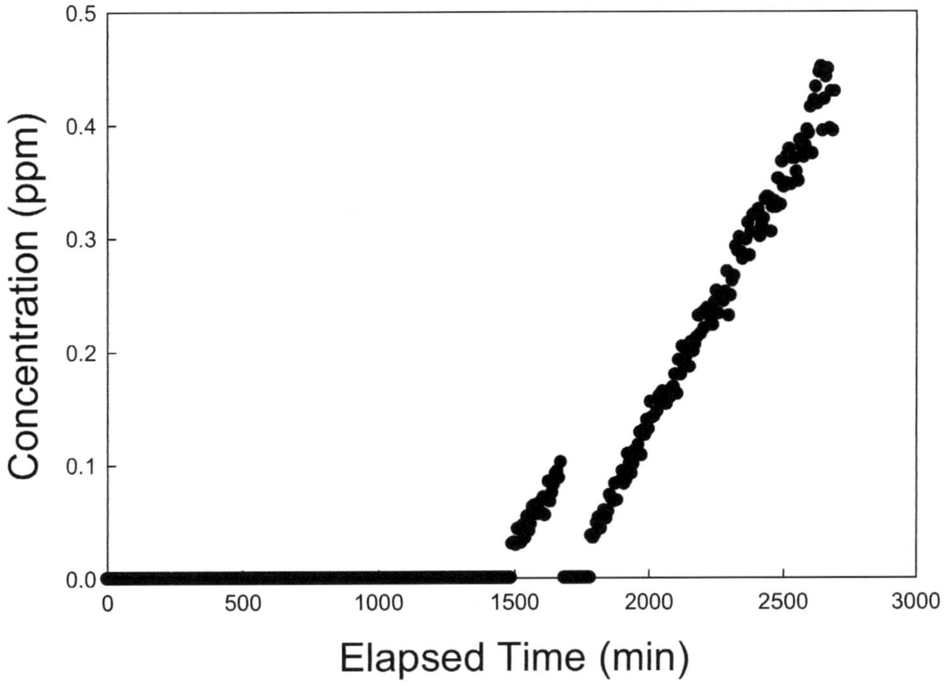

Figure 2: AsH_3 (2.5ppm) breakthrough curve on a 10 mg 20% Pd/Al_2O_3 packed bed at ambient temperature.

Similar results were obtained with PH_3 and AsH_3, although the experimental run time was decreased and breakthrough was never observed. The next set of experiments utilized a bed temperature of 204 °C with 2% Pd/Al_2O_3 remaining in a single contaminant He environment. These choices were made to decrease experimental run time, as the capacity of the 20% Pd/Al_2O_3 was expected to increase with increasing temperature. AsH_3 (2.5 ppm) did not break through the bed after 450 minutes, which corresponded to a total exposure of 0.22 mg AsH_3 for a near 1:1 As:Pd ratio. Experimental run times were unable to be extended beyond this due to safety regulations. PH_3 also did not break through, as the available mass of PH_3 was decreased due to the permeation tube release rate. Hydrogen selenide was available at a much higher concentration (100 ppm) and breakthrough occurred nearly immediately, likely due to overloading the bed.

As a real fuel gas contains multiple contaminants, 2% Pd/Al_2O_3 at 204 °C was exposed to a He stream containing 2.2 ppm AsH_3 and 17.6 ppm H_2Se. Figure 3 shows the results for an experiment running 450 minutes. Under these conditions, H_2Se is removed completely for the entire time. However, AsH_3 breaks through early and increases to about 0.5 ppm after 100 minutes have elapsed, and remains at that concentration for the duration of the experiment. This indicates that there is competitive adsorption between these two contaminants, with H_2Se adsorbing preferentially to AsH_3.

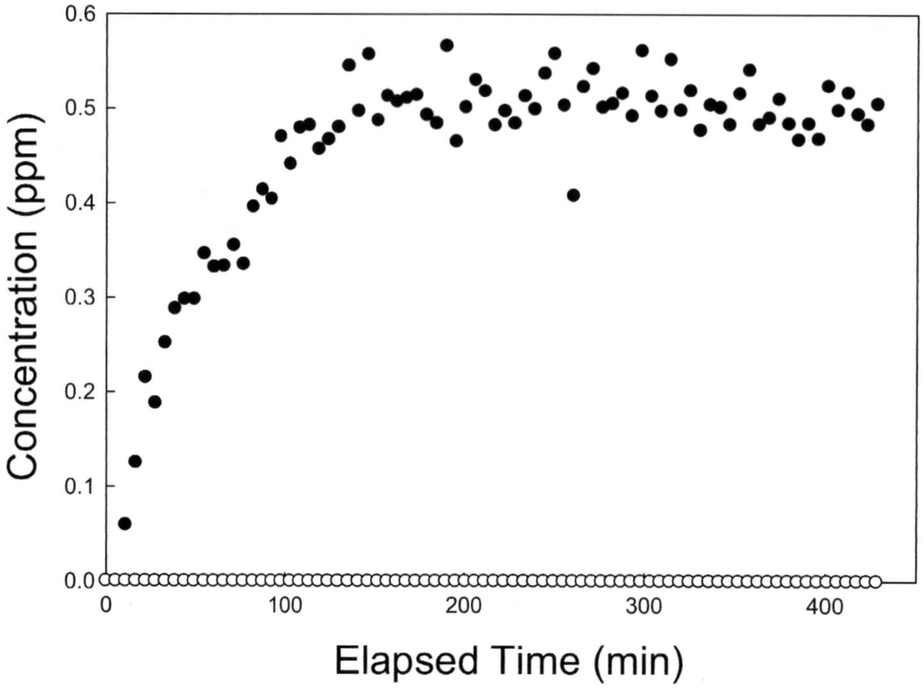

Figure 3: AsH_3 (2.2ppm) and 1 H_2Se (17.6ppm) breakthrough curve on a 10 mg 2% Pd/Al_2O_3 packed bed at 204 °C. ● - AsH_3; ○ - H_2Se.

The ultimate goal of this research is to determine whether Pd/Al_2O_3 can be an effective sorbent for the removal of Hg, AsH_3, H_2Se and PH_3, so it is important to test the sorbent in an as-realistic simulated fuel gas as possible. For this purpose, a simulated fuel gas containing 44% He, 34% N_2, 12% CO, 9% H_2, 1% CO_2, 0.2% H_2S, 21 ppm H_2Se, 4 ppm AsH_3 and 1 ppm PH_3 was passed through 10 mg of a 2% Pd/Al_2O_3 bed at 204 °C. The results can be seen in Figure 4. The three studied contaminants all break through the column nearly upon exposure. This indicates that the sorption characteristics of the Pd/Al_2O_3 sorbent change dramatically in a "realistic" situation. This may be due to the well known affinity of palladium for sulfur, which can block sites and prevent the contaminants from adsorbing. It is also worth noting that the contaminant concentrations are larger than they would be in a real fuel gas.

Viewing the data in Figure 4 on a relative scale can offer additional insight. Figure 5 shows the fractional removal:

$$\text{Fractional Removal} = 1 - \frac{C_{out}}{C_{in}} \qquad (1)$$

of each of the contaminants based on inlet concentration (C_{in}) and outlet concentration (C_{out}). It is evident that H_2Se is removed preferably over AsH_3 and PH_3, which is consistent with the data presented in Figure 3. There is no discernible difference between AsH_3 and PH_3, even with the concentration difference between the two. This could be because of the similar molecular structure of the two contaminants.

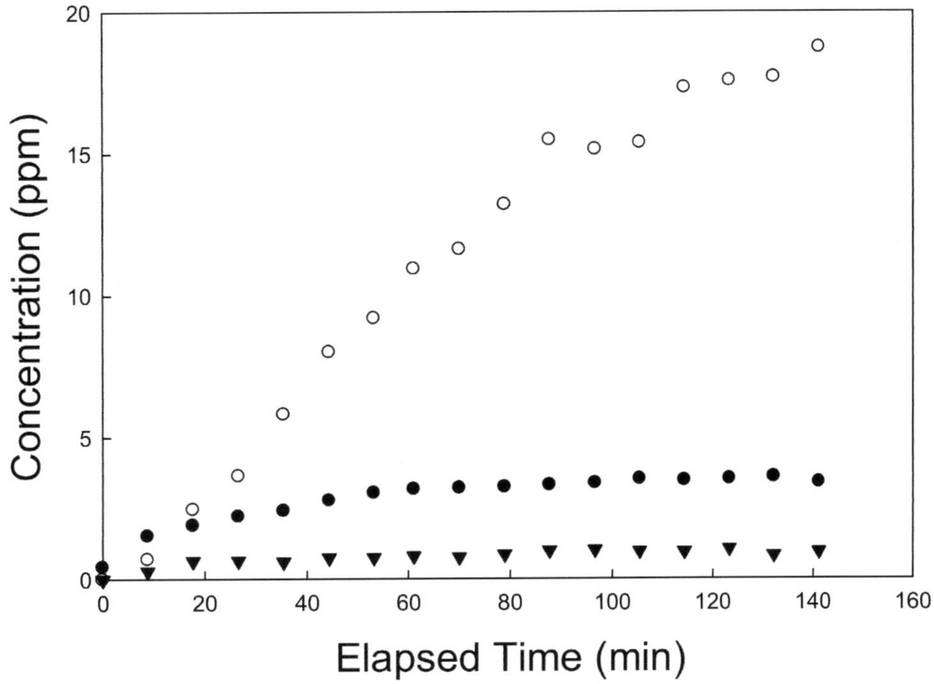

Figure 4: PH_3 (1 ppm),AsH_3 (4 ppm) and H_2Se (21 ppm) breakthrough curve on a 10 mg 2% Pd/Al_2O_3 packed bed at 204 °C in a simulated fuel gas. ● - AsH_3; ○ - H_2Se; ▼ - PH_3.

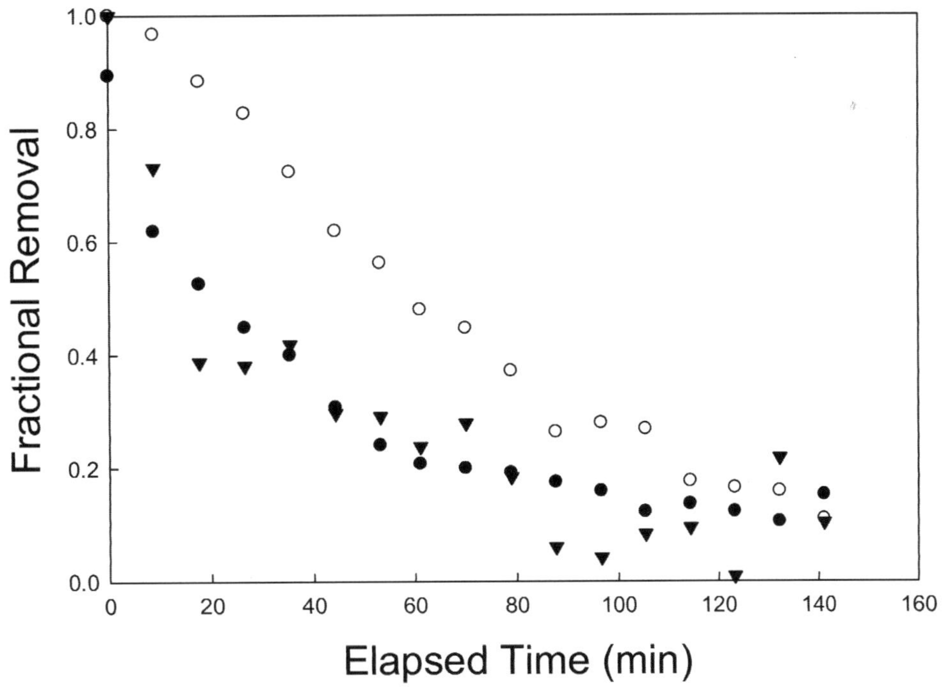

Figure 5: 1 ppm PH_3, 4 ppm AsH_3 and 21 ppm H_2Se removal on a 10 mg 2% Pd/Al_2O_3 packed bed at 204 °C in a simulated fuel gas. ● - AsH_3; ○ - H_2Se; ▼ - PH_3.

Under laboratory conditions, a Pd/Al_2O_3 adsorbent has been proven to be effective in removing a number of gaseous metal contaminants from a gas stream, during these experiments and previous work done in this laboratory. Under single contaminant conditions, the sorbent has a great capacity for the contaminants, often approaching a 1:1 M:Pd ratio before breakthrough is observed. Multiple contaminants will compete against each other for sorption sites, with the presence of additional contaminants such as H_2S in a simulated fuel gas decreasing the adsorption even further. However, given the extreme conditions, such as a very tiny sorbent bed and correspondingly small contact times, under which this experiment was performed, the sorbent still remains a promising technology for the removal of gaseous metal species at high temperatures. It is worth noting that in a recent pilot slipstream test under real coal-derived fuel gas, a palladium sorbent at 288 °F removed nearly 100% of the mercury, arsenic and selenium entering the bed over several hundred hours. Post exposure analysis of this used palladium sorbent bed revealed most of it was still available for additional adsorption of the trace elements, suggesting large capacities for mercury, arsenic and selenium in real syngas.

(36d) Qualitative Prediction of Sorbent Performance for Mercury Capture by Lab-Scale Tests

Evan J. Granite, National Energy Technology Laboratory, Pittsburgh, PA

Erik C. Rupp, National Energy Technology Laboratory, Pittsburgh, PA

Henry W. Pennline, U.S. Department of Energy, National Energy Technology Laboratory, Pittsburgh, PA

It would be desirable to have a simple and inexpensive test for predicting sorbent performance for mercury capture in coal-burning power plants and in integrated gasification-combined cycle (IGCC) facilities. Several lab-scale packed bed reactors have been employed to screen sorbents for the removal of mercury from coal-derived flue and fuel gases. The reactors have used either simulated flue or fuel gas mixtures, or slipstreams from coal-utilizing facilities.

Simple packed bed reactors have the potential to qualitatively predict the performance of sorbents in actual flue or fuel gas. This predictive capability exists despite the fact that the contact time and gas composition may differ significantly from the full-scale application. The performance of different sorbents for mercury removal will be compared under both lab and full-scale conditions. Future research towards the goal of a simple index test to determine full-scale sorbent performance will be discussed.

See more of this Session: Mercury and Other Trace Elements in Fuel: Emissions, Control, and Measurement II
See more of this Group/Topical: Topical 2: Advanced Fossil Energy Utilization

(36e) On the Solubility of Mercury in Liquid Hydrocarbons

Nicolas S. Bloom, Trace Metals Research Group, Columbia Analytical Services, Kelso, WA

D. Gallup, Chevron Energy Technology Company, Houston, TX

Summary. The solubility of Hg^o was determined in 34 pure hydrocarbons, hydrocarbon mixtures, crude oils, and water from -65°C to +65°C (where each solvent remained liquid). The solubility of $HgCl_2$ was also quantified, but in a limited subset of the solvents. The solubility of any Hg species in its respective solvent can be very accurately represented by fitting the data to an exponential function of temperature (E1). A similar volume-based exponential (E2) was best-fit to data for the saturation concentration of Hg^o in air. By converting the Hg^o solubility equations from a mass basis to a volume basis, and then dividing into E2, temperature-dependent unitless "Henry's Law constants" (E3) were also be calculated.

$$[Hg] = k \cdot e^{A \cdot T} \qquad (E1)$$

$$[Hg^o]_{air} = 2.455 \cdot e^{0.0833 \cdot T} \qquad (E2)$$

$$K_{Hf(T)} = [Hg_{air}]_{f(T)} / [Hg_{HC}]_{f(T)} \qquad (E3)$$

For many samples, measurements were made both of Hg^o and total Hg. These data show that previously reported solubility data, made by measuring just total Hg may be biased high by the presence of oxidized Hg compounds, especially if the solvents were not first pre-purified to remove trace amounts of Hg-oxidizing compounds.

Pre-Treatment of Solvents. At the outset of the study, we discovered that some reagent-grade organic solvents rapidly corroded the surface of clean Hg^o, sometimes to the point of leaving only a gritty black powder in its wake. Different brands and grades of the same compound differed widely in corrosivity, suggesting the presence of trace oxidizing compounds. This hypothesis was supported by observations that the corrosion was (a) strongly temperature dependent, and (b) could be eliminated by pre-extracting with a small volume of aqueous $NH_2OH \cdot HCl$. We did not want to risk artifacts later by using solvents that might harbor traces of the reducing agent, however, so, an alternative pretreatment was employed. Fifty grams of Cu granules were added per litre of solvent, which was then shaken for 24 hours. The Cu reduced, but didn't eliminate the Hg^o corrosion. We never saw signs of corrosion on the Cu surfaces.

Obtaining a Clean Droplet of Hg°. The best way to obtain clean droplets of Hg° was to use a 100 mL gas-tight syringe with a fine gauge stainless steel needle to pull liquid from the center of a larger mass, where the mercury is more than a centimeter deep. A 20 mL borosilicate scintillation vial holding 10 mL of Hg° is excellent for this. Since all of the corrosion and dirt particles are forced to the outside of the mass by the strong bouyancy effects, the Hg in the center is extremely clean.

Analytical Methods. Total Hg in hydrocarbons was determined after first digesting ~0.4 grams of sample with 3.0 mL of concentrated HNO_3 using the Anton Paar High Pressure Asher (HPA) (Bloom, et al. 2004). Samples were wet-combusted for two hours in ultra-clean quartz tubes at 300°C and 125 atm pressure. After dilution with reagent water, aliquots were analyzed using $SnCl_2$ reduction, purge and trap on gold, electrothermal desorption of the mercury as Hg°, and quantification via CVAFS. Hg° was independently determined after dilution with oxidizer-free hexane. Diluted samples were purged directly after floating up to 200 ml of sample on the surface of 100 mL pre-purged 0.2% HCl in water contained in a bubbler that has never been in contact with Sn(II). Purging the sample results in the evaporation of the hexane and release of Hg° onto a gold trap (Bloom, 2000). See Table 1 for a QC summary.

Table 1. Quality assurance summary for the lab over the time period that the Hg solubility study was conducted (PQL = practical quantification limit).

parameter	mercury species	
	THg	Hg°
maximum digestion mass (g)	0.40	1.00
dilution volume (mL)	20.0	40.0
max aliquot volume analyzed (mL)	2.0	0.20
max effective mass analyzed (g)	0.04	0.005
MDL (ng/g)	0.5	1.4
PQL (ng/g)	2	6
mean precision at > PQL: (% RPD)	6.1 ± 3.9	5.1 ± 5.9
mean accuracy at > PQL: (% rec)	100.4 ± 6.7	100.5 ± 7.1

Experimental Setup. Samples for analysis were made by completely filling 125 mL borosilicate bottles with solvent, injecting ~1.0 gram of clean liquid mercury, and tightly replacing the Teflon-lined caps. Bottles were incubated at temperatures ranging from -65°C to +65°C with constant gentle agitation or periodic vigorous agitation depending upon the equipment available. Periodically, the agitation was stopped, and the samples allowed to settle at least four hours prior to aliquoting. As a buffer against temperature fluctuations, samples were kept in 400 mL plastic beakers 70% filled with a

temperature-appropriate inert liquid. This beaker was used to carry the sample from the incubator to and from the analytical balance. The entire handling process generally took less than 2 min, resulting in trivial changes in the temperature and so the Hg^o concentration of the sample. The Hg^o concentration in at least one of the samples per temperature was periodically monitored over days to months, until a constant concentration was observed. The actual quantitative data produced for the study were then collected in replicate digestions two or more days after equilibrium was verified. The solubility of $HgCl_2$ was also quantified in a smaller subset of the hydrocarbons. The set-up and procedures were analogous to the case with Hg^o, except that in place of the small amount of Hg^o, surprisingly large amounts of $HgCl_2$ were required to saturate some of the solvents—over 100 grams, in the case of methanol!

A special procedure was employed for the determination of the solubility of Hg^o in water, owing to the ease with which Hg^o is oxidized to Hg(ll) even in extremely carefully prepared water, giving results that can be many orders of magnitude too high. A droplet of clean Hg^o was placed into a bottle filled to the top with ultra-pure argon-purged water to which 100 ppm of stannous citrate (pH 7.0) was added, and the lid sealed on. By adding Sn(ll) to the aqueous sample, the redox potential was forced down to the point that only Hg^o can exist. Aliquots from this sample were then analyzed in the same manner as were the Hg(ll) standards--by $SnCl_2$ reduction, purge and trap, and CVAFS.

Multiple determinations at each of three or more temperatures were measured for each species/solvent pair, and the concentration expressed as an exponential function (E1) of temperature. The solubility of Hg^o in the entire range of hydrocarbons varies over about a factor of five at a given temperature. The lowest solubilities are found in low MW polar compounds such as water and methanol, and highest solubilities are found in aromatic compounds (Table 2). The solubility of $HgCl_2$, is generally much greater than Hg^o in any given solvent, and it varies over a factor of 100,000 across the range of solvents at any given temperature. Conversely to Hg^o, the greatest solubility for $HgCl_2$ is in the most polar in polar solvents (Table 3).

Many hydrocarbons react to form Hg(ll) corrosion products that produce total Hg levels far higher than the true Hg^o concentration (Table 4). Aromatic solvents tended to convert the mercury droplets to a heavy black powder, while most simple alcohols were seen to produce a fine pinkish powder—growing darker with increasing MW. The presence of corrosion products appears sometimes to be due to reaction between Hg^o and the solvent itself, rather than with trace impurities. For example, significant

levels of dissolved Hg(ll) compounds appear to result from almost every oxygen-containing solvent, suggesting that they, themselves, are reacting with the Hg^o, perhaps forming species analogous to Hg(ll)

Table 2. Hg° solubility (ng/g) relationships for some typical hydrocarbons.

$[Hg^{\circ}]$ at 25°C	solvent	Hg° solubility	N	R^2
2,206	*n*-pentane	$396 \cdot e^{0.0687 \cdot T}$	14	0.997
2,139	*n*-hexane	$433 \cdot e^{0.0639 \cdot T}$	10	0.999
1,988	*n*-octane	$437 \cdot e^{0.0606 \cdot T}$	18	0.988
1,884	*n*-dodecane	$404 \cdot e^{0.0616 \cdot T}$	44	0.985
1,313	*iso*-octane	$251 \cdot e^{0.0662 \cdot T}$	18	0.992
1,840	**all alkanes**	$\mathbf{357 \cdot e^{0.0656 \cdot T}}$	**221**	**0.964**
2,718	benzene	$684 \cdot e^{0.0552 \cdot T}$	12	0.964
2,762	toluene	$640 \cdot e^{0.0585 \cdot T}$	23	0.987
2,778	ethylbenzene	$722 \cdot e^{0.0539 \cdot T}$	8	0.993
2,640	*o*-xylene	$598 \cdot e^{0.0594 \cdot T}$	12	0.989
2,708	**all aromatics**	$\mathbf{673 \cdot e^{0.0557 \cdot T}}$	**65**	**0.960**
1,417	methylene chloride	$406 \cdot e^{0.0500 \cdot T}$	14	0.996
640	methanol	$156 \cdot e^{0.0565 \cdot T}$	14	0.979
52.1	water	$17.0 \cdot e^{0.0448 \cdot T}$	13	0.979

Table 3. $HgCl_2$ solubility (ng/g Hg) relationships for several hydrocarbons.

$[HgCl_2]$ at 25°C	solvent	$HgCl_2$ solubility	N	R^2
29,200	*n*-dodecane	$8{,}242 \cdot e^{0.0506 \cdot T}$	7	0.991
4,771,000	toluene	$2{,}599{,}000 \cdot e^{0.0243 \cdot T}$	7	0.943
3,569,000	methylene chloride	$1{,}287{,}000 \cdot e^{0.0408 \cdot T}$	7	0.994
559,600,000	methanol	$191{,}000{,}000 \cdot e^{0.0430 \cdot T}$	7	0.997

Table 4. Total and elemental Hg in solvents equilibrated with an initially clean, shiny droplet of Hg.

solvent	temp (°C)	appearance of mercury droplet	mean conc. (ng/g) Hg_{tot}	Hg^{o}
n-dodecane	21.5	shiny	1,585	1,579
n-pentadecane	21.0	fluffy white ppt	11,120	1,295
n-hexadecane	21.7	shiny	1,629	1,732
ethylbenzene	8.3	dense black powder	1,603	1,106
toluene, washed	21.0	dark surface stain	2,095	2,095
benzene	22.0	shiny	2,415	2,442
methanol	8.3	shiny	728	257
1-butanol	8.3	faint pink ppt	1,747	616
1-hexanol	22.0	reddish ppt	1,646	1,111
1-octanol	21.9	reddish ppt	1,208	905
glycerol	21.5	shiny	9,424	3,609

References

1. Bloom, N.S, Parker J, and Vondergeest, E. 2004. ***RMZ-Materials and Geoenvironment*, 51: 590-593.**

2. Bloom, N.S. 2000. *Fres J Anal Chem* **366**: 438-443.

See more of this Session: Mercury and Other Trace Elements in Fuel: Emissions, Control, and Measurement II
See more of this Group/Topical: Topical 2: Advanced Fossil Energy Utilization

(36f) In-Situ Electrochemical Detection of Trace Metal Vapors at Bismuth-Doped Carbon Screen Printed Electrodes

Donglai Lu, Mechanical Engineerig Department, University of Washington, Seattle, WA

Volatile species of Pb, Cd and Zn were generated at room temperature by the addition of sodium tetrahydroborate (III) to an acidified solution of the analytes. The vapor-phase species were quickly adsorbed into a thin layer of highly porous hydrogel which was dip-coated on screen printed electrode for electrochemical detection using square wave anodic stripping voltammetric method. The squarewave voltammetric stripping response is highly linear over the 10 to 80 ng range of Pb, Cd and Zn examined (2 min deposition), with a detection limit of 1 ng and good precision (RSD.7.4%). The attractive behavior of the new disposable carbon strip electrodes hold great promise for a simple, rapid, sensitive on-site real-time detection of trace metal vapors.

See more of this Session: Mercury and Other Trace Elements in Fuel: Emissions, Control, and Measurement II

See more of this Group/Topical: Topical 2: Advanced Fossil Energy Utilization

(60b) Novel Polymer Membrane Process for Pre-Combustion CO2 Capture From Coal-Fired Syngas

Timothy C. Merkel, Membrane Technology and Research, Inc., Menlo Park, CA

Meijuan Zhou, Membrane Technology and Research, Inc., Menlo Park, CA

Sylvie Thomas, Membrane Technology and Research, Inc., Menlo Park, CA

Haiqing Lin, Membrane Technology and Research, Inc., Menlo Park, CA

Adrain Serbanescu, Membrane Technology and Research, Inc., Menlo Park, CA

Gasification of coal – a low-cost and abundant domestic feedstock – and subsequent conversion to hydrogen and energy with simultaneous CO2 capture can improve national energy security in an environmentally responsible manner. To fulfill the promise of clean energy production through gasification of domestic coal, more efficient and less costly methods of separating the post-shift syngas (hydrogen and carbon dioxide) are needed. Current absorption technologies (Rectisol, Selexol, and PSA) used for this separation are costly, energy intensive, and unlikely to meet DOE CO2 capture targets.

Membranes of various types are currently being considered for H2/CO2 separations in future IGCC power plants. Key membrane challenges identified from previous work in this area include high membrane cost, low flux, instability in a challenging thermal and chemical environment, device reproducibility, scale-up issues, and poor membrane system design understanding. For an Integrated Gasification Combined Cycle (IGCC) membrane process to be successfully implemented, innovations in the membranes and membrane process designs that address these issues are required.

In this talk, novel high-temperature-stable polymer membranes with H2/CO2 selectivities of >10 and hydrogen permeances of >200 gpu at syngas cleanup temperatures of 100–200°C will be discussed. These membranes can be used in an efficient membrane process that will capture >90% of CO2 in post-shift syngas and generate a hydrogen stream with >93% purity at low cost. Results from a recent field test treating real coal-derived syngas will be presented.

Keywords: syngas, membrane, carbon dioxide, hydrogen and IGCC

See more of this Session: Carbon Capture/Separation From Gasification-Based Systems

Abstract: Novel Polymer Membrane Process for Pre-Combustion CO_2 Capture From Coal-Fired Syngas (2010 Spring National Meeting)

See more of this Group/Topical: Topical 2: Advanced Fossil Energy Utilization

(60c) Evaluation of the Technical and Economic Performance of the Calcium Looping Process for CO_2 and H_2s Capture From Coal-Derived Synthesis Gas

Daniel P. Connell, Research & Development, CONSOL Energy Inc., South Park, PA

Shwetha Ramkumar, Chemical and Biomolecular Engineering, The Ohio State University, Columbus, OH

L. S. Fan, Chemical and Biomolecular Engineering, The Ohio State University, Columbus, OH

Robert M. Statnick, Clear Skies Consulting, Cornelius, NC

The Calcium Looping Process (CLP) is being developed to reduce the cost and increase the efficiency of producing hydrogen and electricity from coal-derived synthesis gas. It integrates the water-gas shift (WGS) reaction with in-situ carbon dioxide, sulfur, and halide removal at high temperatures in a single reactor, while eliminating the need for a WGS catalyst. In the CLP, a regenerable limestone-derived sorbent (i.e., CaO or $Ca(OH)_2$) is used to chemically absorb CO_2, sulfur compounds, and halides from the synthesis gas. The removal of CO_2 drives the water-gas shift reaction forward via Le Chatelier's principle, obviating the need for WGS catalyst and enabling the production of high-purity hydrogen. This hydrogen can be sold commercially or combusted in a hydrogen turbine to produce electricity (i.e., in an integrated gasification combined cycle plant). The spent sorbent is heated in a calciner to regenerate CaO for reuse in the process and to release a concentrated CO_2 stream that can be dried and compressed for sequestration. The regenerated CaO is hydrated before being reintroduced into the absorber to prevent deterioration of its performance. Laboratory experiments have shown that the CLP can produce >99% purity hydrogen with less than 1 ppm sulfur impurity. The effectiveness of the hydration step in maintaining the activity of the calcium sorbent also has been experimentally demonstrated.

This paper presents a preliminary techno-economic evaluation of the Calcium Looping Process as applied to a commercial-scale coal-to-hydrogen plant with co-production of electricity. The CLP-based plant is compared with a conventional coal-to-hydrogen plant that includes a water-gas shift reactor, a two-stage Selexol unit for H_2S and CO_2 removal, and a pressure swing adsorber for hydrogen purification. One significant advantage of the CLP over the conventional process is that CO_2 absorption occurs at high temperature (e.g., 600-700 °C), allowing the exothermic heat of reaction to be recovered for use in generating steam. Excess heat from the calciner also contributes to steam production in the CLP, as does the exothermic heat of the lime hydration reaction. As a result, the CLP plant produces significantly more net electricity than the conventional plant for a given hydrogen production rate. CO_2 emissions from the CLP plant are essentially zero, because the tail gas from hydrogen production is combusted in oxygen in the calciner to produce water and sequestrable CO_2. Challenges associated with the development and scale-up of the CLP will be discussed, and the effects of sulfur chemistry on the performance of the process will be explored. Estimated capital costs and fixed and variable operating and maintenance costs for the CLP will also be presented. This preliminary analysis shows that the CLP has the potential to improve the performance and cost of producing hydrogen from coal.

See more of this Session: Carbon Capture/Separation From Gasification-Based Systems
See more of this Group/Topical: Topical 2: Advanced Fossil Energy Utilization

Development of Freeze Granulated Oxygen Carriers for Chemical Looping Combustion of Solid Fuels

Yi Zhang, Aurora Rubel, Kunlei Liu, and James K. Neathery

University of Kentucky, Center for Applied Energy Research, Lexington, KY 40511-8410

yizhang@caer.uky.edu

Chemical looping combustion (CLC) is proving to be a promising technology for the combustion of carboneous fuels producing electricity and a sequestration-ready concentrated CO_2 stream. Recently, more attention is being given to its application to solid fuels. The University of Kentucky, Center for Applied Energy Research (CAER) is presently developing an integrated process to apply CLC to coal combustion. A first step in this work is the development of suitable oxygen carriers (OCs) specifically for the application of CLC to solid fuels.

The materials investigated in this study included Fe_2O_3, NiO and CuO on a support matrix, α-aluminum oxide. The unsupported forms of the metal oxides were examined as well for comparison. All tested OCs were produced by freeze granulation (FG) with subsequent freeze drying and high temperature calcination. After calcination at 1400°C for 6 hours, the particle size distributions were evaluated. Reaction rates were examined by a thermal gravimetric analyzer- mass spectrometer (TG-MS). A mixture of $H_2/CO/CO_2/Ar$ was used as the reducing gas and O_2/Ar was the oxidizing gas mixture. Oxidation/reduction cycles of the OCs were carried out in wet gases to simulate actual conditions in a CLC reactor while experiments from dry gases were also conducted for comparison. Reaction rates, the extent of oxidation/reduction, recycle stability, and the effect of water vapor will be discussed. The data show that the supported OCs on α-alumina all have fewer tendencies to agglomerate after multiple oxidation-reduction cycles than unsupported OCs.

Introduction

Chemical looping combustion (CLC) is a promising technology [1-4] being investigated at the University of Kentucky Center for Applied Energy Research (CAER) for the combustion of solid fuels. CLC's potential is two-fold in that it can provide a high-purity, sequestration-ready stream of CO_2 and theoretically provide a substantial improvement in thermal conversion efficiency for fossil-fuel utilization with CO_2 capture. CLC uses a solid oxygen carrier (OC), usually a metal oxide to transfer the oxygen from the air (Air Reactor) to the fuel in a reducing reactor (Fuel Reactor) without the direct contact between the fuel and air. The oxygen-depleted OC is recycled by using steam or flue gas in the loopseal to the Air Reactor where the carrier is oxidized to its original state. In CLC, air does not directly meet the fuel, and so the product stream from the Fuel Reactor is not diluted with the nitrogen in air. Then nearly pure CO_2 is obtained following condensation of H_2O. The resulting high concentration CO_2 stream is ready for subsequent sequestration without any additional purification [5]. This technology was first proposed primarily for the combustion of gaseous fuels for natural gas turbines and only recently been considered for solid fuels such as coal without gasification of the coal first to syngas [6-9].

Lately, there has been a considerable interest in the application of CLC directly to solid fuels [8, 10-17]. The main challenge with the ability to combust solids in CLC is to have an appropriate OC with fast reactivity, durability, easily separated from the fuel ash, and resistant to any detrimental effect related to exposure to ash. Several OCs have been studied for use with solid fuels [8, 12-14, 17-20]. Previous work at CAER indicated that iron oxide-based powders and catalysts were the best candidates for use as OCs for CLC of solid fuels from the OCs tested. These solids did not agglomerate; are relatively inexpensive; retained reasonable activity; and remained durable through multiple oxidation – reduction cycles. The data suggested that accessibility of reactant gases to the active sites of the OC through adequate porosity was important. It was found that increased H_2, CO concentrations resulted in increased reduction rate for the iron oxides which may prove useful in a CLC process for solid materials as it provides a method to control the reduction rate during processing [8]. Fe_2O_3 powder was used to directly combust a beneficiated gasification char which contained 60% carbon. A mixture of Fe_2O_3: char (90:10) heated at 950°C in Argon resulted in 88% removal of the carbon with the OC as the only source of oxygen. This was encouraging for the application of CLC to the combustion of solid fuels [9]. Iron oxides were also found unaffected by ash concentration up to 75% and in fact some components of the ash provided additional OC activity [21].

This work was to begin development of an appropriate OC for the CLC of a devolatilized coal char in a bench scale fluidized bed (FB) CLC Redox reactor being constructed at the CAER. Freeze granulation was chosen to develop an OC composed of a metal oxide on a solid support of alumina which had suitable reactivity, particle size, porosity, redox recycle ability, and durability for use in the FB reactor. Three metal oxides were chosen for this study, Fe_2O_3, NiO, and CuO. Fe_2O_3 was our main interest with NiO and CuO used for comparison. To help determine the best metal oxide to alumina ratios, the extent and rates of oxidation/reduction, the redox recycle ability, and the effect of moisture were studied by TG -MS.

Experimental

Commercially available metal oxides (Fe_2O_3, NiO, and CuO) and α-alumina (Al_2O_3) powders were used. A slurry, referred to as a slip in this paper, was prepared by ball milling a mixture of solids, additives (dispersant and binder) with deionized water. The slip was pumped via a peristaltic pump to the spray nozzle of the freeze granulator (Figure 1). Pressurized air was also sent to the spray nozzle to combine with the slip creating a uniform spray mist which is sent into a magnetically stirred container of liquid N_2. The spray is regulated by the air pressure and the feed pumping rate of the slip. This results in the production of frozen spherical particles. The size of the spheres (OCs) can be adjusted by changing the air pressure, the feed rate, and composition of the slip. The metal oxide to alumina loadings used to produce freeze granulated (FG) OCs were 100:0, 75:25, 50:50, 25:75 of wt/wt % for each of the three metal oxides studied. Freeze granulation conditions, pump speed and air pressure, were optimized to maximize the particle yield in size range desired. The slips were prepared by ball milling for at least 24 hr. Air pressure was adjusted in the range of 40~60 kPa. The slip flow rate was 23.4~40 ml/min. FG particles were kept in the frozen state until freeze dried. After freeze drying, the FG OCs were calcined at 1400°C for 6 hours to achieve high durability. For CuO samples, the calcination temperature was set at 800 °C. The same freeze drying and calcination procedures were used for all samples. Samples were sieved to 48x150 mesh particle size samples which were tested for extent and rate of oxidation/reduction, redox recycle capability, and agglomeration resistance by thermal analysis-mass spectrometry (TG-MS) under both dry and wet gases.

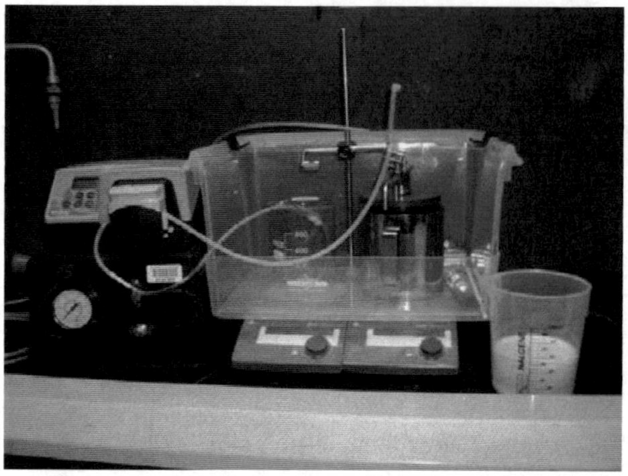

Figure 1. Freeze granulator equipment.

The TG-MS was a Netzsch STA 449 F3 Jupiter equipped with a water vapor furnace and coupled to a Aeolos QMS403 (Figure 2). It is capable of temperature to 1200°C under both steam and dry gases. The MS measured gases leaving the TG to determine the gas constituent in the evolved stream from TG. Conditions for testing are reducing gases of 10% H_2, 15% CO, 20% CO_2 with and without steam and a balance of Ar; oxidizing gases of 20% O_2 in a balance of Ar; and a reaction temperature of 950°C. Oxidations/reductions were carried out to near completion and at least 3-5 redox cycles were done per OC. Gas flow rates and switching was controlled by a set of mass flow controller operated by Lab View software on a dedicated computer. Times for oxidation/reduction and flows could be programmed as needed and the entire operation could be run 24/7 without intervention.

Figure 2. TGMS system used for study.

Results

Figures 3-5 give the weight gain/loss on oxidation/reduction for the different metal oxides loading on alumina for Fe_2O_3, NiO, and CuO respectively. FG Fe_2O_3 showed a slight decrease in the amount of oxygen released (active-O) from the OC after dilution correction as the active metal oxide content decreased. Under wet gases, there was a significant decrease in the weight gain/loss for unsupported (100%) FG Fe_2O_3 OC which did not occur for supported OCs. As contrary, the same amount of oxygen was released during oxidation/reduction of the Fe_2O_3/alumina OCs as in dry gases (Figure 3). The unsupported (100%) FG NiO OC showed a slight decrease in active-O release during the five cycles. However, in previous studies [8, 9] reagent-grade NiO powder agglomerated in one cycle. The unsupported FG NiO OC performance was far better than the pure powder. There was little difference in the weight gain/loss on oxidation/reduction with metal loadings of NiO when run in dry gases. Under wet conditions, the 75% NiO on alumina corrected for dilution gained and lost more active-O than lower loadings (Figure 4). Unsupported FG CuO OC (Figure 5) when run in dry gases agglomerated while the support CuO OC remained active through multiple cycles at 950°C which is substantially higher than the 800°C used for pure CuO powder in a previous work which became inactive quickly due to agglomeration. . There was a decrease in the weight gain/loss on redox for all CuO OCs in wet gases compared to redox in dry gases but there was no evidence of agglomeration or loss of activity with cycles.

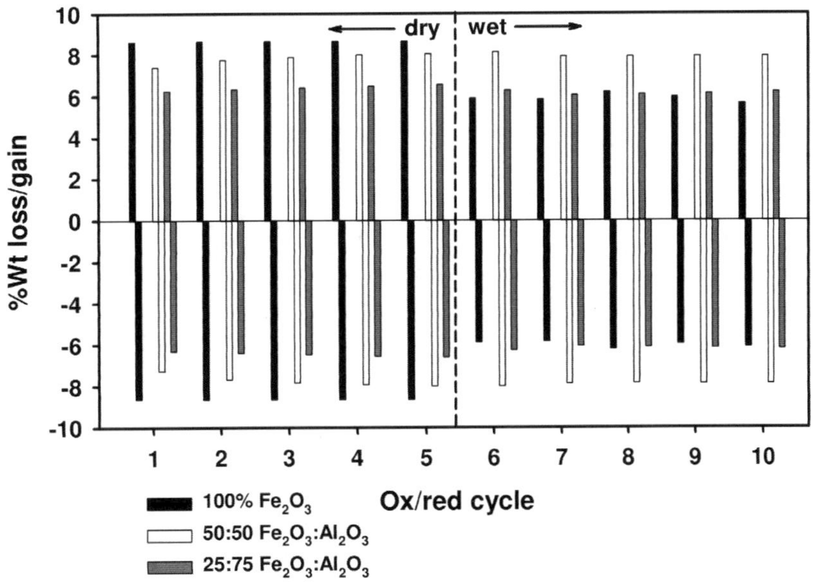

Figure 3. Weight gain/loss on oxidation/reduction of FG Fe_2O_3.

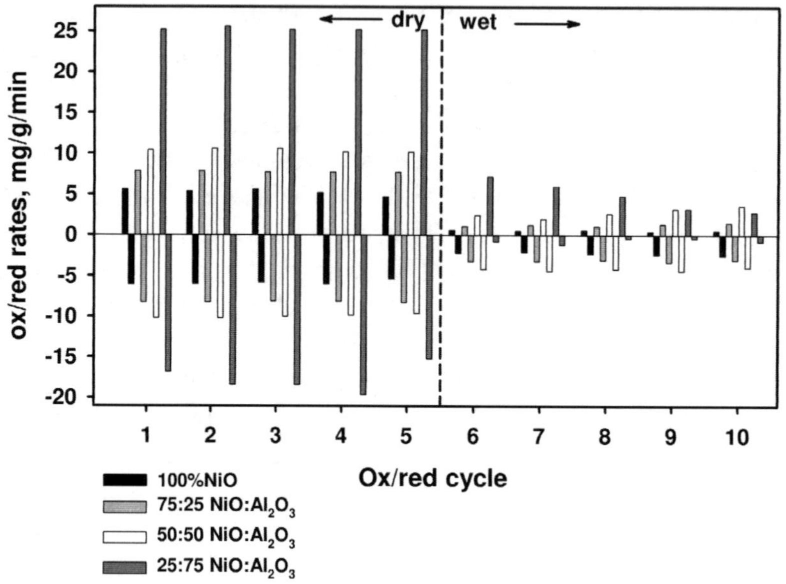

Figure 4. Weight gain/loss on oxidation/reduction of FG NiO.

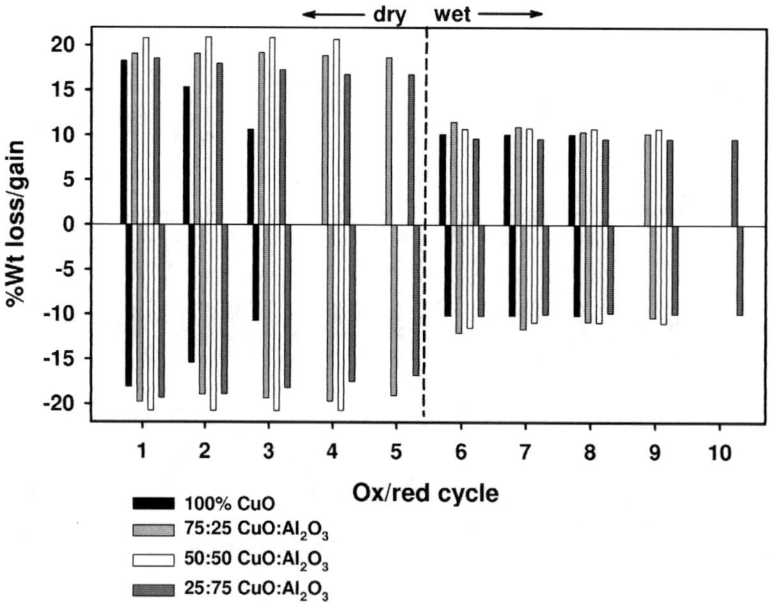

Figure 5. Weight gain/loss on oxidation/reduction of FG CuO.

The loading of active metal oxides on alumina by FG resulted in a significant effect on the rates of oxidation/reduction. For FG Fe_2O_3 cycled between oxidation/reduction under dry gases, there was an increase in the rates as the active metal loading decreased (Figure 6). Under wet conditions rates decreased from comparable dry condition but again there appeared to be a trend toward faster rates with low metal loading especially during reduction though there was considerable variability. One interesting effect of the FG process was that the rates of reduction were greater than oxidation which was different from results with pure metal oxide powders where oxidation rates far exceeded reduction rates [8, 9]. Figure 7 gives the results for the rates of oxidation/reduction for NiO. The trends under dry gases for NiO were the same as for Fe_2O_3 except the increase with lower metal loading was much greater. The increases for supported FG Fe_2O_3 and NiO were 2 and 5 times their unsupported FG OCs respectively. Under wet gas conditions, rates decreased greatly over dry gas conditions, 5 times slower rates for all loadings of NiO. CuO rates (Figure 8) had the same trend as for the other two metal oxides with moist gases even more detrimental with oxidation rates near zero.

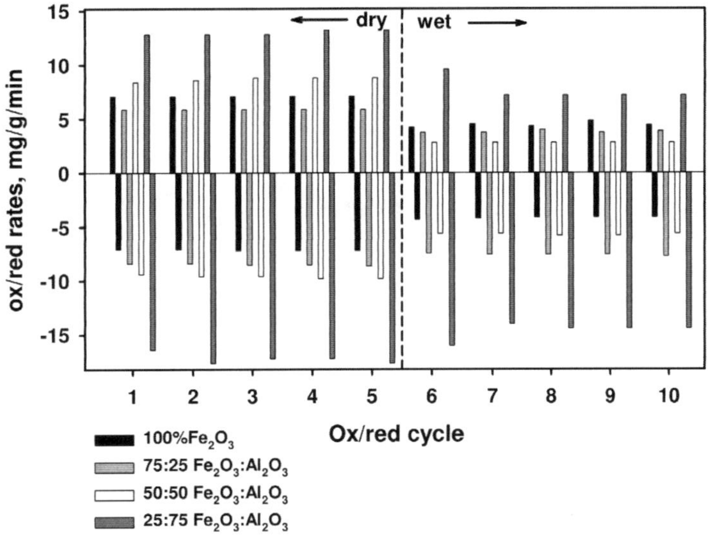

Figure 6. Rate of oxidation/reduction for FG Fe_2O_3 under dry and wet gases.

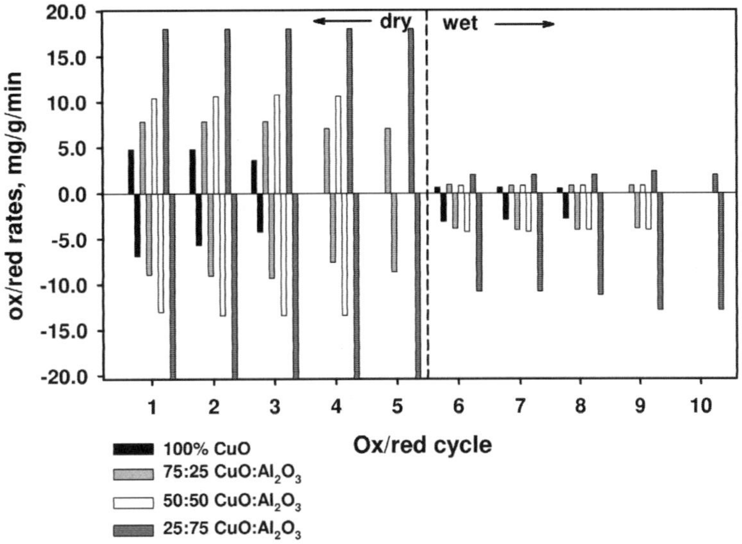

Figure 7. Rates of oxidation/reduction for FG NiO under dry and wet gases.

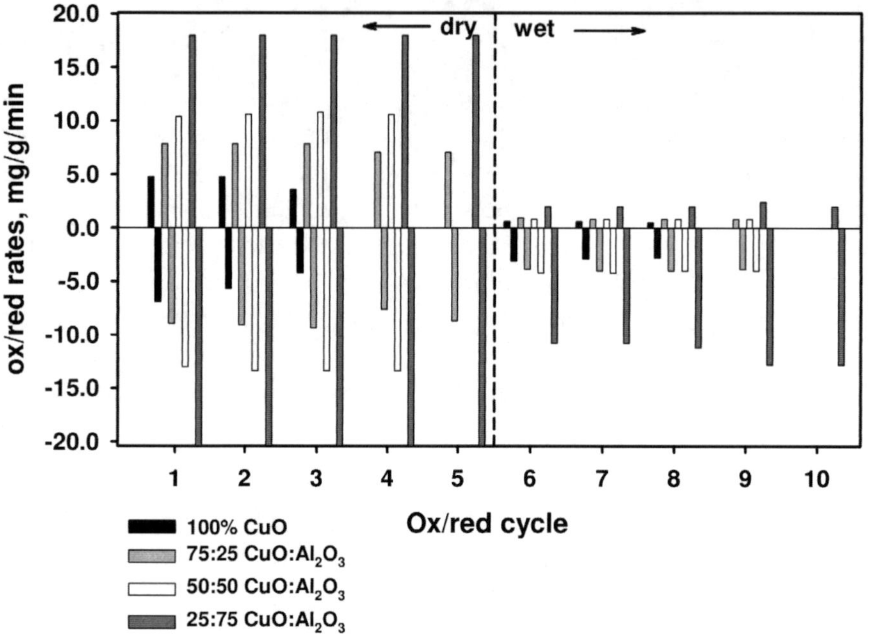

Figure 8. Rates of oxidation/reduction for FG CuO under dry and wet gases.

Discussion

The results suggest the complexity of factors affecting the FG OCs produced in this study. As others have observed loading active metal oxides on a support is beneficial in preventing agglomeration. Our previous work with pure NiO powders [8, 9] agglomerated on the first redox cycle and were no long active after the first cycle. The loading of CuO on a support in this study allowed for the use of CuO at a temperature significantly above that which the pure CuO powders could be used. In this study, CuO/alumina OCs remained active at 950°C where the unsupported (100%) FG CuO OC lost activity due to agglomeration or melting. The spent unsupported FG CuO was a pellet melt of what appeared to be pure Cu.

For all metal oxides in dry gases, the oxidation/reduction rates increased with lowering the active metal oxide concentration in the supported OCs. The definitive explanation for this will undoubtedly involve some combination of pore size, access of gases to reactive sites, spatial proximity of active sites and particle size. All issues which are currently being investigated at the CAER. Each metal oxide will form a spatial arrangement on the alumina support which will depend on its granular size and loading. The reaction rates for the FG metals was significantly lower than for the pure metal oxides again due to limitation of accessibility of redox gases to active oxygen sites which are most likely impacted by particle size, geometry, and accessibility . However, this is probably what protects Ni and Cu from agglomerating as the slower rates reduces temperature extremes and individual metal molecules may not be in close proximity to agglomerate. The slow reaction rates for all the metal oxides in the presence of moisture especially for NiO and CuO is difficult to explain other than the possible blocking of active sites. Water has been shown to react with alumina [22] which could lead to alteration of the geometry and reactivity of the FG OC. However, the better reduction rates over oxidation rates in moist gases could be due to an increase in H_2 via the water gas shift reaction since previous studies on pure metal oxides found that increasing the H_2 content of the reducing gases increased the reduction rate [8, 9].

Conclusions

Though there were benefits (no agglomeration and melting) to the FG of NiO and CuO, the slow rates of oxidation/reduction do not make them viable OC for the CLC of solid fuels. As with our previous studies Fe_2O_3 would be the OC of choice. FG Fe_2O_3 had similar weight gains and losses on oxidation/reduction, were not affected by wet gases when supported on alumina, and maintained better reaction rates in moist gases compare to NiO and CuO. The advantage of FG is that the OC particle size and metal loading can be adjusted to suit the necessary parameters required for separation and reactivity in a CLC reactor.

Acknowledgement

Funding for this work was provided by the Carbon Management Research Group.

References

[1] Croiset, E Thambimuthu, K. Coal Combustion with Flue Gas Recirculation for CO_2 Recovery. In: Riemer, P, Eliasson, BWokauun, A, Editors. Greenhouse Gas Control Technologies, New York: Elsevier; 1999, p. 581-6.
[2] Marion, J, Naskala, Y N, Griffin, T. Controlling Power Plant CO_2 Emissions: a Long Range Approach. In: The First National Conference on Carbon Sequestration, Washington, DC,2001; DOE/NETL-2001/1114
[3] Herzog, H, Eliasson, B, Kaarstad, O. Capturing Greenhouse Gases. Scientific American 2000; 282(2):72-9.
[4] Hatanaka, T, Matsuda, S, Hatano, H. A New-concept Gas–solid Combustion System 'MERIT' for High Combustion Efficiency and Low Emissions. In: Intersociety Energy Conversion Engineering Conference 1997:944-8.
[5] Freund, P. Abatement and Mitigation of Carbon Dioxide Emissions from Power Generation. Powergen 98 Conference. Milan,1998; Powergen 98 Conference
[6] Richter, H Knoche, K. Reversibility of Combustion Processes. In: ACS,1983:71-86.
[7] Anthony, E J. Solid looping cycles: A new technology for coal conversion. Industrial and Engineering Chemistry Research 2008; 47(6):1747-1754.
[8] Rubel, A, Liu, K, Neathery, J, Taulbee, D. Evaluation of oxygen carriers for chemical looping combustion of coal and solid fuels. In: 24th Pittsburgh Coal Conference, Johannesburg, SA,2007:paper 21-1.
[9] Rubel, A, Liu, K, Neathery, J, Taulbee, D. Oxygen carriers for chemical looping combustion of solid fuels. Fuel 2009; 88(5):876-884.
[10] Cao, Y, Casenas, B, Pan, W P. Investigation of chemical looping combustion by solid fuels. 2. Redox reaction kinetics and product characterization with coal, biomass, and solid waste as solid fuels and CuO as an oxygen carrier. Energy and Fuels 2006; 20(5):1845-54.
[11] Cao, Y Pan, W P. Investigation of chemical looping combustion by solid fuels. 1. Process analysis. Energy and Fuels 2006; 20(5):1836-44.

[12] Chuang, S Y, Dennis, J S, Hayhurst, A N, Scott, S A. Development and performance of Cu-based oxygen carriers for chemical-looping combustion. Combustion and Flame 2008; 154(1-2):109-121.
[13] Leion, H, Jerndal, E, Steenari, B-M, Hermansson, S, Israelsson, M, Jansson, E, Johnsson, M, Thunberg, R, Vadenbo, A, Mattisson, T, Lyngfelt, A. Solid fuels in chemical-looping combustion using oxide scale and unprocessed iron ore as oxygen carriers. Fuel 2009; 88(10):1945-1954.
[14] Leion, H, Lyngfelt, A, Mattisson, T. Solid fuels in chemical-looping combustion using a NiO-based oxygen carrier. Chemical Engineering Research and Design 2009; 87(11):1543-1550.
[15] Leion, H, Mattisson, T, Lyngfelt, A. CO_2 capture from direct combustion of solid fuels with Chemical-Looping Combustion. In: Proceedings of the 33rd International Technical Conference on Coal Utilization and Fuel Systems, Clearwater, FL, USA,2008
[16] Leion, H, Mattisson, T, Lyngfelt, A. Combustion of a German lignite using chemical-looping with oxygen uncoupling (CLOU). In: The Proceedings of 33rd International Technical Conference on Coal Utilization & Fuel Systems, Clearwater, FL, USA,2008
[17] Siriwardane, R, Tian, H, Richards, G, Simonyi, T, Poston, J. Chemical-Looping Combustion of Coal with Metal Oxide Oxygen Carriers. Energy & Fuels 2009; 23(8):3885-3892.
[18] Mattisson, T, Leion, H, Lyngfelt, A. Chemical-looping with oxygen uncoupling using CuO/ZrO2 with petroleum coke. Fuel 2009; 88(4):683-690.
[19] Zhao, H, Liming, L, Xu, D, Chuguang, Z. Study on Ni-based Oxygen Carriers for Chemical Looping Combustion by Coal Char. In: 24th International Pittsburgh Coal Conference,2007:1-18.
[20] Zhao, H B, Liu, L M, Wang, B W, Xu, D, Jiang, L L, Zheng, C G. Sol-gel derived $NiO/NiAl_2O_4$ oxygen carriers for chemical-looping combustion by coal char. Energy and Fuels 2008; 22(2):898-905.
[21] Rubel, A, Liu, K, Neathery, J. The Effect of Ash on Oxygen Carriers for Chemical Looping Combustion of Solid Fuels. In: AIChE Extended Abstracts, Nashville, TN,2009
[22] Wittbrodt, J M, Hase, W L, Schlegel, H B. Ab Initio Study of the Interaction of Water with Cluster Models of the Aluminum Terminated (0001) a-Aluminum Oxide Surface. Journal of Physical Chemistry B 1998; 102(34):6539-6548.

(60e) Pulverized Coal Pyrolysis & Gasification in $N_2/O_2/CO_2$ Atmospheres by Thermo-Gravimetric Analysis

Muhammad Faisal Irfan, Living and Environmental Sciences, Fukuoka Women's University, Fukuoka, Japan

The combustion characteristics of Datong coal in the different $N_2/O_2/CO_2$ environments have been extensively studied by using thermo-gravimetric technique. The effects of combustion environment, oxygen and CO_2 concentrations, pyrolysis conditions, particle size and different heating rates were considered. Results showed that, at each temperature zone, the mass loss rate of coal pyrolysis in CO_2 environment is higher than that in N_2 environment. This may be due to the division of CO_2 pyrolysis process into three stages: moisture release, devolatilization and char gasification by CO_2 at higher temperature zone or due to the formation of perfectly hexagonal structure of coal. However, the burning process of coal in O_2/CO_2 environment is almost same compared with that in O_2/N_2 environment for equivalent oxygen concentrations but slightly lesser than O_2/N_2 environment at the end of the process. With the increase of oxygen concentration, the burning rate of coal increases and burnout time is shortened. Among the different pyrolysis temperatures, 500 oC was found to be the best temperature as the ignition point in the O_2/CO_2 atmosphere lies near or just after this temperature. Moreover, the surface area produced by this char at this temperature was much higher in CO_2 atmosphere (400 m2/g) than N_2 (52 m2/g). This simultaneous process of pyrolysis and gasification seems to be advantageous because of production of the nascent char in one process and avoiding of the excess use of gasifying medium from room temperature to 500 oC. As the heating rate increases, coal particles are faster heated in a short period of time and burnt in a higher temperature region, but the increase in heating rate has almost no obvious effect on the combustion mechanism of coal. At different heating rates the values of the Arrhenius activation energy (E) and pre-exponentional factor (A) were estimated from the TGA data by applying different methods such as Freeman-Carroll, Doyle and Coats-Redfern. A trend of decrease of activation energy values were observed with the increase of different heating rates.

See more of this Session: Carbon Capture/Separation From Gasification-Based Systems
See more of this Group/Topical: Topical 2: Advanced Fossil Energy Utilization

FUTURE FEED-STOCKS FOR PETROLEUM AND PETROCHEMICAL INDUSTRIES

Dr D P Lal

1. **Indroduction**

. The phenomenal rise in crude oil price, observed sometime back, has increased the urgency to explore the alternate feed-stocks which the petroleum and petrochemical industries have to adopt in future. Though there is a temporary drop in price, but once the world economic conditions become normal, such up-ward price trend, cannot be ruled out. This is due to the fact that the world crude oil reserve explored so far (approx. 1230 billion barrels) is expected to last only 70-80 years, current consumption level being approx. 30 million barrels per day. Even if the reserve under the arctic (which is estimated at around 90 billion barrels) is explored, this additional reserve will meet world demand for another 9 years only. Further, it is estimated that there will be 35% increase in worldwide petroleum demand by the year 2015. This rising crude oil demand against falling reserve would eventually push up the crude oil price to uneconomic level. So the world is faced with the daunting question;- how this rising demand is going to be met economically?

2. **Need to Search for Alternate Feed-stocks**

For reasons mentioned above, it has now become imperative to find out alternate feed-stocks for petroleum and petrochemical industries on a priority basis. It is predicted that after 20-30 years, the price of many of the alternate feed-stocks now being explored and described later in the paper, would become competitive compared to price of crude oil based feed-stocks. Hence, most of the developed countries in the world, as well as many developing countries are now engaged in this phenomenal task of finding out appropriate alternate feed-stocks.

The other compelling factor which has forced the world community to look for alternate feed-stocks and fuels is the harmful contribution of the fossil fuels to global warming. While the present international search is mainly aimed at finding alternate resources to meet the energy requirement of the world without contributing to or preferably reducing the global warming; but the search will also lead to finding out liquid products containing linear and aromatic hydrocarbons from sources other than crude oil which could be recovered to produce feed-stocks for refineries and petrochemical plants.

In this paper, however, we will concentrate on those areas of research and development which are going to lead to finding out alternate feed-stocks for refineries and petrochemical plants.

3. New Feed-stocks in the Initial Phase

It need not be emphasized that techno-economic criteria will ultimately determine how different feed-stocks are going to be adopted in the concerned industries as the time passes. Hence, one of the deciding criteria for selecting the new feed-stocks will be how best and how long the existing investments made in this field (namely the existing petroleum refineries) can be utilized with modifications wherever necessary. From this consideration; it becomes obvious that residues, heavy crude oil and high sulphur crude oil which are presently relatively uneconomic to process, will be the first group of feed-stocks to be considered. The technologies required for processing these feed-stocks (namely, hydro-cracking process, coker technology, sulphur removal process etc.) are well-proven and are already in operation in certain refineries and can be easily adopted in other refineries. But profitable and environmentally safe outlets of the by-products produced in bulk scale from these feed-stocks (i.e petroleum coke and sulphur) should be found out before these feed-stocks are adopted in any refinery.

The transportation of these feed-stocks to any refinery would not be a problem, because, the existing means of transportation (i.e LCC's and VLCC's) can be used for this purpose.

The next group of feed-stocks which could be considered for processing in the existing refineries are the oil stocks recovered from tar-sands and tar-shale. New technologies required for this oil recovery are in advanced stage of development mainly due to extensive R&D work done under the sponsorship of DOE, USA. While, number of pilot plants are in operation within USA for oil recovery from tar-sand; ENI is already planning to invest approx. US $ 3 billions to recover oil from tar-sand in Congo. From the scale of investment, it seems that this is going to be a commercial plant; which is an indication that the technology has sufficiently developed, so that a commercial scale plant could be built based on the latest advances made. However, since transportation of this raw material (i.e tar sand) over long distance would be uneconomic, hence these plants should be located near the source of tar-sand. But the oil recovered could be transported over long distance to far off refineries using conventional modes of transportation.

The liquid products produced from the refineries by processing the above-mentioned feed-stocks would be substantially similar to those produced from crude oil, available at present. And hence the present types of products (i.e, naphtha, BTX, kerosene etc) would continue to be available for the downstream petrochemical plants. Thus, no major change is anticipated in the technologies presently being used in the petrochemical plants for treating the feed-stocks supplied from refineries processing residue, heavy crude oil etc..

4. New Feed-stocks in the Second Phase

Many of the countries in the world have got huge reserves of coal and natural gas. With price of fossil fuels going up, these countries will endeavor to produce liquid hydrocarbons based on these raw materials. In fact, the world coal reserve is estimated at around 100,000 quadrillion btu; and is expected to last for approx. 300 years. Thus countries like India and Australia, which are rich in coal reserve, could try to make liquid fuels from coal, based on some of the established process. South Africa, another coal-rich country is already producing various liquid hydrocarbons from coal.

There are two process steps to convert coal to liquid hydrocarbons. At first, gasification of coal is done with a mixture of steam and air to produce synthesis gas (a mixture of hydrogen and carbon-monoxide) by using Kopper-Totzek,, Shell or BGL coal gasification process. The synthesis gas so produced is converted to liquid hydrocarbons through Fischer Tropche process. In addition, in the steel industries, coke required in blast furnace is produced through pyrolysis of raw coal, which releases number of aromatic hydrocarbons (benzene, toluene, xylene, napthene etc.) which could be recovered in pure form and used as feed-stocks of chemical and petrochemical industries.

The Sasol coal-based complex in South Africa is the fore-runner in this field. Besides fertilizers and fuels, Sasol is producing numerous liquid hydrocarbons from coal, which are used as feed-stocks in various chemical and petrochemical industries. Syntroleum, a USA company is producing jet-fuel and diesel from synthesis gas. There is some initiative to set up coal based liquid hydrocarbon production plants in India in collaboration with Sasol.

The interesting feature of the Fischer-Tropsch process is that the product profile can be adjusted by using appropriate type of catalyst. In fact, a patent has been taken recently where the catalyst can yield olefinic naphtha and hydrogenated distillate suitable for use as naphtha cracker feed-stock.

Like coal, natural gas is likely to be an important feed-stock for production of liquid hydrocarbons. The world natural gas reserve is estimated at around 6100 trillion cubic feet and is expected to last for around 250 years. Approx. 40% of the reserve lies in the Middle East and 25% in Russia. Rest of the reserve is spread over mostly in Asia and North & South Americas. Conversion of natural gas to synthesis gas is done through steam reforming process widely used in the fertilizer plants all over the world. Thereafter, conversion of synthesis gas to various hydrocarbons could be done using the Fischer-Tropsch process or its modified versions. Shell has developed successfully a very clean and efficient process to produce diesel from natural gas.

At this point, it is worth mentioning that compositions of liquid products produced from synthesis gas as discussed above would be substantially different from those of fossil fuels. Hence, the scope for utilizing the existing refineries to recover selected cuts from the liquid products produced in Fischer-Tropsch process based plants, would be rather limited. It would be a chemical engineering challenge at that point of time as how best to utilize the investments made in the existing refineries to achieve this objective. However, feed-socks for the petrochemical plants as used presently would be available from the Fischer-tropsch process based plants and hence the existing petrochemical plants would remain suitable to treat new feed-stocks. This is obvious from the experience of Sasol.

5. Renewable Feed-stocks like Bio-fuels

The general consensus of the experts is that bio-fuels are the answers to meet the energy need of the future generations, because of the duel benefits offered by this type of fuels. These benefits are;- 1) the feed-stocks for bio-fuels are renewable in nature and would be available indefinitely unlike fossil fuels, coal and natural gas which have limited reserves; and 2) there is no net release of carbon-di-oxide when bio-fuels are used, because during growth of the plants supplying raw materials for production of the bio-fuels, nearly as much carbon-di-oxide is consumed by the plants, as is released during combustion of the corresponding quantity of bio-fuels. Hence, it is an elegant solution to the problem of global warming by limiting release of carbon-di-oxide to the atmosphere.

Extensive research and development work has been going on in different countries all over the world for finding out the most appropriate process for production of bio-fuels which will not only meet the fuel requirement and arrest the global warming, but also will not have any adverse impact on food crops production. The research work and field studies done with these objectives have already led to development of three generations of bio-fuels, which are reviewed in the following paragraphs.

In fact most of the developed and developing countries have made it obligatory to blend fossil fuels with bio-fuels (Renewable Transport Fuel Obligation – RTFO). Consequently, world bio-fuel demand has approached nearly 100 million tonnes per year. Hence, well known oil companies like Shell, Chevron, BP, ENI, Petrobas, Virgin Atlantic etc. are also deeply involved in developing bio-fuels from various feed-stocks.

5.1 First Generation of Bio-fuels

The first generation of bio-fuel started with production of alcohols from sugarcane and corn. In this development, Brazil took the lead with extensive sugarcane cultivation followed by USA producing alcohol from corn. Brazil now generates nearly 47% of its electricity requirement from renewable resources;-mainly alcohol and hydro-electricity.

Dry land crop sorghum is also an attractive crop for alcohol production, because the juice extracted from it contains 15-20% sugar compared to 10-15% in that extracted from sugarcane; and comparatively much less water is required for cultivation of sorghum.

But there are some major adverse impacts of producing bio-fuel like ethanol from sugarcane and corn. Because, diversion of land area from food crop cultivation to corn and sugarcane cultivation is leading to scarcity of land area available for growing food crops and hence rise in food crop price. According to a study by IMF, approximately 10% rise in food crop price is attributable to this factor. In addition, heavy use of fertilizers and pesticides for cultivation of these crops pollutes ground and river water. The extensive deforestation which is taking place particularly in Amazon rain forest area and Bolivia is attributed to this diversion of land area for sugarcane and corn cultivation. This is also leading to global warming because of reduction in carbon-di-oxide sinks. Because of these adverse impacts; USA, Japan and the European countries are discouraging any further blending of alcohol with petrol. However, this is not applicable to all countries. In any country, if waste dry land is available, sorghum could be grown there to produce alcohol without any adverse impact on food production.

The other type of bio-fuel, namely bio-diesel is being developed and produced in semi-commercial scale in different countries from wide variety of feed-stocks, i.e, palm oil, coconut oil, castor oil, soybean oil, jatropha seed oil etc. It may be of interest to know that Rudolf Diesel, the inventor of diesel engine first tried out his engine with peanut oil and he expected that vegetable oils would be used for driving diesel engines. This, however, did not materialize because of wide availability of diesel from petroleum refineries. But the financial viability of production of bio-diesel from feed-stocks like palm oil, coconut oil, castor oil etc. depends to a great extent on the price of the feed-stocks, which is quite volatile. Depending upon international availability, the price of any of these feed-stocks could be quite low or un-economically high. Hence for long term economic viability of these diesel production units; these should be so designed that these are capable of operation with any of these feed-stocks without sacrificing operating efficiency.

Asiatic Centre of Genome Technology in Malaysia has completed sequencing study of palm genome and jatropha genome to develop high oil yield varieties of these plants. Malaysia and Indonesia are targeting to replace 5% of their fuel consumption by bio-diesel within 2-3 years. India is targeting to replace 10% of transport fuel by bio-fuel by 2017.

A study funded by DOE, USA states that enough bio-diesel could be produced to meet total demand of diesel in USA (approx. 140 billion gallons) from oil growing plants planted over a land area of 9.5 million acres which is only a small fraction of total crop land area (approx. 450 million acres) now being used in USA. There are more than 20 companies in USA producing bio-fuels (both alcohol and bio-diesel) to meet the obligatory blending requirement of petrol and diesel with respective bio-fuels.

A British Company, Sun Bio-fuels PLC has invested \$20 million in Tanzania to plant jatropha oil seeds over a land area of 9000 hectares to produce bio--diesel. Besides bio-diesel, they intend to produce glycerine (a by-product of the process for recovery and purification of bio-diesel from jatropha seed oil), varnishes, soaps, pesticides and medicines for skin disease. A Chennai (India) based Company, DI Mohan Biofuels Ltd is planning to invest \$ 17 million to process 24000 tonnes of jatropha seeds annually to produce 8000 tonnes of bio-diesel along with by-products like glycerne, varnish etc. Indian Railway is considering cultivation of jatropha over 90,000 acres of land area available to them, to meet their diesel requirement.

Countries like India which have large area of dry waste land are in a unique position to produce bio-diesel from jatropha seeds. Because, these drought tolerant plants can be grown on dry wasteland. And hence cultivation of jatropha would not compete with production of food crops. If whole of 60 million hectares of wasteland available in India is used for jatropha cultivation, then approx. 30 million tonnes of bio-diesel could be produced from the jatropha seeds, besides large volumes of other by-products like glycerine, soap, varnish etc. It goes without saying that this would give a big boost to rural industrialization and employment. Such large area covered with green plants would act as a large sink of carbon-di-oxide and contribute to arresting global warming.

It is worth-mentioning that Botanical Survey of India has identified 400 species of plants and trees which can yield oil. But jatropha has emerged as the preferred option. The other plants which have good potential are pangamia and ponamia pinnala. Pangamia is a leguminous tree and is helpful for nitrogen fixing. Its leaves are used as organic manure.

Relative oil yield of various types of plant products per acre of plantation is furnished in Table -1.

5.2 Second Generation of Bio-fuels

Research and development on second generation of bio-fuels started with the objective of producing this product from waste vegetable items like corn stover, cane bagasse, wheat straw, hay, soft wood, stems, leaves, wood chips, pulp, various types of grass and other similar cellulosic waste materials. Ligno-cellulosic compounds in these materials are broken down with enzymes to free the sugar molecules which are then separated and fermented to produce alcohol. Number of patents have been taken to produce ethanol from lingo-cellulosic feed-stocks mentioned above. USA, Canada, Denmark, China, Japan and Brazil have already set up semi-commercial scale cellulosic ethanol plants to produce bio-ethanol from the second generation feed-stocks

Direct conversion of lingo-cellulosic feed-stocks to gasoline like products is the latest development in this field. A research team in MIT has patented a process which involves rapid heating of this type of feed-stocks in a liquid medium in presence of certain catalyst followed by rapid cooling which yields a liquid mass containing many of the components of gasoline. It seems second generation of bio-fuels i.e cellulosic ethanol has got a distinct commercial possibility and is likely to compete with the first generation bio-fuels.

5.3 Third Generation of Bio- fuels

Third generation of bio-fuels have been developed using various types of algae as feed-stocks. There are number of advantages of using algae as raw material for production of bio-fuels. These are;-

1) Oil-yield from algae could be up to 15 times higher compared to other raw materials mentioned above per acre of land area required for growing these feed-stocks..

2) By developing appropriate culture of algae, either diesel or petrol can be produced from it, as well as the oil yield can be improved.

3) Algae can be grown either in fresh water, or in salt/brackish water, or in contaminated water/ effluent, in pond or in back-waters along sea coast. Thus algae can help treatment of effluent from a locality and there is minimum constrain regarding medium of its growth.

4) Growth of algae can be accelerated by passing carbon-di-oxide through the algae containing mass. Thus it could solve the problem of greenhouse gas emission from a power plant by absorbing carbon-di-oxide from the flue gas emitted from the power plant.

With all these significant advantages, no doubt algae has become one of the most preferred feed-stock for producing bio-fuels. Hence, extensive research and development work has been taken up in most of the developed countries and in some of the developing countries, to find out the most appropriate type of high oil yield variety of algae. Exxon, Mobil, Shell and Department of Defense (USA) are making serious efforts to be the commercial producer of bio-fuel from algae. DOE funded a project of quick growing algae for sequestering carbon-di-oxide emission from coal-fired power plant. Some of these algae are of high oil yield variety. USA is also collaborating with China for developing high oil yield variety of algae. As a result of persistent R&D work, 15 algae based bio-fuel plants are in the process of construction in USA. The largest among these is the one by Petrosun Ltd to be built in Rio Hando, Texas; to produce 4.4 million gallons of bio-fuel from algae. While most of the development so far has taken place with algae grown in ponds; but research work is in progress to grow it inside bio-reactors either exposed to sun-light or to artificial light. A Dutch Company has developed photo-bioreactors which can produce 6-times more algae even in cold climate. Arizona Public Service Company uses algae bioreactors to capture carbon-di-oxide from flue gas of their power plant. Oil recovered from algae thus grown is used as fuel in the same power plant.

The bio-reactors are getting more popular since these occupy much less space compared to large water body required for algae growth and relatively oil yield being much higher. Oil recovery process from algae is very simple. Oil is extracted by expeller followed by solvent extraction for recovery of residual oil from algae. The algae cake can be used as fuel.

The process of producing bio-fuel from algae has got very high potential. Because, there is immense scope of improving this process on various fronts. Through genetic study of algae strains, its oil yield could be increased manifold. Already some researchers have succeeded, in laboratory scale, to produce algae strains which yield nearly 10 times more oil compared to the present varieties. Through systematic bio-chemical engineering study, the design of the bio-reactors used for algae growth could be improved to make these;- high throughput, non-clogging, continuously operated ones. This would improve plant productivity and reduce plant cost; and at the same time occupy minimum land area.

In spite of numerous advantages of third generation of bio-fuels, first and second generation bio-fuels might be worth producing in some areas because of some local advantages. For example, countries producing sugar would find it advantageous to produce alcohol through fermentation of molasses, since this is a profitable way of disposing off this effluently. Similarly, countries having large areas covered by wasteland would find it beneficial to plant jatropha trees in the wasteland, thereby producing bio-diesel from jatropha seeds; and at the same time increasing green coverage thereby reducing global warming and improving climatic conditions. In the same way, where large volume of cellulosic wastes (hay, wheat straw, wood chips, soft wood etc.) are available, it might be a profitable way to dispose off these materials by producing bio-fuels from these waste products. Hence, research and development work is going on in different countries on all these three generations of process for bio-fuel production.

6. Scope of Using Bio-fuels as Feed-stocks for Petrochemical Industries

As is obvious from above description, so far only two types of bio-fuels have been produced in semi-commercial scale. These are;- bio-alcohol and bio-diesel. Only one study has been reported where it was possible to produce a mixture of hydro-carbons like fossil fuels in laboratory scale from feed-stocks used in developing second generation of bio-fuels. Techno-commercial feasibility of producing this product is yet to be proven.

However, from bio-alcohol, ethylene could be produced using the well-established alcohol de-hydration process. The process of conversion of ethylene to higher olefins like propylene, butene, butadiene etc. have been already developed. Thus, there could be a mother plant using bio-alcohol as the feed-stock and producing a range of olefins which could be used as feed-stock for various polymer plants and other olefin-based chemical plants.

Since production of benzene from ethylene is till now an expensive process, hence there could be a problem to meet economically the feed-stock requirement for the aromatics-based chemicals and polymers plants based on alcohol. However, benzene and higher aromatics should be available from coal through pyrolysis process (as is used for producing coking coal in steel plants) for another two centuries or so. By that time, hopefully it would be possible to develop aromatic products ranging from benzene to higher aromatics based on renewable natural resources as feed-stocks, as has been developed in laboratory scale now.

7. Conclusions

Based on the facts and figures presented above, it could be concluded that after reserves of fossil fuels are exhausted, a range of alternate feed-stocks would appear in the market to meet the feed-stock requirement of petroleum and petrochemical industries. Alternate feed-stocks which are likely to be used initially are heavy and high-sulphur crude oil, followed by oil recovered from tar-sand and tar-shale. Simultaneously coal-based feed-stocks would be developed particularly in the coal-rich countries, as has already happened in South Africa. Countries rich in natural gas reserve would then start marketing the liquid hydrocarbon products synthesized from natural gas following the process described earlier.

Bio-fuels like bio-alcohol and bio-diesel are already used in both developed and developing countries as partial replacement of fossil fuels. Though presently this replacement is mainly aimed at reducing greenhouse gas concentration in the atmosphere to arrest global warming and thereby improve climatic conditions; but the proportions of the bio-fuels are likely to increase further because of the price competitiveness of bio-fuels against fossil fuels and also because of the over-riding concern to control global warming. Three generations of bio-fuels are likely to be developed simultaneously in different countries depending upon local advantages.

However, success of bulk production of bio-fuels on a sustained basis in future, would depend to a great extent on how much we succeed in maintaining climatic conditions congenial for growth of plants and algae used for bio-fuel production. To create favorable climatic conditions, it is imperative that the global warming which has attained an alarming proportion, must be first contained and then reversed at least to the level prevailing 50 years back. This would help restoration of coverage of the glaciers to the original heights and thereby sustain adequate water flow-rate through the rivers which in turn would replenish the ground water reserve and ensure supply of adequate quantity of water for cultivation. Simultaneously, strategic afforestation work should be taken up to stop soil erosion, preferably by planting trees which would supply feed-stocks for bio-fuel production.

Fortunately, the process of bio-fuels production itself should help to control global warming by reducing generation of green house gases; and thereby sustain growth of renewable resources for bio-fuel production. Thus the process of bio-fuel production and controlling global warming are mutually complementary and conform to the basic principle of sustainable growth.

TABLE - 1

RELATIVE OIL-YIELD FROM VARIOUS RENEWABLE RESOURCES

PLANT SPECIES	OIL-YIELD, Gallons/year/acre
CORN	15 – 18 (Alcohol)
SUGARCANE	20 – 25 (Alcohol)
SOYBEAN OIL	45 – 50 (Diesel)
SUNFLOWER OIL	100 – 105 (Diesel)
PALM OIL	600 - 650 (Diesel)
COCONUT OIL	280 – 300 (Diesel)
CASTOR OIL	150 – 155 (Diesel)
OLIVE OIL	120 – 130 (Diesel)
RAPE SEED OIL	125 – 130 (Diesel)
MICRO-ALGAE	1850 – 5000 (Alcohol/ Diesel) (Per acre of pond}

Hydrogen Production from Residual Heavy Oil

G. Prieto[1,2*], O. Prieto[1], C. R. Gay[1], S. Mededovic[2], K. Takashima[2], A. Mizuno[2]
[1] Departamento de Ingeniería Química – Facultad de Ciencias Exactas y Tecnología – Universidad Nacional de Tucumán - República Argentina
[2] Department of Ecological Engineering - Toyohashi University of Technology – Toyohashi - Japan

Abstract. The predominant chemical reactions conducted in a cold plasma reactor are essentially dehydrogenations when hydrocarbons are the reactive. Taking advantage of this characteristic, the present study is focused on the production of hydrogen from residual heavy oil, using argon as carrier gas. The main goal is to consider the process as an alternative method for production of a gas rich in hydrogen and free of carbon monoxide.

Keywords: Cold Plasma, Nonthermal Plasma, Heavy Oil, Hydrogen Production.

1. Introduction

Various technologies have been developed for hydrogen production in view of the increasing use of fuel cells in the automotive industry. In essence, a fuel cell is an electrochemical reactor that generates energy from the catalytic conversion of hydrogen. The majority of the present technologies for hydrogen production are based in the water reforming reaction or in reactions of partial oxidation of methanol, ethanol, and/or propane. In addition to hydrogen, these processes also produce CO which has to be removed from gas mixture due to its poisonous effect on the catalyst, even at concentrations of just few parts per million.

Thus, it is necessary to find an alternative method for the production of a gas rich in hydrogen and free of carbon monoxide.

During the last decade, the main focus of the non-thermal plasma technologies have been oriented towards the destruction of harmful compounds in the air, however, when hydrocarbons are introduced into plasma reactors dehydrogenation takes place and hydrogen is the main byproduct. In the present study, the plate-to-plate non-thermal plasma reactor shown the best performance to convert heavy oil, a residue in the refineries, into a mixture of hydrogen and light olefins. The plasma reactor was driven by the AC high voltage and argon was used as a carrier gas. The parameters Energy Efficiency and Hydrogen Selectivity were evaluated using the methodology of the Statistical Experimental Design.

The present paper analyzes the production of hydrogen when a residual heavy oil is processed in nonthermal plasma chemical reactors. The products distribution exhibited high concentration of hydrogen as a consequence of light olefins production. The high concentrations of olefin compounds of low molecular weight can be drawn taking into account that dehydrogenation is a common reaction occurring in plasmas. Compounds such as paraffin normally form a variety of products containing considerable amount of olefins, in general, in a high unselective behavior (Boenig, 1988). The mayor byproducts obtained were hydrogen and hydrocarbon compounds with two, three and four atoms of carbon, being ethylene the compound in highest concentration.

The main goal is to consider the process as an alternative method for production of a gas rich in hydrogen and free of carbon monoxide. Furthermore, light olefin

---

[*] To whom all correspondence should be addressed.
Address: Departamento de Ingeniería Química - Facultad de Ciencias Exactas y Tecnología, Ave. Independencia 1800, (4000) San Miguel de Tucumán, República Argentina.
E-mail: gprieto@herrera.unt.edu.ar

hydrocarbons such as ethylene and propylene are commonly employed as additives in the NOx removal conversion systems for power plants and vehicles exhaust gases.

The physical characterization of the electrical discharges in the field of the nonthermal plasma technologies has been discussed in many papers (Szekely, 1983; Penentrante and Schultheis, 1993; Benthe and Chang, 1995) and the general fundaments of the nonthermal plasma chemistry has also been discussed (Boenig, 1988). Nevertheless, the mechanisms of these processes have received less emphases because of the complexity derived from the interaction among the structure of the chemical compounds, their reactivity and the physical characteristics of reactors, such as their geometry, configuration and nature of electrodes, type of carrier gas, treatment time and applied power.

The experimental runs were planned and carried out according to the techniques provided by the Statistical Experimental Design. One possible way is to use the experimental information to construct mathematical models capable to represent the actual state of the process. These models must correlate target parameters, such as efficiency or yield, with those variables that may exert influence over them. Then, the study is oriented to acquire experimental data, to select the appropriate model and to estimate the parameters involved in the model. Empirical models are very useful, and many times the only way of modeling, especially when the phenomenon under study is complicated and not sufficiently understood. Frequently, complete theoretical models are not available and sometimes any attempt to develop one could be not justified from an economic point of view (Box et al., 1978).

A methodology that involves mathematics and statistic techniques appropriate to model and to analyze how a target response (selected response of interest) is influenced by many variables, is the Response Surface Methodology. When the mathematics form of the function is unknown, the relation between the variables under study and the selected response can be adjusted to a polynomial function. These polynomials can be best adjusted by the "central composite design". This design lends by itself to a sequential experimentation. The fundamentals and underlying philosophy of these techniques are discussed in many papers and textbooks (Montgomery, 1991; Davies, 1979; Myers, 1971).

The present research was initiated in the Department of Ecological Engineering, Toyohashi University of Technology, Japan, in 1998 with the purpose to investigate a nonthermal plasma plate-to-plate reactor to convert a residual heavy oil (high molecular weight hydrocarbons, essentially paraffins in the range of n-C10 to n-C25, into light hydrocarbons (mainly olefins) to be employed, for example, as gasoline components for the new generation of engines in the so-called "hybrid process"; or to be added in small amounts to certain catalytic processes for reduction of nitrous oxides compounds in the power plants and vehicles exhaust gas treatments.

2. Experimental

Figure 1 shows the schematic of the nonthermal plasma reactor of the type plate to plate, designed for the present study. It was constructed with an acrylic resin tube of 50 mm in diameter, with perforated plates made of stainless steel as electrodes. Voltage was provided by a transformer of 9KV (peak to peak), 220 V, 60 Hz. The selected carrier gas was argon. The lower electrode, connected to the ground, was covered by a glass fiber sheet and imbibed with heavy oil for each experimental run.

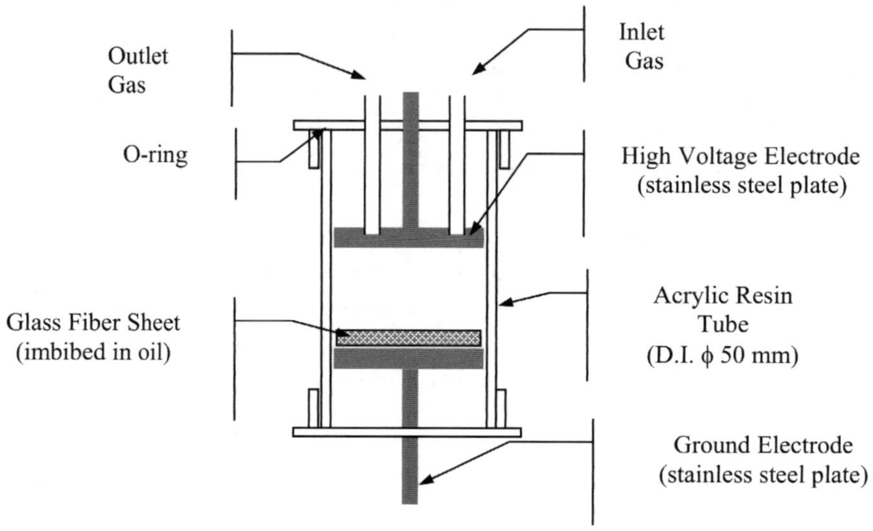

Figure 1: Schematic of the plate to plate reactor

Heavy oil (n-C10 to n-C25), from the bottom of the refining unit was used as raw material. The hydrocarbons produced were analyzed by gas chromatography, FID detector (Shimadsu GC-14B, stainless steel packed column, 4 mm in diameter and 3 m in length: Porapack Q, 60/80).

Hydrogen was analyzed by a gas chromatograph Shimadsu GC-8A, TCD detector and stainless steel packed column, 4 mm in diameter and 2 m in length: Molecular Sieve 13-X, 60/80.

The selected input variables and their respective ranges were:
Gap Distance: 0.3 – 1.0 (cm)
Input Power: 3.0 – 12.0 (W)
Gas Flow: 0.3 – 1.0 (mL/min)

Two other non-thermal plasma reactors of similar characteristics: Packed-Bed, and Needle-Plate were evaluated to understand which of them is the most effective in carrying out the conversion of heavy oils to produce Hydrogen. The parameters examined were gap distance between electrodes, carrier gas flow rate and applied power.

3. Results and Discussion

A statistical experimental design technique, the complete factorial experiments at two levels, defined the number of experiments to carry out into the selected range for each variable under study (Box et al., 1978; Davies, 1979). The purpose was to evaluate the efficiency of the cold plasma reactor in converting residual heavy oil into a gas mixture with hydrogen as the main product and low molecular weight olefins in minor concentrations (Prieto et al., 1999; 2003; 2005).

The products obtained were hydrogen, methane, ethane, ethylene, hydrocarbons with 3 and 4 carbon atoms and only traces of high molecular weight hydrocarbons. At the end of each experimental run a deposit of a carbonaceous compounds was observed on the surface of the glass fiber, previously imbibed in oil and placed on the ground electrode before each run.

Regarding the performance of the process, among all the reactors assayed, the plate-plate nonthermal plasma reactor showed the best efficiency in converting heavy oil into hydrogen and light olefin hydrocarbons as well as the best selectivity towards hydrogen production.

The selected output variables were:

Efficiency of the process in producing hydrogen

Efficiency (μL/Joule) = Hydrogen produced / Input Power

Selectivity of the process in producing hydrogen

Selectivity (vol. %) = Efficiency (H_2) / Efficiency (total C compounds)

Figures 2, 3, and 4 display a convenient way for a comparative graphical representation of the experimental results.

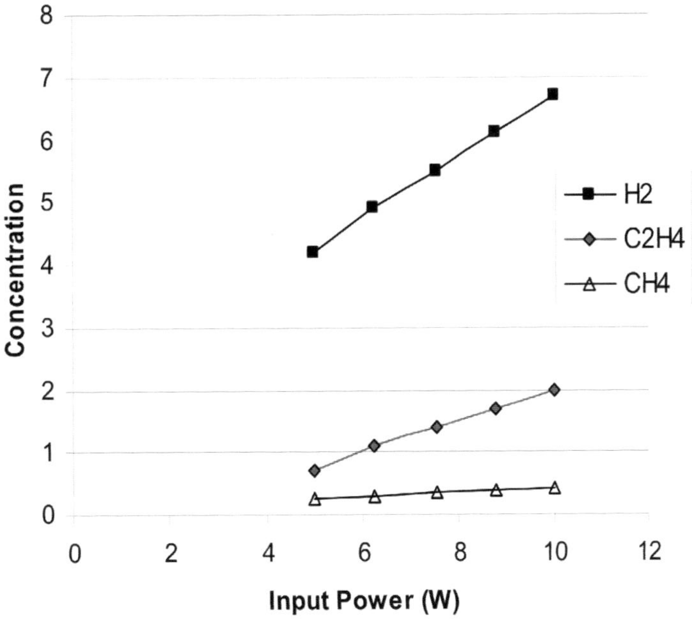

Figure 2: Effect of the Input Power on the concentrations of hydrogen, ethylene and methane, at the maximum residence time (minimum Gas Flow = 120 mL/min). Gap Distance between electrodes 0.65 cm (mean value of the experimental range). Concentrations are expressed in %.

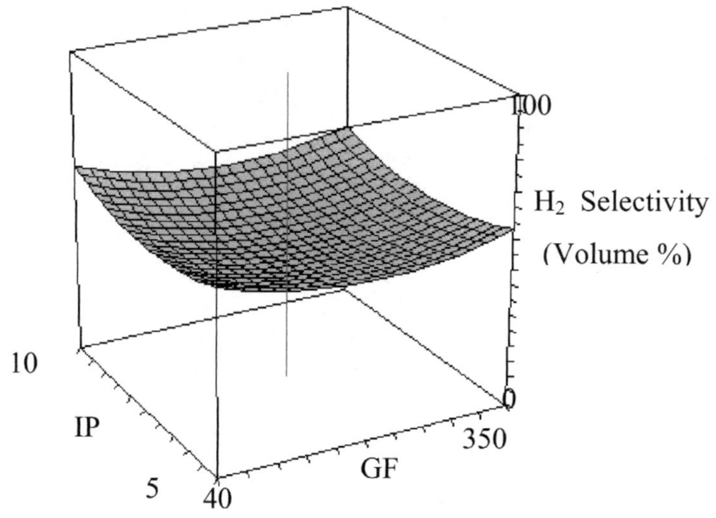

Figure 3: Response surface for the selectivity (volume %) towards hydrogen formation as a function of the Input Power (W), and the Gas Flow (mL/min). Gap Distance between electrodes 0.65 cm (mean value of the experimental range).

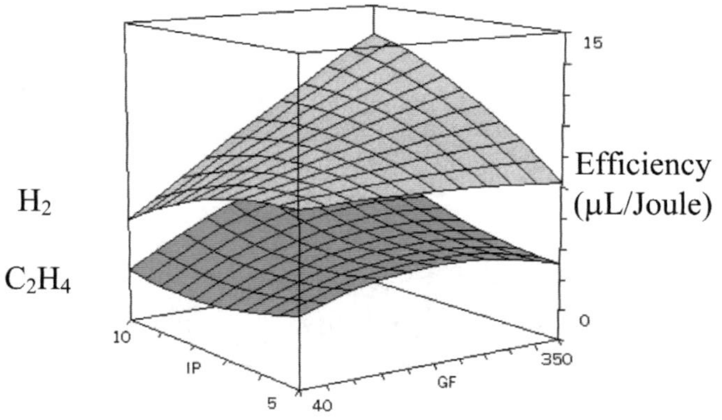

Figure 4: Response surface for the efficiency (μL/Joule) of hydrogen and ethylene production as a function of the Input Power (W), and the Gas Flow (mL/min). Gap Distance between electrodes 0.65 cm (mean value of the experimental range).

The Analysis of Variance for the experimental results demonstrated that there exists a strong interaction between the variables Input Power and Gap Distance. This means, we cannot infer about how the Efficiency is going to change if the variable Input Power is modified, because the response of the system will depend on the particular value of the variable Gap Distance.

The complexity derives from the fact that all the factors above mentioned present interaction among them. This means that the effect of all and each one of the variables cannot be evaluated independently. The effect that each one may exert on the system will depend on the particular conditions established for the rest of the variables.

These conclusions are guaranteed by the application of the statistical experimental design techniques, which are fundamental tools for such highly complex systems.

Table I shows the conditions and the experimental results obtained in processing heavy oil (n-C10 a n-C25) with a cold plasma reactor of the type plate-to-plate.

Table I: Experimental conditions and experimental results obtained by processing a residual heavy oil with a nonthermal plasma plate-to-plate reactor.

Processing variables			Products Concentration (%)						Selectivity: S (vol%) Efficiency: Eff (μL/joule)			
Gap Distance (GD) cm	Gas Flow (GF) mL/min	Input Power (IP) W	CH_4	C_2H_4	C_2H_6	C3	C4	H_2	S C_2H_4	S H_2	Eff H_2	Eff C_2H_4
0,44	120	4,8	0,139	0,624	0,024	0,055	0,084	2,067	20,84	69,07	8,61	2,6
0,86	120	4,8	0,110	0,660	0,014	0,054	0,074	1,786	24,46	66,18	7,44	2,75
0,44	330	4,8	0,040	0,168	0,014	0,039	0,031	0,588	19,02	66,78	6,74	1,92
0,86	330	4,8	0,047	0,205	0,012	0,028	0,038	0,616	21,67	64,98	7,06	2,35
0,44	120	10,2	0,298	1,901	0,029	0,113	0,210	4,840	25,72	65,49	9,49	3,73
0,86	120	10,2	0,091	0,640	0,011	0,046	0,068	1,536	26,74	64,2	3,01	1,25
0,44	330	10,2	0,172	1,145	0,029	0,099	0,145	2,410	28,63	60,25	13	6,18
0,86	330	10,2	0,100	0,610	0,017	0,044	0,082	1,914	22,06	69,18	10,32	3,29
0,3	225	7,5	0,120	0,672	0,020	0,049	0,072	1,779	24,77	65,6	13,56	3,36
1	225	7,5	0,082	0,427	0,019	0,034	0,053	1,280	22,52	67,54	6,4	2,13
0,65	49	7,5	0,296	1,933	0,021	0,112	0,174	5,343	24,53	67,81	5,82	2,1
0,65	401	7,5	0,077	0,495	0,015	0,048	0,081	1,230	25,46	63,22	10,96	4,41
0,65	225	3	0,029	0,088	0,011	0,017	0,008	0,382	16,42	71,45	4,78	1,1
0,65	225	12	0,201	1,458	0,023	0,083	0,145	3,505	26,92	64,74	10,95	4,56
0,65	225	7,5	0,154	0,784	0,020	0,056	0,092	1,934	25,77	63,6	15,21	3,92
0,65	225	7,5	0,147	1,124	0,018	0,070	0,132	2,410	28,80	61,76	12,05	5,62
0,65	225	7,5	0,123	0,839	0,020	0,059	0,102	1,945	27,17	63	9,73	4,19
0,65	225	7,5	0,121	0,737	0,022	0,051	0,090	1,862	25,58	64,59	14,41	3,69
0,65	225	7,5	0,110	0,800	0,017	0,049	0,094	1,950	26,51	64,58	9,75	4
0,65	225	7,5	0,135	0,920	0,019	0,060	0,110	2,296	25,99	64,87	11,48	4,6

Previous to these set of experiments shown in Table I, another experimental design was planned to convert the residual heavy oil, essentially paraffins in the range of n-C10 to n-C25, into light hydrocarbons (mainly olefins). The purpose was to add the olefins in small amounts to certain catalytic processes for reduction of nitrous oxides compounds in the power plants and vehicles exhaust gas treatments. In this set of experiments hydrogen was not analyzed, only hydrocarbons were analyzed by Gas Chromatography. Figure 5 represents a three dimensional model of the Efficiency in producing the total low molecular weight hydrocarbons.

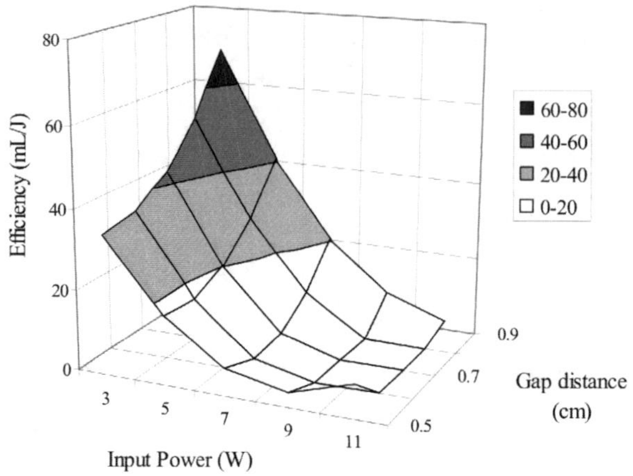

Figure 5: Surface response for the efficiency as a function of Gap Distance (cm) and Input Power (W) using a plate-to-plate reactor. Gas Flow rate: 225 mL/min.

The analysis of the experimental results showed that there exists strong interactions between the variables Input Power and Gap Distance which makes no possible the individual analysis for each one of both variables. The interaction is evidenced because the same increment in the Gap Distance causes an increase of the Efficiency duplicating its value, if the Input Power is at the lowest level, but if the Input Power is at the highest level, the Efficiency exhibits almost no changes. There is also an interaction between the variables Gas Flow and Gap Distance. Clearly, the strong interactions between the variables evidence the high complexity of these systems and can explain results of Efficiencies extremely high, such as 70 (μL/J). This particular high value for the efficiency evidenced the existence of a parametric sensibility of the system. It is worth to put the emphasis in the importance of the statistical experimental design techniques, because they lead us to find the parametric sensibility of the system, if there exists, among all the variables studied into the selected experimental ranges. The importance of a high process efficiency in producing light olefins implies a high efficiency in producing hydrogen gas.

These promising results encouraging us to try some other conditions in view to reproduce such high efficiencies and to increase the hydrogen production. Then, we planned and carried out other experiments: a) increasing the volume of liquid: filling the reactor with heavy oil from the bottom up to the surface of the electrode, with and without the quartz micro fiber sheet imbibed in heavy oil; b) similar conditions, but replacing the ground electrode by another of the same size and material (stainless steel) but with a bigger mesh in size; c) similar conditions, but replacing the ground electrode by a stainless steel perforated plate. None of these experiments could reach nor improve the efficiency of 70 (μL/J).

Another set of experiments were carried out with the same reactor filled with pellets of a dielectric material $TiBaO_3$ in between the electrodes (plates): Packed-Bed reactor, expecting higher efficiencies for increasing the micro discharges in the pellets contact points and expecting perhaps a catalytic effect, but the highest efficiency reached was 12 times lower than for the best efficiency obtained with the plate-to plate reactor. The experiments carried out with the same reactor but replacing the high voltage plate electrode by a needle, were also unsuccessful.

Then, for the plate-to plate reactor with the ground electrode covered with a quartz microfiber sheet imbibed in heavy oil, the intense electric field initiates multiple streamers along its length creating the reactive chemical species. For the variables, The optimal set of values to get efficiencies in the order of 70 microliters per Joule, corresponded to a Gap Distance of about 0.9 cm with stainless steel plate electrodes in an acrylic tube of 50 mm inner diameter, applying a low Input Power around 5 W.

4. Conclusions

Experimental results show the nonthermal plasma process to convert a residual Heavy Oil into a gas rich in Hydrogen in a tubular reactor type plate-to-plate is very promising reaching a selectivity of about 70% with the important advantage to be generated free of carbon monoxide formation when using Argon as carrier gas. This method for producing a gas with high content of hydrogen free of carbon monoxide results very interesting and the development of an appropriate technology for the implementation of this process, appears as worth whiling. It is worth to put the emphasis in the importance of the statistical experimental design techniques, because they lead us to find the parametric sensibility of the system, if there exists, among the variables studied into the selected experimental ranges.

References

Benthe, T.G.,.Chang, J.S. (1995). Handbook of Electrostatic Processes, Chapter 9: Gas Discharge Phenomena, pp.147. Editors J.S.Chang, A.J.Kelly and J.M.Crowley. Publisher Marcel Dekker, Inc., New York, 1995.

Boenig, H.V. (1988). Fundamentals of Plasma Chemistry and Technology. Technomic Publishing Co. Inc., Lancaster, Pennsylvania, U.S.A. (ISBN 87762-538-7).

Box, G.E.P., Hunter, W.G., Hunter, J.S. (1978). Statistics for Experimenters. Editorial John Wiley and Sons.

Dascalescu, L. (1993). An Introduction to Ionized Gases. Theory and Applications, Toyohashi University of Technology, Aichi, Japan.

Davies, O.L. (1979). The Design and Analysis of Industrial Experiments. 2nd Edition, Imperial Chemical Industries Limited, New York. (ISBN 0-582-46053-0).

Montgomery, D. C., Editor (1991). Design and Analysis of Experiments, Third Edition. Editorial John Wiley & Sons.

Myers, R. H. (1971). Response Surface Methodology, Editorial Allyn and Bacon, Boston.

Penetrante, B.M., Schultheis, S.E., Editors (1993). Non-Thermal Plasma Techniques for Pollution Control. Part A: Overview, Fundamentals and Supporting Technologies. NATO ASI Series G: Ecological Sciences, Vol. 34, Part A. Springer-Verlag, Berlin. (ISBN 0-387-57174-4).

Prieto, G., Okumoto, M., Shimano, K., Takashima, K., Katsura, S., Mizuno, A., Prieto, O., Gay, C.R. (1999). IEEE-IAS Annual Meeting, Phoenix, Arizona, USA.

Prieto, G., Okumoto, M., Takashima, K., Katsura, S., Mizuno, A., Prieto, O., Gay, C.R. (2003). Brazilian Journal of Chemical Engineering, Vol. 20, N° 1, pp. 1-10.

Prieto, G., Prieto, O., Gay, C.R., , K., Takashima, K., Katsura, S., Mizuno, A., (2005). ENPROMER 2005 Congress, 2nd Congress on Chemical Engineering, and 4th Mercosur Congress on Process System Engineering, Rio de Janeiro, Brazil.

Szekely, J. (1984). Plasma Processing and Synthesis of Materials, Materials Research Society Symposia Proceedings, Vol. 30, pp. 1-13, 1983, Boston, Massachusetts, U.S.A. Editorial North-Holland, Publisher Elsevier Science Publishing Co. Inc., 1984. (ISBN 0-444-00895-0).

(95b) Hydrogen Production From Ammonia Decomposition: Study On the Synergetic Effect of DBD Plasma and Cheap Metal Catalysts

Li Wang, Department of Catalytic Chemistry and Engineering & National Key Laboratory of Fine Chemicals, Dalian University of Technology, Dalian, China

Yue Zhao, Department of Catalytic Chemistry and Engineering & National Key Laboratory of Fine Chemicals, Dalian University of Technology, Dalian, China

Hongchen Guo, Department of Catalytic Chemistry and Engineering & National Key Laboratory of Fine Chemicals, Dalian University of Technology, Dalian, China

Weimin Gong, Department of Catalytic Chemistry and Engineering & National Key Laboratory of Fine Chemicals, Dalian University of Technology, Dalian, China

Introduction

Hydrogen has been considered as an ideal candidate to solve the urgent energy and environmental issues, especially on-site H2 production for fuel-cell vehicles [1,2]. Ammonia decomposition is an appealing hydrogen production route for fuel cell application [3]. Since 2004, studies related to the in-site generation of COX-free hydrogen via the decomposition of ammonia have increased very rapidly, almost all of them used tradditional catalytic route which was characterized by using supported transitional metal catalysts and fixed-bed reactor. The emphases of these studies were given to the effects of the different active components, supports and promotors of the catalysts on the decomposition activity of ammonia. According to the published works, (1) noble metal Ru is the most active catalyst, carbon nanotubes (CNTs) are the most effective support, and the alkaline substances like KOH is the best promotor. At 350 °C, a kind of K-Ru/MgO-CNTs catalyst, the most active catalyst ever prepared, has reached a hydrogen yield of 585 ml/min.g-catal [4]; (2) recombinative desorption of surface-adsorbed nitrogen atoms is a rate-determining step in ammonia decomposition [5]. The recombinative desorption rates of surface-adsorbed nitrogen atoms of non-noble metals like Fe, Co and Ni are much lower than that of the noble metal Ru. As a result, the catalytic activities of the non-noble metals for ammonia decomposition are much slower than that of Ru; (3) although CNTs supported Ru catalysts already possess excellent catalytic activity, improvements must be made for them to overcome problems including the shortage of Ru resource, the high cost of CNTs, and the instability of the catalyst caused by the methanation of CNTs in H2 atmosphere during long-term usage [6].

Here, we would discuss a novel plasma-catalysis method for ammonia decomposition. Our motive is to apply non-equilibrium plasma to ammonia decopmposition catalyzed by non-noble metal catalysts, so as to accelerate the recombinative desorption rates of surface-adsorbed nitrogen atoms. Based on systematic explorative experiments, strong activity-enhancing synergetic effects have been seen between dielectric-barrier discharge (DBD) plasma and supported non-noble metal catalysts including Fe, Cu, Mo, Ni and Co.

By employing XRD, FT-IR, XRF, UV-Raman, XPS, TEM and TPD, the nature of the activity-enhancing synergetic effect between DBD plasma and supported metal catalysts was investigated. The promotion of the recombinative desorption rates of surface-adsorbed nitrogen atoms by DBD plasma has been confirmed by plasma desorption (PD) technique.

Results and Discussion

Based on systematic explorative experiments, strong activity-enhancing synergetic effects were seen between dielectric-barrier discharge (DBD) plasma and supported non-noble metal catalysts, including Fe, Cu, Mo, Ni and Co. (1) In the case of being used separately at 410℃, the conversion of NH3 obtained with Fe2-3N catalyst and DBD plasma was only 7.4% and 7.8%, respectively. However, in the same conditions the combination of the DBD plasma with Fe2-3N catalyst dramatically increased the conversion of NH3 to more than 99%. In other words, in the absence of DBD plasma, reaction temperature of 550 °C was needed for Fe2-3N catalyst, to reach more than 99% ammonia conversion. In the presence of DBD plasma, however, the reaction temperature for Fe2-3N catalyst to reach more than 99% ammonia conversion decreased by 140 °C. (2) Among transitional catalysts investigated, the activity-enhancing synergetic effect of DBD plasma and supported metal catalysts increased in Cu < Fe <mo<ni<co plasma-catalysis="plasma-catalysis" has="has" that="that" reaction="reaction" not="not" catalysts="catalysts" on="on" reasonable="reasonable" desorption="desorption" molecules, ="molecules," study="study" co="Co" also="also" technique..="technique.." employing="employing" decomposition="decomposition" plasma="plasma" cu,="Cu," lower="lower" therefore,="Therefore," but="but" with="with" and="and" of="of" energetic="energetic" characterization="characterization" is="is" we="we" to="to" will="will" have="have" ni="Ni" dbd="DBD" cleave="cleave" various="various" it="it" recombinative="recombinative" can="can" directly="directly" mo,="Mo," promotion="promotion" surface-adsorbed="surface-adsorbed" the="the" non-equilibrium="non-equilibrium" inelastic="inelastic" activity="activity" sequence.="sequence." n-h="N-H"

Acknowledgements

The authors acknowledge the financial supports from the Natural Science Foundation of China (NSFC: Grant No. 20473016 and 20673018)

Reference:

[1] J. E. Mason, World energy analysis: H2 now or later, Energy Policy, 2007, 35, 1315-1329. [2] M. Z. Jacobson, W. G. Colella, D. M. Golden, Cleaning the Air and Improving Health with Hydrogen Fuel-

Cell Vehicles, Science, 2005, 308, 1901-1905. [3] A. Klerke, C. H. Christensen, J. K. Nørskovb, T. Vegge, Ammonia for hydrogen storage: challenges and opportunities, J. Mater. Chem. 2008, 18, 2304-2310. [4] S. F. Yin, B. Q. Xu, S. J. Wang, C. F. Ng, C. T. Au, Magnesia–carbon nanotubes (MgO–CNTs) nanocomposite: novel support of Ru catalyst for the generation of COx-free hydrogen from ammonia, Catal. Lett. 2004, 96, 113-116. [5] S. F. Yin, B. Q. Xu, X. P. Zhou, C. T. Au, A mini-review on ammonia decomposition catalysts for on-site generation of hydrogen for fuel cell applications, Appl. Catal. A: Gen. 2004, 277, 1-9. [6] S. J. Wang, S. F. Yin, L. Li, B. Q. Xu, C. F. Ng, C. T. Au, Investigation on modification of Ru/CNTs catalyst for the generation of COx-free hydrogen from ammonia, Appl. Catal. B: Environ. 2004, 52, 287-299.

See more of this Session: Plasma for Energy and Environmental Applications
See more of this Group/Topical: Topical 2: Advanced Fossil Energy Utilization

(95c) Plasma Reforming of E85 for Hydrogen Rich Gases Production

Xinli Zhu, School of Chemical, Biological and Materials Engineering, Center for Biomass Refining, University of Oklahoma, Norman, OK

Trung Hoang, School of Chemical, Biological and Materials Engineering, Center for Biomass Refining, University of Oklahoma, Norman, OK

Lance L. Lobban, School of Chemical, Biological and Materials Engineering, Center for Biomass Refining, University of Oklahoma, Norman, OK

R.G. Mallinson, Chemical, Biological, and Materials Engineering, University of Oklahoma, Norman, OK

E85 (85 vol.% ethanol and 15 vol.% gasoline) is a partly renewable fuel, which is commercially available in gas stations. Hydrogen rich gas production from E85 for fuel cell or internal combustion engine applications is being considered as a potential way to reduce CO2 emissions with increasing interest. The complex composition of E85 requests the catalyst to be active for two kinds of molecules (ethanol and gasoline) and must be resistant to sulfur for a catalytic process. Thus this process typically relies on noble metal catalyst (Rh) that is operated at high temperature with high H2O/Carbon ratio. In this work, we demonstrate the possibility of reforming of E85 using a nonthermal plasma reactor, which avoids using an expensive catalyst and is operated at lower temperature. The results show that hydrogen rich gases (62-66% H2 and ~27% CO, with small amounts of CO2, C2 hydrocarbons) could be achieved. Compared to ethanol plasma reforming at identical condition, E85 reforming produces slightly higher C2 hydrocarbons.

See more of this Session: Plasma for Energy and Environmental Applications
See more of this Group/Topical: Topical 2: Advanced Fossil Energy Utilization

(95d) Non-Thermal Plasma Reforming of Refractory Tars and Oils Generated by Biomass Gasification

Lyman Frost, Ceramatec, Inc, Salt Lake City, UT

S. Elangovan, Electrochemical, Ceramatec, Inc, Salt Lake City, UT

J. Hartvigsen, Electrochemical, Ceramatec, Inc, Salt Lake City, UT

One of the barriers to wider use of many kinds of biomass is the fact that they generate refractory tars and oils when gasified. These compounds cause difficulty when they are passed to downstream components in the system by causing plugging and drop in performance. The compounds can be condensed and removed but this involves extensive heat exchange, increased cost, and removal to a hazardous waste landfill.

Working as a subcontractor to Emery Energy, a Salt Lake City company with expertise in biomass gasification, Ceramatec will be applying its non-thermal plasma reformer technology to eliminate any refractory tars and oils that exit the Emery gasifier. Preliminary laboratory work indicates that it is possible for the plasma reformer to operate on the gas exiting the gasifier and convert these tars and oils to additional hydrogen and carbon monoxide. This work is being done at Western Research Institute in Laramie, Wyoming under a US Department of Energy award to Emery Energy.

This paper will present an overall process structure of the gasification process and the non-thermal plasma reformer in particular. Operational characteristics and laboratory results related to the reforming of the refractory tars and oils will be presented. The larger pilot sized facility at Laramie will also be described. Data may be available from the larger pilot sized facility in Laramie by the time for the conference and if available will be presented.

See more of this Session: Plasma for Energy and Environmental Applications
See more of this Group/Topical: Topical 2: Advanced Fossil Energy Utilization

Non-thermal Plasma Logistic Fuel Reformer for TARDEC

L. Frost, J. Hartvigsen, S. Elangovan; Ceramatec, Inc

Presently, the reforming of sulfur-containing, liquid hydrocarbon fuels, such as JP-8 and NATO F76, is generally handled through catalytic processes. Typical systems use either autothermal approaches or catalytic partial oxidation. The weaknesses of the catalytic approaches are well known, and those systems are not necessarily compatible with the operational characteristics desired for many fuel cell systems. Many catalysts are sensitive to the presence of sulfur. In addition, the mass of the catalyst must be heated to an even temperature to prevent the formation of coke and other undesirable byproducts, thus, impacting the ability of the reformer to start quickly and respond to load changes. The mass of the catalyst operates against both rapid startup and high volumetric energy density. Ceramatec recently completed a project for the United States Army (TARDEC) using a non-thermal plasma to catalyze the reforming of JP-8.

Reforming of Liquid Hydrocarbons

Operation on liquid hydrocarbon logistic fuels drawn from any available infrastructure, or from onboard the vehicle, is an absolute requirement for a truly useful and practical fuel cell power system for many applications. Unfortunately, catalytic steam reforming of these fuels has proven difficult due to adverse effects that the high aromatics content and deeply-bound sulfur have on catalyst activity.

In partial oxidation, catalytic partial oxidation (CPOX), and autothermal reforming (ATR) processes, a substoichiometric amount of air is combined with the fuel, and the mixture is burned, with or without a catalyst, to generate the synthesis gas. Conventional partial oxidation processes have been plagued with soot and carbon deposition issues, and in some cases, the soot management system may be larger and more complex than the reformer. The large size, the thermal mass of the catalyst in the case of CPOX or ATR operation, the hydrogen required in a hydro-desulfurization system, and the operational difficulties associated with the partial oxidation process, make each of these processes difficult to use for compact synthesis gas production.

Non-thermal plasma reforming is an approach that solves many of the above problems. This approach uses a low-amperage, high-voltage plasma arc to catalyze the reforming of the heavy hydrocarbons. The energy needed to complete the reformation process can be introduced by burning a portion of the fuel or by heat integration with the rest of the system. Much as would be accomplished by a solid catalyst, the non-thermal plasma creates radicals that promote the reforming process. Unlike the solid catalyst, the non-thermal plasma approach is not sensitive to sulfur since it is constantly renewed with each plasma arc generated and since there are no catalyst sites for sulfur to deactivate. Likewise, without the solid catalyst bed there is no thermal mass that must be heated to temperature and thus start-up time is substantially reduced. The unit may be started in a partial oxidation mode and then switched to steam reforming as temperatures sufficient to produce steam are reached. This provides substantial flexibility in operations by going from exothermic to endothermic operations dependent only on control algorithms and the mode that makes the most sense at a point in time. The system is flexible, sulfur tolerant, and consumes a minimum of the energy value of the synthesis gas produced. The system does need some electrical energy to produce the plasma (i.e. ~ 1-3% of the energy in the input fuel) and does not eliminate the sulfur from the synthesis gas but as stated it is not sensitive to this sulfur.

GlidArc ® Plasma Reformer

Ceramatec has developed a very promising alternative approach to the logistic fuel reforming challenge. The approach employs thermal and process integration of a low-power, non-thermal, *plasma-catalyzed* reformer for use with heavy hydrocarbon fuels. The plasma reforming technology is an innovative, integrated process that retains the thermodynamic advantages of steam reforming and combines it with the robust, highly energetic reaction zone characteristic of partial oxidation. The process is easily scaleable for use in compact power systems, ranging from a few kilowatts to hundreds of kilowatts. The laboratory-scale reformer (Figure 1) is compact, with an internal volume of 0.6 L and a reformation capacity of 5 kWt fuel rate. The plasma reformer generates a highly active, non-thermal plasma source of non-equilibrium electric discharges that act as an activation source for plasma chemical reactions. The operating temperature is set by the thermodynamics of the reforming reaction, which typically means that temperatures in excess of 800°C are required for high conversion of hydrocarbons to synthesis gas. Reforming reactions are highly endothermic, consuming up to 30% of the heat of combustion of the fuel. This reformation heat is not supplied by the low-energy plasma, but may be supplied by either partial oxidation of the fuel within the reformer or process integration with other temperature matched heat sources

The device has at least two diverging knife-shaped electrodes along which multiple electric discharges glide. These electrodes are immersed in a fast-flowing gas. The progress of an electric discharge is shown in Figure 1 (right). A high-voltage, single- or poly-phase, relatively low-current discharge is generated across a gas flow between the electrodes. The discharge forms at the closest point, spreads along the electrodes, and disappears; another discharge immediately forms at the initial spot. The geometry of the electrodes, flow conditions, and characteristics of the power supply determine the path of the arc. The electrodes do not need to be cooled, so the electrical energy is directly and totally transferred to the processed gas. The voltage can be as high as 20,000 volts. Multiple discharge systems can be installed for large gas flow rates. Any gas or vapor can be directly processed. Droplets, mist, and powders can be present without the risk of soot formation.

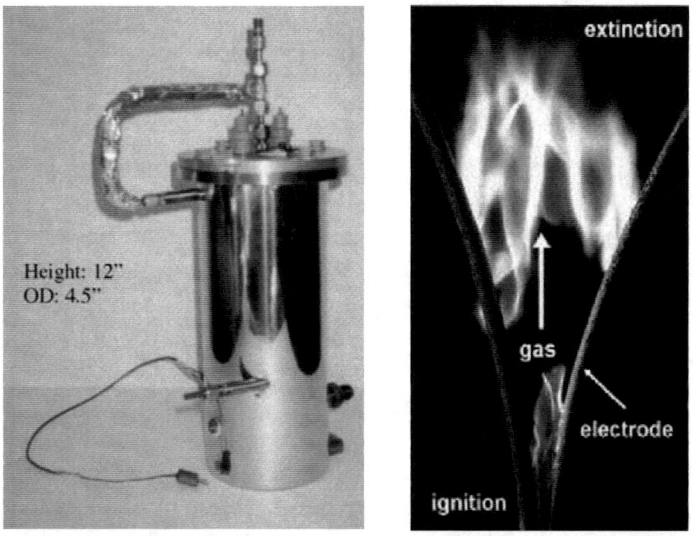

Figure 1. GlidArc plasma reformer: compact 5-kWt JP-8 reformer (left); inside view of the GlidArc (right).

JP-8 fuel obtained from a Chevron refinery was reformed using the reformer. The analysis provided by the refinery indicates that the fuel had a total sulfur of 430 ppmw and an aromatic content of 15.3 vol%. A reformate gas analyses using air for the oxygen supply performed over a 16-hour operation is shown in Figure 2. The H_2 + CO content of the reformate was around 30 vol%. Even with the 15% aromatic content of the JP-8, there was only about 1.5 vol% of methane in the reformate and negligible higher hydrocarbons.

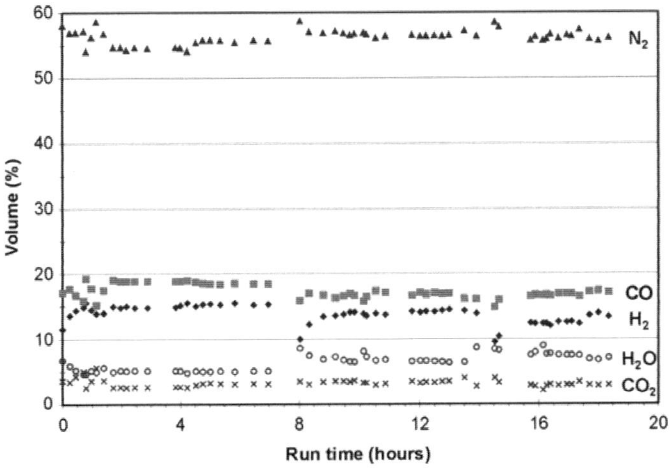

Figure 2. Gas composition of JP-8 reformate.

The plasma reformers are compact, easily scalable, and the electrical consumption is low—estimated to be in the 1–3% range of a fuel cell's full load output, enabling high overall system efficiency. Tests have successfully reformed pipeline-grade natural gas, commercial propane, gasoline, and diesel fuel at atmospheric pressure, and with electrical power as low as 90 W for a 5-kWe fuel rate (about ~2% of the fuel cell output). The reformer continues to function well at reduced fuel flow rates; however, electric power input remains relatively constant and, thus, consumes a proportionally greater fraction of the input energy. Also, if the reformer is not integrated within the SOFC thermal enclosure, heat losses from the reformer will represent a greater fraction of the fuel heating value.

Plasma Catalyzed Steam Reformer Design and Performance

The partial oxidation mode GlidArc ® reformer is designed to minimize heat loss from the reaction zone, and to recover some of the heat loss in preheating the reactant air stream. In contrast, an endothermic steam reformer, whether a conventional catalytic or plasma catalyzed must be designed to provide the required heat transfer duty of 25-30% of the feed stream heat of combustion in order to drive the endothermic reforming reaction. In the case of the GlidArc® reformer, it is desirable to preheat the steam/air oxidant mixture to near reforming temperature since heat transfer area in the plasma zone is very limited. A large post plasma zone heat transfer area is also provided to allow further conversion and residence time. A first implementation of the endothermic plasma reformer design is illustrated in Figure 3 below. The design of Figure 3 has been fabricated and tested using JP-8, and other fuels as previously noted. The reformer installation in the test furnace is shown in Figure 4 below.

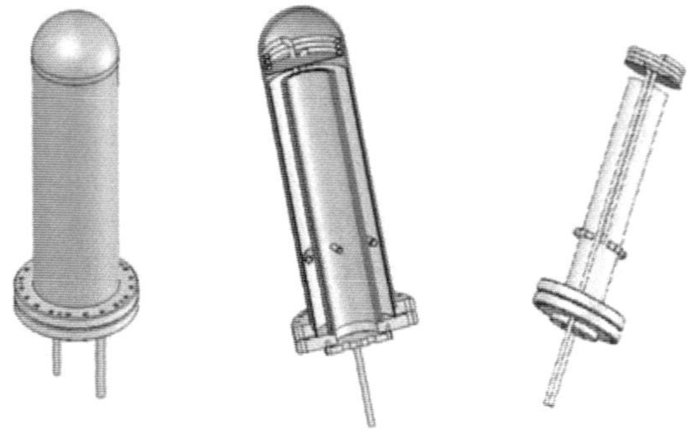

Figure 3. Solid model of endothermic plasma reformer design

Figure 4. Endothermic plasma steam reformer installation in electric test furnace.

Initial results of JP-8 testing using the endothermic plasma catalyzed steam reformer design show the improved performance with respect to reformate composition (Figure 5) and reformate heating value

(Figure 6) relative to operation in partial oxidation mode. In both cases (i.e. partial oxidation and endothermic steam gasification) the plasma zone geometry and electric power input were fixed.

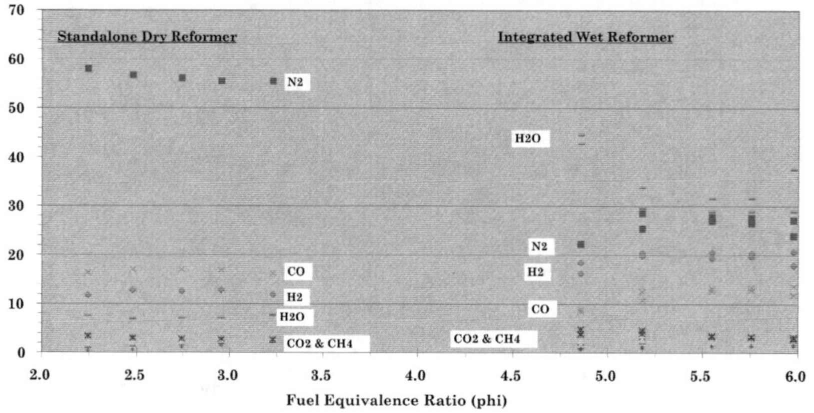

Figure 5. Reformate composition (%) from partial oxidation and steam reforming modes.

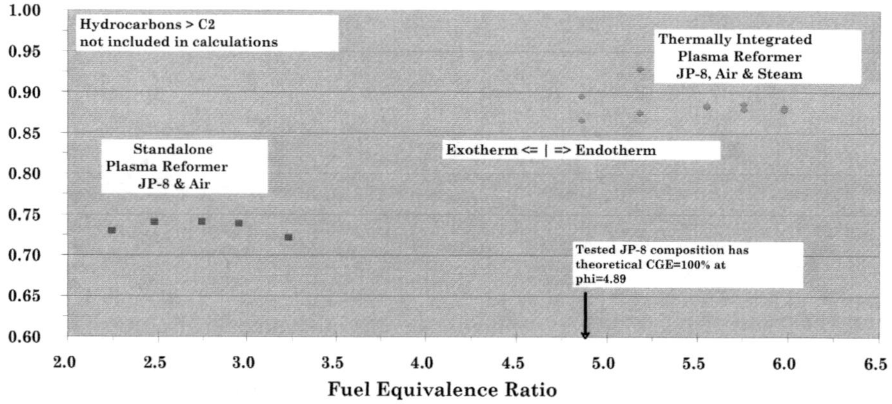

Figure 6. Reformate heating value of partial oxidation and steam reforming modes

The compact plasma reformer requires no cooling, uses simple power supplies, does not exhibit significant electrode corrosion, and can be easily integrated into other processes. The system may be modularized and scaled up using multiple-electrode configurations. Based on the experience with 5-kWt fuel reformer, Ceramatec has supplied several 60-kWe oxidation mode units to a commercial customer. These reactors used six electrodes to increase size of the plasma zone for larger throughput. Ceramatec has successfully proven that the plasma reformer can produce soot-free JP-8 reformate at temperatures that match well with a SOFC system. Ceramatec has utilized its non-thermal plasma reformer to reform a number of different fuels as illustrated in the following table (table 1).

Table 1

Steam Reforming of Heavy Hydrocarbons

INPUT	H2	O2	>CH4	CH4	CO
JP-8	33-36%	.1-.4%	~.7	3-4%	14-15%
Synthetic	45-46%	.01-.04	.7-1.1%	.6-.8%	18-20%
Synthetic + Naphthalene	44-46%	.03-.1%	~0%	.5-.7%	19-20%
JP-8	12-16%	.6-1%	5-6%	7-8%	14-15%
Synthetic	26-31%	.01-.08%	4-5%	~7%	~18%
Synthetic + Naphthalene	26-30%	.01-.07%	3-4%	~7%	~19%

The numbers shown in the lower part of the table (yellow) are at the initiation of the reaction and obtained from samples in the reaction zone. The numbers in the upper part of the table (blue) are from the exit from the non-thermal plasma reformer. It can be observed that as expected the plasma zone numbers show a lower percentage of hydrogen than shown at the exit from the reformer. The numbers also show a higher percentage of hydrogen for the synthetic fuel since this fuel is a straight chain hydrocarbon and enters with essentially zero percent sulfur. The small percentages of methane at the exit of the reformer are probably caused by recombination of the carbon and hydrogen in the presence of the iron based reformer piping. No double bond carbon compounds were detected indicating a good reforming process.

TARDEC Plasma Reformer

Ceramatec recently completed a contract for TARDEC that involved design, fabrication and delivery of a non-thermal plasma reformer that is capable of producing enough synthesis gas to fuel 10 kWe of solid oxide fuel cells. The unit included a sulfur removal unit fabricated by IntraMicron that was capable of on-line regeneration of the sulfur capture material. The unit did not include the solid oxide fuel cells but was designed to be relatively compact and have the potential to be integrated with solid oxide fuel cells in the future. The design is shown in Figure 7 (a) below. The unit consisted of two M-shaped plasma units with a heat exchanger between them. The plasma unit is located in the center leg of the M shaped unit. If fuel cells were to be added, the enclosure would be expanded slightly and the fuel cells placed between the heat exchanger and reformer for heat integration.

Figure 7 (b) shows the as-built unit. The unit is approximately 28" x 26" x 34" and delivers enough reformed fuel for 10 kWe at an expected efficiency of 30%. The unit was operated successfully on JP-8 with approximately 400 ppmw sulfur and will be undergoing additional testing during 2010. The tests in 2010 are expected to include feeding the reformate to ~ 2.5 kWe of solid oxide fuel cells.

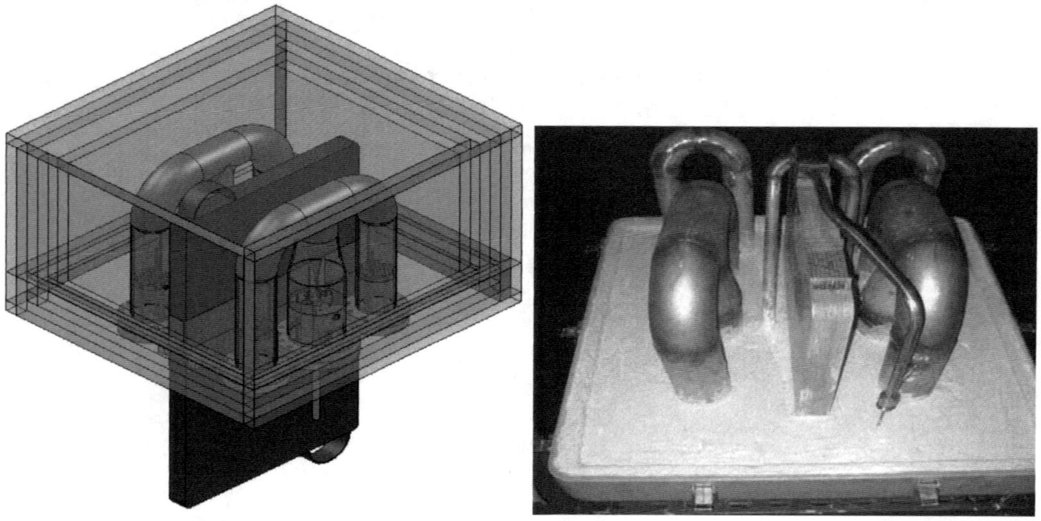

Figure 7: (a) Design of 10 kWe plasma reformer; (b) internals as built

Summary

In summary, the plasma-catalyzed reformer has been demonstrated to be sulfur-tolerant in reforming JP-8 fuel containing over 400 ppmw total sulfur and in excess of 15 v% aromatics. Non-methane hydrocarbon slip is negligible. The reformer operation is soot free and the power consumption for the plasma was of the order of 1 – 3% of fuel LHV.

(95f) Fluidized Bed Plasmas Reactor for Catalyst Preparation. Application for SCR NOx by Hydrocarbons in Mobile Sources

Marjorie Foix, Université Pierre et Marie Curie et Ecole Nationale Supérieure de Chimie de Paris, Paris, France

Patrick da Costa, Université Pierre et Marie Curie, Paris, France

Cedric Guyon, Ecole Nationale Supérieure de Chimie de Paris, Paris, France

Michael Tatoulian, Ecole Nationale Supérieure de Chimie de Paris, Paris, France

Nowadays, almost all major chemicals are produced by catalytic processes. Among them, heterogeneous catalysis plays a very active role, because of environmental concerns. The use of plasmas for catalysis is now well developed1. Plasma treatment in a low pressure systems has been already used to replace the thermal calcinations of catalysts. Fluidized bed reactors offer the possibility to lead to homogenous treatment and have the additional advantage of excellent heat transfer rates between the gas and the particles2, so that it can modify the catalytic properties of the prepared materials3. The aim of our work is to prepare plasma-prepared catalysts dedicated to NOx abatement in mobile sources. In such applications, silver based catalysts could be used4 since ethanol was chosen as reducing agent for NOx removal in lean burn conditions. After typical catalysts preparation, cold fluidized bed plasma reactors were used to activate alumina supported silver catalysts in order to obtain a sustainable synthesis process. For the sake of comparison, a classical calcination in air at 500°C was performed. Concerning the plasma process, we have studied the influence of the plasma exposure time as well as the nature of the gas treatment (Ar and Ar/O2 plasmas). First, the alumina was sieved between 200 and 350 μm in order to obtain a homogeneous fluidization, and to avoid the loss of powders in the pumping system. The 2.5 wt% Ag/ alumina was prepared by an excess impregnation method with AgNO3 as precursors in reduced pressure. Then, the catalyst was dried at 120°C overnight, prior to the two different processes, i.e. (i) calcination in air at 500°C for 2 h (10°C.min-1) or (ii) plasma treatment in the fluidized bed reactor. The fluidized bed reactor (FBR) consisted in a cylindrical Pyrex glass tube. The impregnated powders were fluidized by an Ar gas or a mixture of Ar/O2 gas passing through a porous glass plate. Two external electrodes were connected to the radio frequency generator in order to create the plasma in the fluidized bed region. The excited species formed in the plasma were characterized by in-situ diagnostic such as optical emission spectrometry (OES). Different samples were treated in Ar for different plasma treatment times (10 to 60 min) and one sample was prepared in O2/Ar mixture during 30 min. In order to understand the effect of the plasma treatment on silver based catalysts, a bench of surface characterizations such as TEM, UV-VIS, TPO were performed. In order to select the more efficient plasma treatment, SCR NOx by ethanol was performed, with a GHSV= 190,000h-1, under 500ppm NO, 10%O2 and 2500ppm C2H5OH in argon. Alumina supported silver based catalysts prepared under plasma conditions followed the same trends than the one obtained with the classical calcination (Fig 1.). A volcano type curve was always obtained due to the competition between ethanol oxidation and NOx reduction. The maximum of NOx conversion was found at 410°C on calcined catalyst whereas this maximum was shifted to lower temperatures depending on the plasma treatment. The higher activity was found for catalyst treated in argon during 30 min (RF-2). These catalysts were characterized by different surface analyses in order to understand the difference in terms of catalytic activity and in order to indentify the silver species. The first results in TPO showed that nitrates species were still present on the surface of catalysts after plasma treatment (RF-2), whereas these species completely disappeared when the catalysts were calcined. Furthermore, UV-VIS measurements revealed the presence of the same ionic silver species (Ag+, Agn+) on both plasma (RF-2) and calcined samples, whereas the metallic silver species were only observed on calcined catalysts. These preliminary tests could explain the difference observed in deNOx mostly at low temperatures. Conclusions Other characterizations are in progress in order to confirm these results. Moreover, we plan to replace the R.F generator by a microwave source, which presents the advantage to create plasma with higher density of energy. We would wish to try to test another catalyst in order to validate these first on another catalyst systems in the field of the stationary sources for example the palladium based catalysts were chosen since the main reducing agent is methane5. References 1.A. Brockhaus, D. Korzec, F. Werner, Y. Yuan, J. Engemann - Characterization of a microwave plasma by in situ diagnostics - Surface and Coatings Technology, Part 1 Vol. 74-75 p. 431-442 (1995). 2.R. Jafari, M.Tatoulian, W. Morscheidt, F. Arefi-Khonsari - High Stable Polymerised Acrylic Acid coating deposited on Polyethylene (PE) films in a low frequency discharge. Reactive & Functional Polymers, Vol.66 p. 1757–1765 (2006). 3.W. Xia, O. F.-K. Schlüter, C. Liang, M. W.E. van den Berg, M. Guraya, M. Muhler - The synthesis of structured Pd/C hydrogenation catalysts by the chemical vapor deposition of Pd(allyl)Cp onto functionalized carbon nanotubes anchored to vapor grown carbon microfibers - Catalysis Today, Vol. 102-103 p. 34-39 (2005). 4.Y. Hoon Yeom, M. Li, Wolfgang M.H. Sachtler, E. Weitz - Low-temperature NOx reduction with ethanol over Ag/Y: A comparison with Ag/γ-Al2O3 and BaNa/Y- Journal of Catalysis, Vol. 246 p. 413-427 (2007). 5.P. Da Costa, R. Marques, S. Da Costa - Plasma catalytic oxidation of methane on alumina-supported noble metal catalysts – Applied Catalysis B: Environmental, Vol. 84 p. 214-222 (2008).

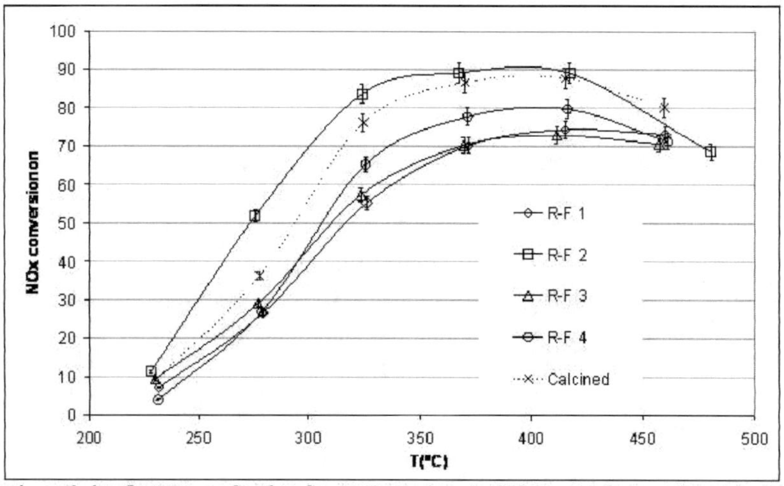

Fig. 1: Catalytic activity for NO_x reduction by C_2H_6O over Ag/Al_2O_3 catalysts. Plasma treated catalysts: in Ar RF-1(10 min), RF-2(30 min), RF-3(60 min); in Ar/O_2 RF-4(30 min), Reaction conditions: 500 ppm NO, 2500 ppm C_2H_6O, 10 vol.% O_2, GHSV=~190000 h^{-1}, Ar balance, total flow rate = 250 $ml.min^{-1}$

See more of this Session: Plasma for Energy and Environmental Applications
See more of this Group/Topical: Topical 2: Advanced Fossil Energy Utilization

(95g) Production of Hydrogen From Sugar by a Liquid Phase Electrical Discharge

Selma Mededovic Thagard, Department of Chemistry, Colorado State University, Fort Collins, CO

Graciela Prieto, Department of Chemical Engineering, National University of Tucuman, San Miguel de Tucuman, Argentina

Ellen R. Fisher, Department of Chemistry, Colorado State University, Fort Collins, CO

Kazunori Takashima, Department of Ecological Engineering, Toyohashi University of Technology, Toyohashi, Japan

Akira Mizuno, Department of Ecological Engineering, Toyohashi University of Technology, Toyohashi, Japan

In this study, the authors tested the possibility of producing hydrogen by electrical discharge in a sugar (sucrose) solution. The results have shown that after fifteen minutes, the hydrogen concentration measured from the electrical discharge in sugar solution using gas chromatography is four times higher than that measured from the electrical discharge in tap water. Moreover, when the initial solution conductivity of the tap water and sugar solution was raised by a factor of twenty, the hydrogen concentration from the sugar solution was increased twenty seven times compared to the hydrogen concentration measured from tap water. We hypothesize that the mechanism of hydrogen production is that the electrical discharge in water converts sugar to ethanol, which then further decomposes to give hydrogen. Hydrogen production from the sugar solution was also compared to that from an ethanol-water solution. Overall, these results demonstrate that the electrical discharge in a sugar solution converts sugar into carbon dioxide and hydrogen (ethanol) within minutes, whereas the timescale required for the same reaction under the action of yeast is measured in days. This makes electrical discharge a competitive new technology for the hydrogen production from renewable sources.

See more of this Session: Plasma for Energy and Environmental Applications

See more of this Group/Topical: Topical 2: Advanced Fossil Energy Utilization

(95h) Forming Small Organic Molecules From CO_2 Using a Dielectric Barrier Discharge (DBD) Plasma

Michael A. Lindon, Physics, West Virginia University, Morgantown, WV

Dushyant Shekhawat, National Energy Technology Laboratory, Morgantown, WV

Earl Scime, Physics, West Virginia University, Morgantown, WV

Michael Gallagher, National Energy Technology Laboratory, Morgantown, WV

Matt Galante, Physics, West Virginia University, Morgantown, WV

Jeffrey Atkinson, Physics, West Virginia University, Morgantown, WV

Bruce Tepke, Physics, West Virginia University, Morgantown, WV

We have designed a dielectric barrier discharge plasma source to explore the use of DBDs in CO_2 sequestration. DBDs operate at atmospheric pressure and employ high voltage pulses across an anode-cathode pair to create a high density, low temperature plasma. Typical commercial DBDs operate at 60 Hz, but there are significant advantages to operating DBDs at much higher frequencies. Here the goal is to accelerate electrons to energies of several eV while leaving the background gas at the inlet temperature. These energetic electrons will electronically excite CO_2 molecules and act as a catalyst for CO_2 breakdown into CO and O, which can be collected and used for future chemical processes. Theory suggests that this non-thermal process will be several times more efficient than thermal systems (DBD 48kJ/mol versus thermal 340kJ/mol)*. The DBD design is highly flexible allowing for easy addition of catalysts and control over pressure, electrode distances, voltages (0 kV to over 10 kV) and operating frequencies (1 Hz to over 10 kHz).

* K. Jogan, et al "The Effect of Residence Time on the CO_2 Reduction from Combustion Flue Gasses by an AC Ferroelectric Packed Bed Reactor" IEEE transactions on industry applications, VOL. 29, NO. 5, (1993)

See more of this Session: Plasma for Energy and Environmental Applications
See more of this Group/Topical: Topical 2: Advanced Fossil Energy Utilization

Measurement of OH Radical in DBD Reactor with/without Catalyst by HPLC for Toluene Removal

Yufang Guo [1,*], Xiaobin Liao [1,2], Daiqi Ye [2]

(1. College of Environmental Science and Engineering, Guangzhou University, Guangzhou, Guangdong 510006, China.2.College of Environmental Science and Engineering, South China University of Technology, Guangzhou, Guangdong 510006, China.)

Abstract: In this paper, a new kind of way was used to detect the concentration of OH radical in DBD reaction. A film which was impregnated with salicylic acid was used to catch OH radical in plasma reaction at room temperature and atmospheric pressure. Manganese oxide was used as a catalyst. The combination effect of plasma and manganese oxide catalyst was studied on the toluene removal efficiency and hydroxyl radical formation. In addition, in order to investigate the performance of MnO_X catalyst and the formation of other radicals, a characterization of the catalyst was performed by BET, XPS, TPR, and FT-IR. It was found that the amount of OH radical increases with the increase of specific energy density (SED), and the concentration of HO* in gas stream with catalyst is much lower than that without catalyst. The concentrations of HO* without catalyst were in the range of (5.9~23.6) $\times 10^{13}$ radicals/cm^3. But with catalyst, they just varied with (0.67~4.12) $\times 10^{13}$ radicals/cm^3. In addition, OH radical concentration was detected under different relative humidity including 0, 20%, 40%, 60% and 80%. The results showed that the highest OH yield was obtained at a RH of 20%. The results of XPS and FT-IR suggested that various oxygen species and some compounds containing N element deposited on the catalyst after plasma reaction. And the spectra of FT-IR also verified the formation of OH on the surface of catalyst.

Keywords: Dielectric barrier discharge; Catalyst; Hydroxyl radical; Toluene removal

1. Introduction

The emission of volatile organic compounds (VOCs) from various industrial processes is one of the most important sources for air pollution. VOCs are damaging for both human health and environment, either directly for their toxicity and malodorous nature or indirectly as ozone and smog precursors [1]. Non-thermal plasma (NTP) was thought to be an available way to decompose diluted VOCs. While NTP alone has many disadvantages, such as low energy efficiencies, poor selectivity to carbon dioxide, and byproduct formation, an attempt to overcome these limitations is a combination of NTP method with catalysts [2-4]. Various catalysts such as noble metals, metal oxides or mixture were investigated in NTP reaction [4-6].

Hydroxyl radical (HO*) was supposed to be a very important reactive species because it was responsible for the oxidation of VOCs and affected their removal efficiency in plasma system [7-9]. But seldom researches were concerned about the formation of HO* in non-thermal plasma system with/without catalyst. In addition, due to the lives of plasma radicals are very short (some less than 10^{-9} seconds), the detecting work becomes difficult. In the last few years, some researchers investigated several ways to study the radicals in plasma reaction. For example, laser induced fluorescence (LIF) has been already employed by authors [10-15] for studying NO removal process in various non-thermal plasma reactors. Recently, several researchers succeeded

* Corresponding author. E-mail address: yufanguo@hotmail.com (Y.-F. Guo).

in measuring the OH radicals using LIF technique in the pulsed corona discharges [16-20]. Z. Machala et al demonstrated several ways of use of the UV-vis optical emission spectroscopy (OES) for the diagnostics of atmospheric pressure air and nitrogen plasmas. However, in these articles the real amount of radicals had not been affirmed.

In this study, a new technology was introduced to detect the concentration of OH radical in dielectric barrier discharge(DBD) reaction. A film impregnated with salicylicacid was used to collect OH radical in the DBD reaction at room temperature and atmospheric pressure. When salicylic acid reacts with OH radical, 2,3-dihydroxybenzoic acid (2,3-DHBA) and 2,5-dihydroxybenzoic acid (2,5-DHBA) are produced. Salicylic acid,2,3-DHBA and 2,5-DHBA could be detected by HPLC. Therefore, the concentration of OH radical could be calculated.

Manganese oxide was used as a catalyst and it was placed in the discharge area of a wire-plate dielectric barrier discharge reactor. The combination effect of plasma and manganese oxide catalyst was studied on the toluene removal efficiency, byproducts inhibition and hydroxyl radical formation. In addition, a characterization of the catalyst was performed by BET, XPS, TPR, and FT-IR.

2. Experimental

2.1 *Experimental set up*

The experiments were performed at room temperature and atmospheric pressure. The apparatus mainly includes a gas supplying and regulating system, a self-made DBD reactor with power supply system and a gas analysis system. The initial toluene concentrations were controlled from 200mg/m^3 to 800mg/m^3. The gaseous toluene was regulated by controlling the air flow rate from gas cylinder through pure toluene liquid (>99.5%) which was kept in a water bath (T =25±1℃). After passing through a mixing chamber, the feeding gases entered the DBD reactor. And the gas flow rates were controlled at a range of 150 ml/min to 450ml/min. The feeding gases with different humidities, which were measured by a humidity meter (Center 310, China), were simulated by controlling water bubbling system. Gas samples were obtained from the effluent by an auto-sampler and were analyzed on-line. Toluene and other volatile organic compounds were analyzed by a gas chromatograph (GC-7890II, China) equipped with a hydrogen flame ionization detector (FID) and a DB-5MS capillary column (30m*0.25mm*0.25mm) heated at 80℃.

2.2 DBD reactor and power supply system

Fig.1 shows the structure of a wire-plate DBD reactor in the reaction. Two epoxy resin boards (200mm*45mm*0.8mm) were used to form the reactor walls and acted as dielectric barriers (the dielectric constant ε = 3.6). A high voltage electrode was made of brass wire with 0.6mm in diameter. This electrode was shaped into a spiral in order to increase the energy density in the reaction volume. The wire-to-wire distance was 8mm. Two grounded electrodes were made of copper nets and fixed on the two sides of a middle epoxy resin board. The total length of the reactor was 200 mm, while the effective length was 150 mm. When no catalyst was used, the gap between the high-voltage electrode and the grounded electrode was 8 mm, resulting in a cross sectional area for the flow channel of 400 mm^2 and a reaction volume of 60 cm^3. To add a catalyst, it was supported by two pieces of nickel foam (150 mm ∗ 25 mm ∗ 2 mm) fixed to the grounded electrodes. This produced a gap of 6 mm, a cross sectional area for the flow channel of 300 mm^2 and a reaction volume of 45 cm^3.

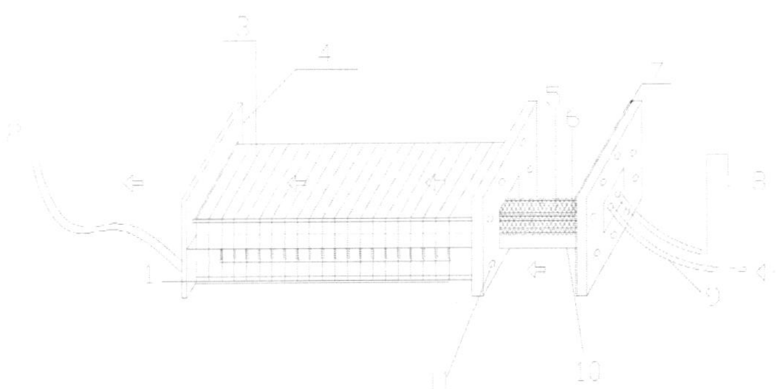

Fig.1. Structure of the DBD reactor with catalyst.

(1) positive electrode, (2) polyethylene tube (gas out), (3) brass wire, (4) epoxy resin board (1mm), (5) nickel foam, (6) copper net, (7) silicone pad, (8) ground electrode, (9) polyethylene tube(gas in), (10) epoxy resin board (2mm), (11) drilled hole for screw

High voltage power was supplied by a booster (0～250V) cooperated with a high voltage AC transformer (5～25KHz, 30kV) in series. In this experiment the frequency was controlled at 9.512 KHz. The applied voltage was measured by a high voltage probe (Tektronix P6015, America) and a digital oscillogragh (Tektronix TDS1002, America) was used to measure the voltage and current deposited to the reactor. The input power included the discharge power (the power deposited to the reactor) and the power consumption of the transformer and the circuit. The discharge power was calculated from the applied voltage and the reactor current. In this experiment, the discharge power was proportional to the input power by a factor of 0.5~0.63 in the DBD–catalyst system and by a factor of approximately 0.35 in the DBD system.

2.3 Preparation of a sampling film for HO detection*

The sampling method is similar to the previous article [21]. To collect HO* radical produced in the plasma reaction, one piece of filter paper was cut to a calculated size (300 mm × 45 mm), then a solution, which was made by 0.3 g salicylic acid (SAL) dissolving in 10 ml anhydrous ethanol (99.9%), was dripped equably by a injector on this filter paper. After that, the paper was dried and weighed until all the solution was loaded on. The filter film was folded as shown in Fig.2. Though the reaction of SAL and HO* was a gas-solid interface reaction, sampling film collection efficiency could achieve approximately 100%. The coarse surface of film could easily absorb HO* in the gas stream, and the reaction kinetics between HO* and salicylic acid belonged to the type of diffusion, hence its reaction rate was large [22]. In addition, the method to place the film also favored the effective trapping for its short-range collision.

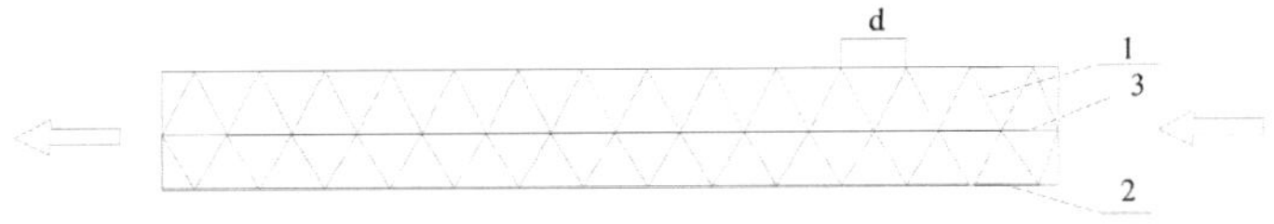

Fig.2. Position of filter film in the DBD reactor. 1. filter film; 2. epoxy board; 3.copper net.

2.4 HO analysis system*

The analysis method for OH radical was similar to the reference [22]. A HPLC (Ultimate 3000, USA), which was performed using a C_{18} reversed-phase column (5 m, 250 mm × 4.6mm) with the column temperature of 35°C, was used to separate and determine the amount of products and residual SAL. HPLC was run isocritically with a mobile phase consisting of 40% (v/v) methanol, with 60% deionized water at a flow rate of 1 ml/min. The HPLC system coupled with an UV detector was used to monitor SAL and hydroxylated derivatives at 300 nm at the condition above. In the HPLC chromatograph the retention time for 2,5-DHBA, 2,3-DHBA and SAL is 5.604 min, 7.525min and 17.686 min, respectively.

The concentration of HO* can be calculated by Eq.(1).

$$C_{OH}(Radicals/cm^3) = \frac{C_{Total} \times V_L \times N}{F_{Total} \times F_g \times t \times \eta \times (1-\alpha)} \quad (1)$$

C_{Total}——Hydroxylated derivatives concentration formed during sampling (mol/l);

V_L—— Scrubbing solution volume after sampling (l);

N——Avogadro's number (6.02×10^{23});

F_{Total}——Fraction of main hydroxylated derivatives in total products (assumed to be 100%);

F_g——Air flow rate (ml/min);

t——Sampling period (min);

η——Extraction efficiency (supposed to be 100%);

α——Loss rate of HO* passing the sampling films (set to zero).

2.5 Catalyst preparation

Impregnation with precursors of catalytically active phases constitutes a classical method to prepare heterogeneous catalysts. In this paper, catalyst of manganese oxide/alumina/nickel foam was prepared with a mass ratio of 50%:50%:100%. And the catalyst was prepared by an incipient wetness impregnation method. First a sol of alumina was loaded on nickel foam until the mass ratio was up to 50%:100%, then precursors of $Mn(NO3)_2$ were deposited on the support. After that, it was dried at 393K for 2 h, and then calcined at 723K for 4h.

2.6 Catalyst characterization

Specific surface area analysis of BET for the catalyst was measured by a surface area analyzer (ASAP 2020M, USA). The sample (0.1g) was pretreated with a temperature of 200 ℃ in a vacuum system for 4 h. Then, N_2 was used as an adsorbate to allow the temperature down to -196 ℃when the surface of catalyst was measured.

TPR analysis was carried out by using a micromeritics TPR/TPD 2900 instrument with a self-assembly thermal conductivity detector (TCD). Sample was pretreated at 400℃ using high-purity Ar (30ml/min) for 1 h. Then H_2 (100% by volume) was used as a reducing gas with a flow rate of 1 ml/ s. The catalyst was heatinged up to 800 ℃ at a rate of 10℃/min. Hydrogen consumption was measured approximately by TCD.

The elemental compositions and their valences on the surface of catalysts were analyzed with XPS using VG Multilab 2000 spectrometer(Germany) equipped with a hemispherical electron analyzer and Mg Ka radiation source (hν=1253.6eV). All binding energies are referenced to the C 1s line at 284.6 eV, which provided an accuracy of ±0.48 eV within full scanning of 0~1400 eV. A XPS Peak 4.1 software was used for fitting.

The FT-IR spectra of catalysts after plasma reaction were recorded by using an FT-IR spectrometer (Thermo Nicolet 6700, USA) at room temperature. Selfsupported wafers (13 mm diameter, 15±1 mg) were formed by pressing the catalyst powder scratched from the catalyst. Sample A was a catalyst loading with salicylic acid while sample B did not load.

3. Results and discussion

3.1 Effect of specific energy density on toluene removal efficiency and HO formation*

Fig.3(a) shows that toluene removal efficiency increases significantly when MnO_X catalyst is introduced into the discharge area. For example, with the catalyst in situ, the toluene removal efficiency (η) was 64.8% at a SED of 180 J/L, which was two times more than that of 31.8% without catalyst. It also illustrates that the toluene removal efficiency increased with the increasing of SED whether catalyst exists or not. When SED was regulated from 140 J/L to 320 J/L, η varied from 48.7% to 99.8% with MnO_X catalyst while η of 25.6% ~ 95.8% existed without catalyst.

Fig.3(b) shows that the HO* yield increases with the increase of SED, and the concentration of HO* in gas stream with catalyst is much lower than that without catalyst. The concentrations of HO* without catalyst were in the range of (5.9~23.6) $\times 10^{13}$ radicals/cm^3. But with catalyst, they just varied with (0.67~4.12) $\times 10^{13}$ radicals/cm^3.

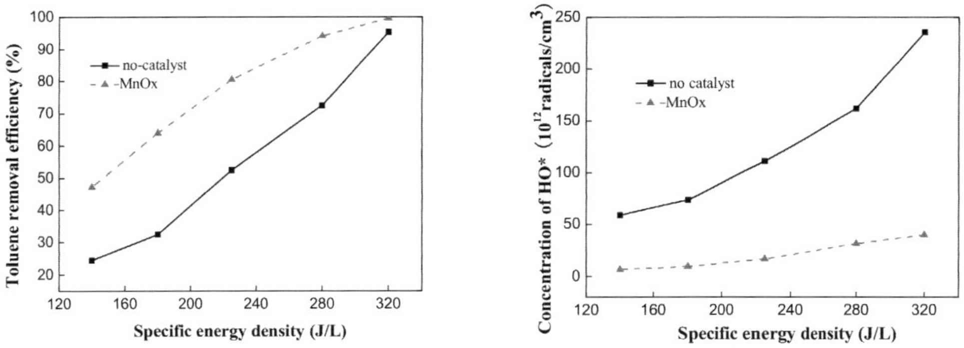

(a)Toluene removal efficiency (b) Hydroxyl radicals

Fig.3 Effect of specific energy density on the toluene removal efficiency and HO* formation

(Relative humidity: 20%; Toluene initial concentration: 400mg/m^3; Gas flow rate: 450ml/min)

3.2 Effect of relative humidity on toluene removal efficiency and hydroxyl radicals

From Fig.4 (a) and (b), it could be seen that there was a parallel trend between the HO* density and toluene removal efficiency without catalyst in gas stream at various relative humidities (RH). The highest HO* yield and toluene removal efficiency were achieved with a RH of 20%. However, when water was added to the DBD reactor with catalyst in situ, there was no direct relationship between the quantities of HO* in the gas stream and the toluene removal efficiency. The highest HO* yield was also obtained at a RH of 20%, and there was a fine distinction of HO* yield among various humidities. But the toluene removal efficiency decreased with the increase of humidity when catalyst added. It is possible that water covers on the surface of the catalyst and results in a decrease in the reaction probability.

The highest HO* yield and toluene removal efficiency were achieved with a RH of 20% because of the electronegative characteristic of water. When the humidities ranged from 0 to 20%, with the increasing of RH, more H_2O molecules collide with high energy electrons and form OH radicals, resulting in a higher removal efficiency. However, water also has an adverse effect due to its electronegative characteristics. Increasing the humidity limits the electron density in the system and quenches the activated chemical species [23]. So the HO* yield decreased when RH exceeded to 20%.

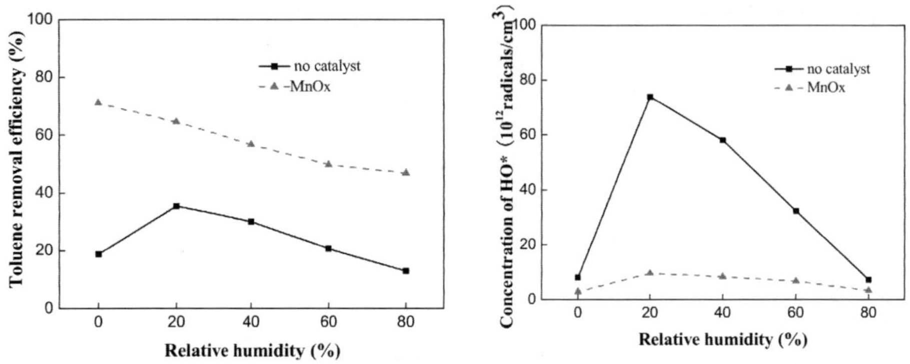

(a) Toluene removal efficiency (b) Hydroxyl radicals

Fig.4 Effect of relative humidity on (a) toluene removal and (b) hydroxyl radicals

(Specific energy density: 180J/L; Toluene initial concentration: 400 mg/m^3; Gas flow rate: 450 ml/min.)

3.3 Characterization of catalyst

3.3.1 BET

The BET surface area of the MnO_X catalyst was measured at 84.44 m^2/g. Compared this data with other metal oxide catalysts in the former article [2], it was found that manganese oxide owns a much larger BET surface area. Commonly speaking, a higher BET surface area means a higher dispersion of the catalyst and a bigger contact area between active component and gas, which leads to a better catalytic activity. Therefore, manganese oxide catalyst has a good catalytic activity, and this is in good consistent with the results of the plasma-catalytic performances test on toluene decomposition.

3.3.2 XPS

XPS technique is suitable for identifying the various states of species deposited on the catalyst. Fig. 5 shows the complex chemical composition of the MnO_X on the catalyst surface through XPS spectrum. The C1s region exhibited aromatic/Alkanes, aliphatic, and oxygen-bearing carbon.

The survey spectrum was recorded in the binding energy (BE) ranging from 0 to 1400 eV (Fig. 9(a)). The peak positions corresponding to Mn2p, O1s, N1s and C1s have been searched in the BE range of 635-660 eV, 528–534 eV, 395-402 eV and 280-288 eV, respectively. The curves obtained have been shown in Fig. 9(b), (c) and (d). The peak position corresponding to Mn 2p 1/2 and Mn 2p 3/2 are respectively 654.8 and 642.6 eV, and this result is in good agreement with that given in the article [24].

According to the XPS data-base, there are three oxygen species on the catalyst surface. A BE at 528.8 ~ 529.7eV was a characteristic peak of lattice oxygen O^{2-}(OL). A weak adsorption of oxygen O^{-}(OC, A) on the surface falls at a BE of 531.0 ± 0.5eV. And BE at 532.6 ± 0.5eV belongs to the adsorbed oxygen molecule O^{2-}(OC, A). The presence of the peak in a BE curve at 654.8 eV is assigned to Mn^{2+}, and 529.0 eV represents lattice oxygen. Therefore XPS studies indicate Mn^{2+} and oxygen in O^{2-} chemical states, and confirms the formation of MnO_X when $Mn(NO_3)_2$ was calcined at 450 ℃. In addition, from Fig. 9(d) it could be seen that a small amount of N elements has also appeared on the catalyst surface. The source of N on the catalyst comes from the adsorption of NOx or N radicals dissociated from N_2 in background gas.

The BE for C1S was at 282 -285 eV and 285-288 eV. According to lasurface data-base, the BE of 282 -285 eV due to the existence of Al-C, C-O-Al, C-H, and 285-288 eV ascribes to C-N,C=O，C-O,C-N or the OH in the air plasma.

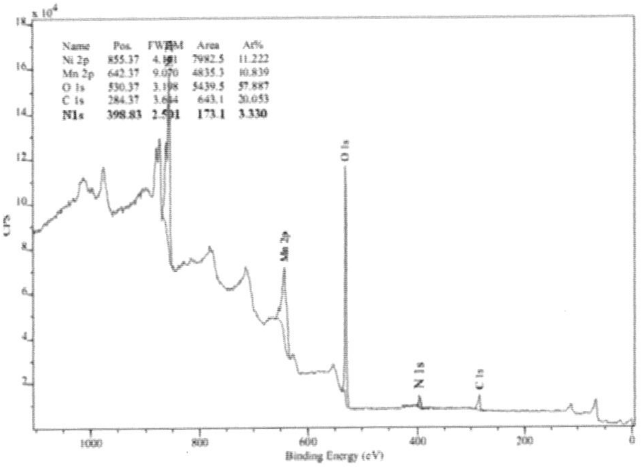

(a) XPS survey spectrum for the catalyst after reaction

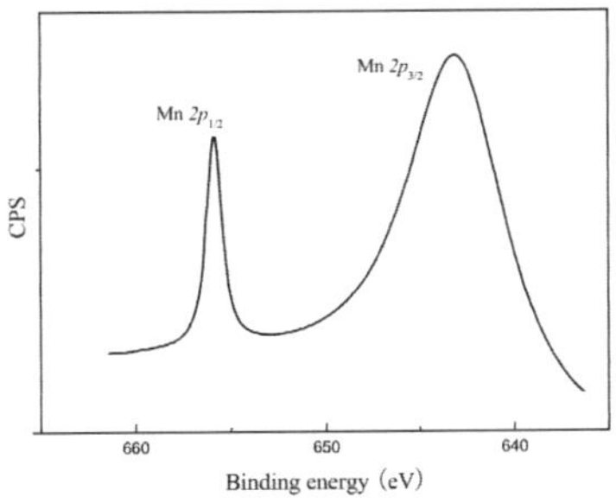

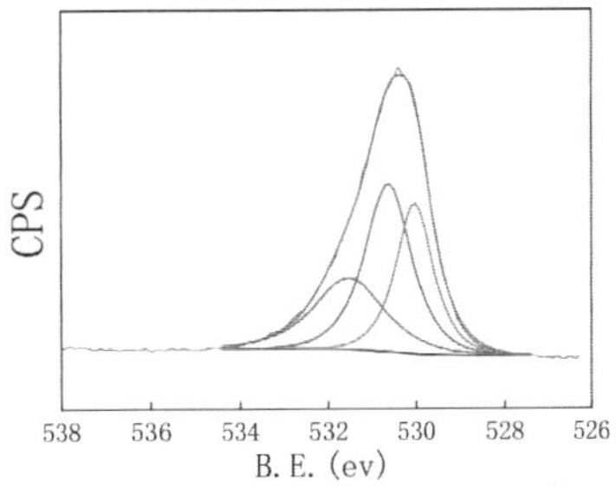

(b) XPS spectra of $Mn^{4+}2p$ (c) XPS spectra of O^1s

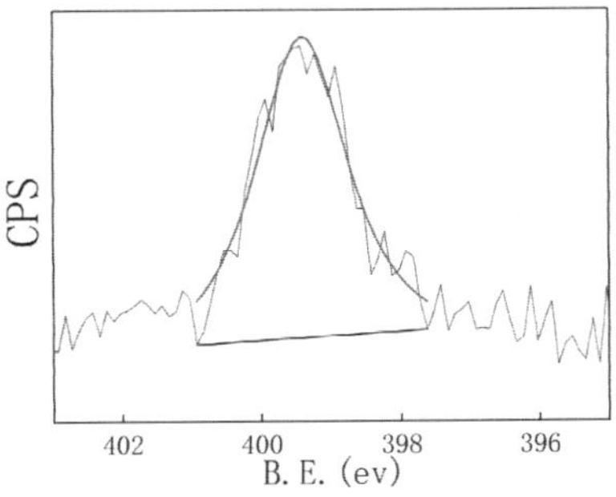

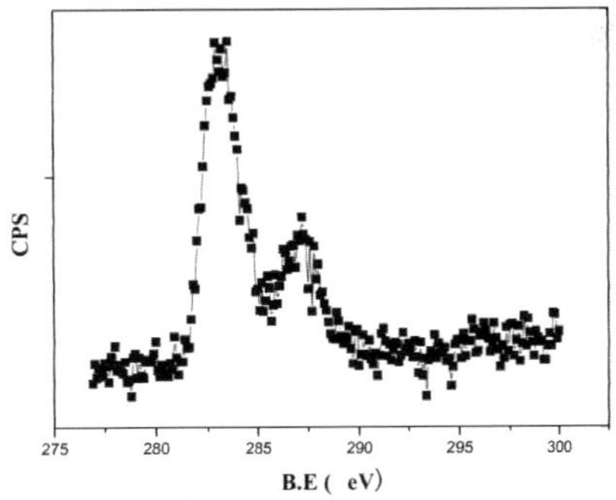

(d) XPS spectra of N (e) XPS spectra of C

Fig. 5. (a) XPS survey spectrum for the catalyst after reaction, (b) XPS spectrum for Mn, (c) XPS spectrum for O.(d) XPS spectrum for N element (e) XPS spectrum for C element

3.3.3 TPR

Fig.6 shows the TPR curves for MnO_X catalysts. Under the experimental conditions，there are two distinct reduction peaks of α (the former) and β (the latter). And the reduction temperature was 350 ℃and 460 ℃, respectively. This result is in good agreement with the H_2-TPR results of MnO_X reported in the literature [25].

These two peaks could be attributed to the two-step reduction of MnO_X. The first step corresponds to the reduction of MnO_2 to Mn_3O_4 and the second step is due to the further reduction of Mn_3O_4 to MnO. And the reduction of $Mn_3O_4+H_2 \rightarrow 3MnO+H_2O$ consumes more hydrogen than that of $3MnO_2+H_2 \rightarrow 2Mn_3O_4+H_2O$.

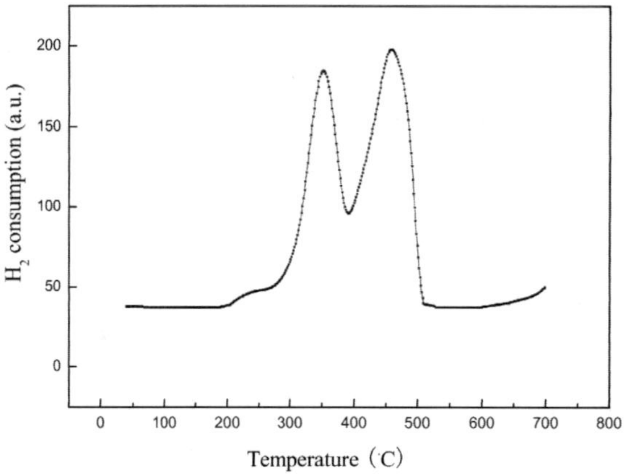

Fig.6 TPR curves for the MnO_X catalysts

3.3.4 FT-IR

Fig.7 shows the FT-IR survey spectrum of sample A and sample B for MnO_X after reaction. Sample A was made of MnO_X catalyst after reaction loaded with SAL, and sample B did not load with SAL. The result of FT-IR proved that hydroxyl existed in the spectrum of sample A, but not existed in sample B. For example, the band of 1660 cm^{-1}, which appears in sample A but not in sample B, presents C=O stretching band of DHBA. The band of 4000–3000 cm^{-1} was the hydrogen bond networks for DHBA, and 3500-3200 cm^{-1} belongs to the peak of OH. The 1580 cm^{-1} bands related to benzene ring. And the 1450 and 1300 cm^{-1} bands was the peak of C=C stretching of benzene nucleus. The 1250 cm^{-1} band represents C-O stretching of COOH atom. 1157 and 1034 cm^{-1} peaks are also assigned to COH and other vibrations. The 895 cm^{-1}, 700 cm^{-1} and 661 cm^{-1} band related to COH, OCO and other vibrations. Compared the spectrum with SAL in the reference [26], there are two groups of COH, and the band of 3000-2800 cm^{-1} belongs to the stretching vibration of saturated C-H of liquid paraffin.

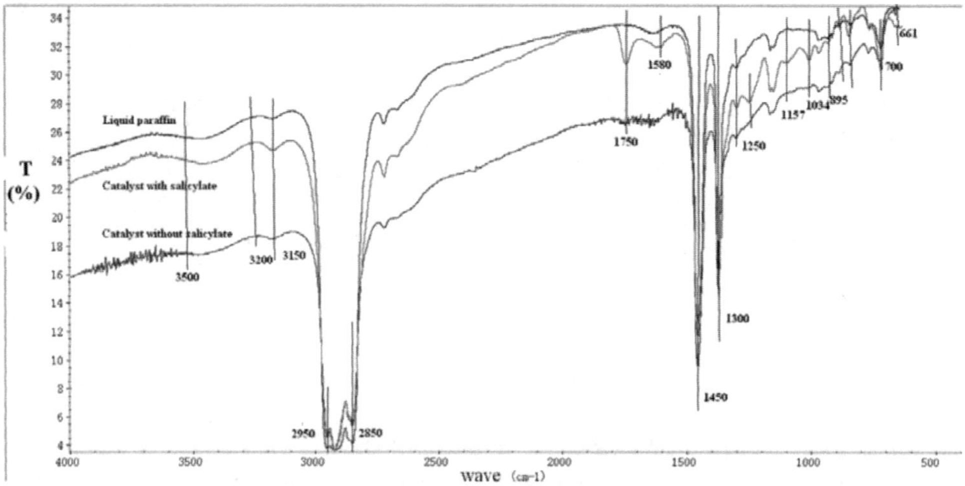

Fig.7 FT-IR survey spectrum of sample A and sample B. Sample A: catalyst with salicylate, Sample B: catalyst without salicylate

4. Conclusions

Toluene removal efficiency increased significantly when MnO_X catalyst was introduced into the discharge area. And the concentration of HO* in gas stream with catalyst is much lower than that without catalyst. The quantities of HO* radicals in the gas phase and on the surface of catalyst have been described in a plasma-catalysis system. The quantities of HO* radicals on the surface of catalyst was five times more than that in the gas phase. It proves that HO* would be attached on the surface of catalyst. In addition, the results of XPS and FT-IR also proved that various oxygen species, hydroxyl and some compounds containing N element deposited on the catalyst after plasma reaction. Therefore, combination of non-thermal plasma and catalysts could control the distribution of active oxygen species in situ such as ozone (O_3), oxygen atoms, hydroxyl radical (HO*), and final oxidation products such as CO and CO_2.

Acknowledgement

This work was supported by the National Natural Science Foundation of China (No.50708021), the Guangzhou University Xinmiao Project (No. gyfl-1001), and the Guangzhou University Scientific Research Startup Project (No. gyfl-1002). In addition, key laboratory of environmental protection and eco-remediation of Guangdong regular higher education institutions was gratefully acknowledged.

References

[1] H.L. Chen, H.M. Lee, S.H. Chen, Y. Chao, M.B. Chang, Appl. Catal. B: Environ.85 (2008) 1.

[2] Y.F Guo, D.Q Ye,K.F Chen, J.C He, Catal. Today 126 (2007) 328

[3] S.Fuamura, A. Zhang, H.Einaga, H. Kabashima, Cataly. Today 72 (2002) 259

[4] W. Mista , R. Kacprzyk， Catal. Today 137 (2008) 345

[5] M. B. Chang, H. M. Lee, Catal. Today 89 (2004) 109.

[6]H.-H. Kim, S.-M. Oh, A. Ogata, S. Futamura, Appl. Catal. B Environ. 56 (2005) 213.

[7] T. Oda, J. Electro. 57 (2003) 293.

[8] R. D. Bach, O. Dmitrenko, J.Am.Chem.Soc. 128(2006)1474.

[9] W. J. Bian, X. L.Ying, J. W. Shi, J. Hazard. Mater.162 (2009) 906.

[10] S. Kanazawa, T. Ito, Y. Shuto, T. Ohkubo, Y. Nomoto, J. Mizeraczyk, IEEE Trans. Ind. Appl. 37 (2001) 1663.

[11] S. Kanazawa, Y. Shuto, N. Sato, T. Ohkubo, Y. Nomoto, J. Mizeraczyk, J.S. Chang, IEEE Trans. Ind. Appl. 39 (2003) 333.

[12] G.J. Roth, M.A. Gundersen, IEEE Trans. Plasma Sci. 27 (1999) 28.

[13] H. Hazama, M. Fujiwara, M. Tanimoto, Chem. Phys. Lett. (2000) 542.

[14] F. Fresnet, G. Baravian, S. Pasquiers, C. Postel, V. Puech, A. Rousseau, M. Rozoy, J. Phys. D: Appl. Phys. 33 (2000) 1315.

[15] F. Tochikubo, T. Watanabe, Hakone VII, 1(2000) 219.

[16] A. Ershov, J. Borysow, J. Phys. D: Appl. Phys. 28 (1995) 68.

[17] R. Ono, T. Oda, J. Electrost. 55 (2002) 333.

[18] F. Tochikubo, S. Uchida, T. Watanabe, Jpn. J. Appl. Phys 43 (2004) 315.

[19] S. Kanazawa, H. Tanaka, A. Kajiwara, T. Ohkubo, Y. Nomoto, M. Kocik, J. Mizeraczyk, J.-S. Chang, Thin Solid Films 515 (2007) 4266

[20] L. Magne, S. Pasquiers, C. R. Physique 6 (2005) 908

[21] Y.F. Guo, X.B. Liao, D.Q. Ye, J Environ Sci-China 20 (2008)1429
[22] X. Ren, K. Shao, G. Miao, China Environ. Sci. 21(2001) 115.
[23] Y.F. Guo, D.Q. Ye, K.F. Chen, Y.F.Tian, Plasma Chem Plasma Process 26(2006) 237
[24] L.J.Yuan, Z.C.Li, J.T.Sun, K.L.Zhang,Y.H.Zhou, Mater. Letters 57 (2003) 1945
[25] A.Chen, H.L Xu, Y.H Yue, W.M Hua, W.Shen, Z.Gao, J.Mole.Cataly.A: Chem. 203 (2003) 299
[26] L.M Yang, Y.Z.Xu, Y.L. Su , J.G Wu,K.Zhao, J.Chen, M.KWang，Spectroch.Acta Part A 62 (2005) 1209

Hydrogen production by reforming of logistic diesel fuel in supercritical water

Carsten Cremers, Dominik Müller, Karsten Pinkwart, Jens Tübke
Fraunhofer Institute for Chemical Technology (ICT), Pfinztal, Germany

Introduction

Compared to other fuels, diesel fuels offer significant advantages with respect to logistic properties such as energy density, availability and ease of handling. That is why they are still the preferred fuel for all larger scale military applications. However, the only way to produce electrical energy from diesel available today is the use of internal combustion engines (ICE) driven generators. The use of such ICE generators has some disadvantages. For example they have a strong IR signature and produce a lot of noise. Fuel cells and in particular low temperature fuel cells do not show these disadvantages. However, they require hydrogen as fuel in most cases. In principle it should be possible to produce hydrogen from diesel via a reforming reaction, but common hydrocarbon reformers operate at very high temperatures eliminating one important advantage of using a low temperature fuel cell. Furthermore, the evaporation of diesel often leads to diesel decomposition and soot formation clogging the tubing downstream of the actual reactor. Special measures need to be taken to avoid this reaction. In case that the reforming reaction is done via catalytic partial oxidation (CPO), a process has been suggested by the group of Aicher, where a part of the reaction is performed during heating-up and evaporation of the fuel[1]. It is reported that in this way diesel can be evaporated without soot formation. However, the entire process has some disadvantages. First the hydrogen yield from CPO is much lower compared to steam reforming reaction. The same holds true for the hydrogen content in the product gas. Furthermore, the CPO reaction also runs at very high temperatures and is strongly exothermic so that heat needs to be released at these elevated temperatures. The group of Peters has reported on an ATR diesel and jet fuel reformer explicitly developed to feed a HT-PEMFC. For this approach also a lot of effort has to be made in fuel evaporation and mixing of fuel with vapor and oxygen in order to avoid either soot formation or the evolution of hot spots[2]. A disadvantage of both systems CPO and ATR is that the reforming process does not offer much possibility to use chemical energy in the anode off-gas or reject gas of a membrane separation unit. Only in case that the reformer is used to feed an SOFC anode, off-gas may be used to heat up the cathode air supply as described by Apfel et al.[3]

Alternatively steam reforming has been investigated for hydrogen production from diesel. Beside the problem of evaporating the diesel fuel, here in addition heat needs to be delivered to this highly endothermic reaction. Best success has for been found using micro-structure technology with integrated combustors. Such reactors for hydro carbon fuels have been developed e.g. by Forschungszentrum Karlsruhe[4], the institute for Mikrotechnik Mainz (IMM)[5] and by PNNL/ Batelle[6]. However most of these reformers are built for smaller power classes and all of them are not able to operate on fuels with high sulfur contents as defined in most specifications for logistic diesel fuels.

The approach described here is to do the reforming in supercritical water. The chemistry of the actual reforming reaction is the same as for the steam reforming reaction. So the reactor can exploit all benefits of the SRF like high yield of hydrogen and offers a heat sink for reject gas or anode off-gas because of the endothermic reaction. However, the process differs in many points from the usual steam reforming which renders it beneficial in comparison. Some of these advantages are the possibility to dissolve the diesel fuel in supercritical water instead of evaporating it, the high pressure facilitating gas clean-up via a separation membrane and the high excess of water in the process allowing for reduced reaction temperatures as well as the transport of heat within the supercritical water stream. All of these advantages help to design a simple reactor suitable to deliver ultrapure hydrogen for PEMFC from diesel fuels

Basic information on the reactant and reactions

Supercritical water is an environmentally friendly medium and it is well known in industry and science. Thermo-physical properties of supercritical water such as density, viscosity, relative permittivity and hydrogen bonding are quite different from those of steam or liquid water[7]. In contrast to the liquid state, mixing of supercritical water with non-polar gases and organic compounds containing large, non-polar groups is generally possible[8]. Gasification of hydrocarbons in supercritical water is based on the characteristics that supercritical water acts not only as a solvent but also as a reagent[9]. Table 1 gives a compared overview for density ρ, dynamical viscosity η and static dielectricity ε of the liquid, gaseous and supercritical aggregate state.

Table 1: Density, dynamical viscosity and static dielectricity of water in different aggregate states

aggregate state	liquid (25 °C, 0.1 MPa)	supercritical (400 °C, 25 MPa)	gaseous (400 °C, 0.1 MPa)
ρ / g cm^{-3}	1	0.17	0.0003
η / mPa s	0.89	0.03	0.02
ε	78	5.9	1

The reference diesel which was investigated had a chemical composition corresponding to the molecular formula $C_{12.95}H_{24.38}$. No data concerning its critical behavior were available. Therefore it was replaced in parts of the experiment by n-decane $C_{10}H_{22}$. To the latter thiophene could be added in variable amounts to simulate sulfurous impurities. Physical data of n-decane at its critical point are given in table 2.

Table 2: Physical data of n-decane at its critical point

$T_{critical}$ / °C	$p_{critical}$ / bar	$\rho_{critical}$ / g cm^{-3}
345	22.99	0.228

Steam reforming of n-decane would proceed according to equation (1) forming CO which could be further converted to hydrogen via water gas shift reaction (2).

$$C_{10}H_{22} + 10\,H_2O \rightarrow 10\,CO + 21\,H_2 \qquad (1)$$

$$CO + H_2O \rightarrow CO_2 + H_2 \qquad (2)$$

Standard heat of reactions for theses reaction can be found in literature and are given by e.g. Dreyer et al.[10] with ΔH^0_{SRF} = 1563 kJ mol^{-1} and ΔH^0_{WGS} = -41 kJ mol^{-1}.

Taking into account the high excess of water which will be present in the supercritical reaction and the lower reaction temperature it can be assumed that CO_2 will be the main product of the reaction, leading to reaction equation (3)

$$C_{10}H_{22} + 20H_2O \rightarrow 10CO_2 + 31H_2 \quad (3)$$

The hypothetical standard enthalpy of reaction calculates to ΔH^0_{SCRF} = 1153 kJ mol^{-1}.

Experimental set-up

The basic set-up for the tests of the catalyst for supercritical reforming of diesel has been described elsewhere in more details[11]. Basically it comprises two reactor sections each fed independently with a HPLC pump. In the first section water is heated to the desired temperature above the critical temperature. The reactor is built as capillary coiled onto an electrically heated cylinder and water is lead through the capillary. The capillary is then connected to the second downstream part of the reactor via a T-fitting. The fuel is pressed into the system through the third connection. The second part of the reactor consists of a tubular fixed-bed reactor. A scheme of the set-up is given in figure 1.

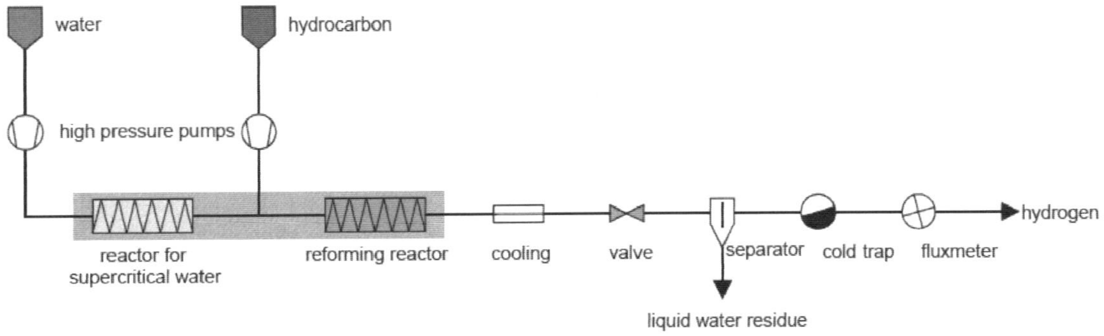

Figure 1: Schematic of the catalyst test-bed for supercritical reforming

In a modification of the experiments it was tested, whether the supercritical conditions could be used in the desulphurization of fuels with high sulfur content like NATO F76 fuel. For this purpose a two stage analytics was added to set-up depicted in figure 1. The first stage is applied directly after the first separator; here the gas phase is tested for its content of sulfurous components using gas chromatography. The sulfur content in the liquid phase is measured by electrochemical methods. After the separator an alkaline scrubber is placed to adsorb acidic sulfurous components from the gas phase which left the reactor. After the scrubber gas phase and liquid phase are again tested for remaining sulfurous components either via gas chromatography or electrochemically. A flow chart of this modified set-up is given in figure 2.

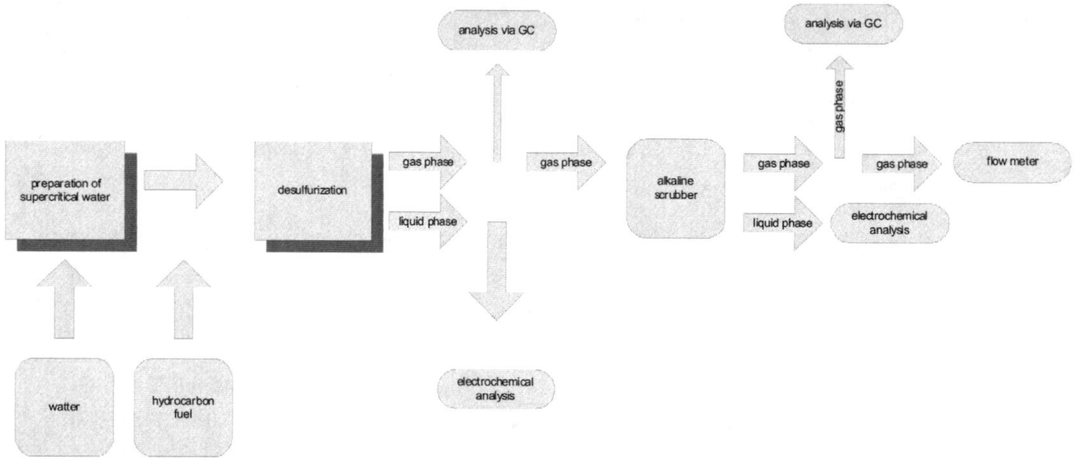

Figure 2: Flow chart for test with fuels with high sulfur content

Finally a demonstration unit with a calculated capacity for a hydrogen production of about 210 W thermal power, sufficient to feed a 100 W PEMFC, was designed, built and tested. The piping and instrumentation diagram of the demonstration unit is depicted in figure 3.

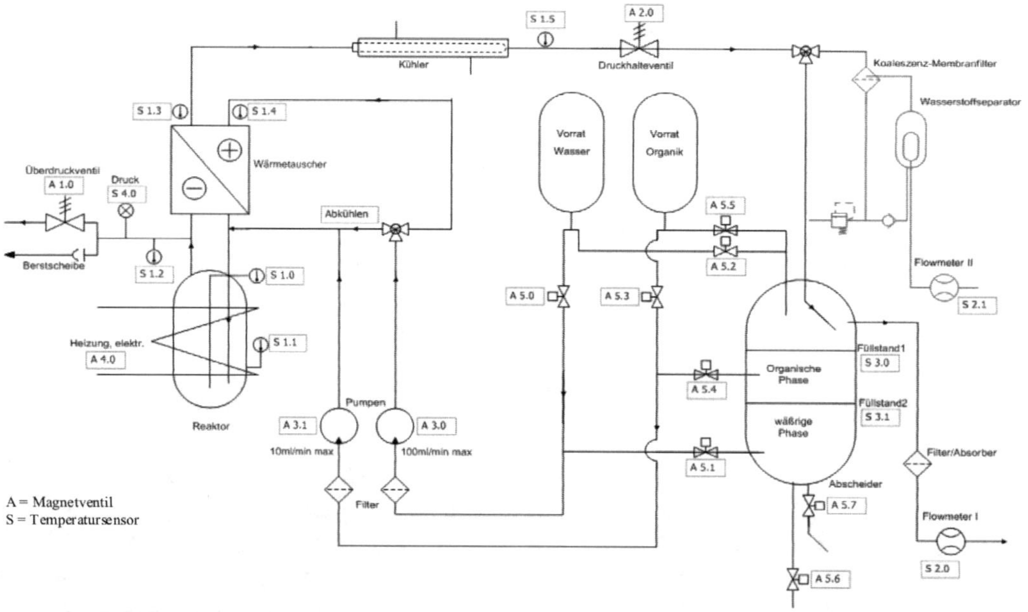

Figure 3: P&ID of the demonstration unit

Results

As there are no particular catalysts for reforming under supercritical conditions available today, a number of catalysts for the reforming of different substrates were tested at the beginning. Table 3 lists the tested catalysts.

Table 3: Specification of tested catalysts

	catalyst A	catalyst B	catalyst C	catalyst D
typically used for	naphtha, natural gas, LPG	natural gas	naphtha	methanation
Component		weight fraction / wt%		
NiO – Nickel oxide	45.2	>1	20	32.4
MgO – Magnesium oxide	12	-	12-16	-
K_2O – Potassium oxide		-	8.5	-
CaO – Calcium oxide			8-10	4
SiO_2 - Silicon oxide	10		14-16	
Al_2O_3 - Aluminium oxide	10		30-40	variable
carrier		$CaAl_{12}O_{19}$ – Calcium-aluminate	Calcium-aluminate	
shape	tablets	ring	ring multi hole	spheres
catalyst-volume / mL	0.072			0.047
catalyst-surface / mm^2	95			64
bulk density / kg/L	0.95-1.0	0.9	1.0	1.0
mass of catalyst in the reactor / g	15.0	15.9	14.8	13.4

Tests were performed using n-decane as substrate with a fraction of 10% (v/v) in supercritical water. The reaction temperature was set to 550 °C and the pressure was set to 250 bar. The fluid hourly space velocity was 360 h^{-1} corresponding to an average residence time of 10 s. In general with all the catalysts hydrogen, methane and carbon dioxide were the main components in the reactor effluent. The yields of these products differ however. With respect to hydrogen production yield the activity decreases in the order D >> B > A >> C (cf. figure 4). Of these catalysts D and A achieve the highest conversion of about 80% of the n-decane. The achieved conversion with catalyst B is about 45% whereas with catalyst C only 15% conversion is achieved. Thus, catalyst D is the most suitable for the reforming reaction. This is assigned to the content of alkaline and alkaline earth metals in the catalyst[11], a positive effect which has also been found earlier for the reforming via catalytic partial oxidation[9]. The high yield of carbon dioxide as compared to carbon monoxide is expected as a thermodynamic consequence of the high excess of water in the reaction mixture. The high yield of methane should be due to the comparatively low reaction temperature which favors the Methanation reaction.

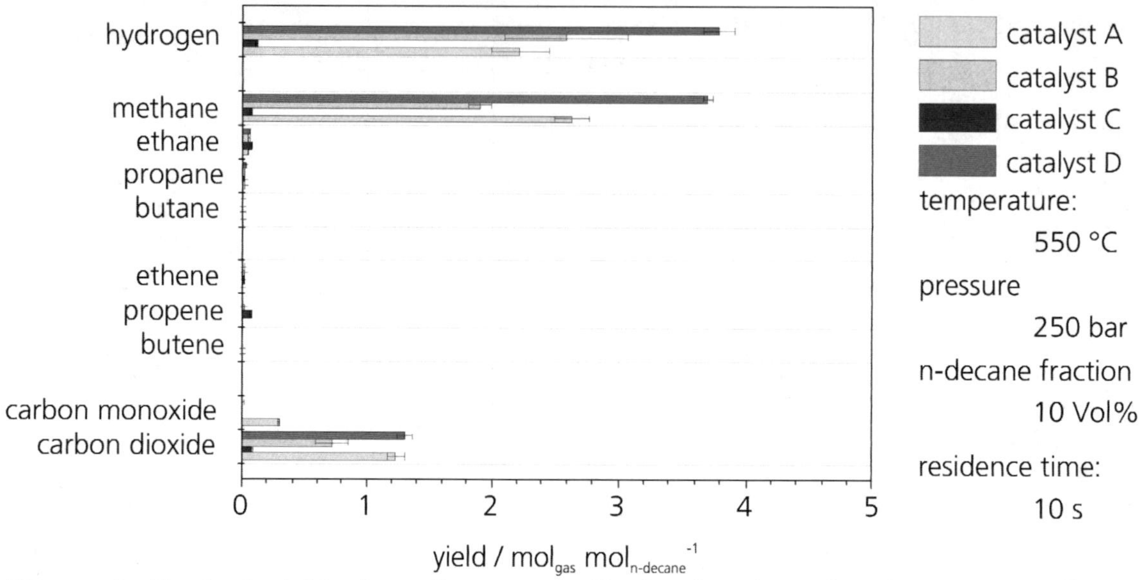

Figure 4: Product yields from the supercritical reforming of n-decane at different catalysts

Catalyst D was therefore used for the investigation of low sulfur diesel reforming. It was found that the reforming is possible with similar yields and conversion rates when the space velocity was further reduced to values of about 90 h^{-1}.

In a next step the use of NATO diesel with higher sulfur content was investigated. Here it was found that the base metal catalyst used successfully for the low sulfur content fuel reformation does not work as properly when the sulfur content is increased. Therefore, alternative noble metal catalysts were employed. Using real world NATO diesel with up to 2 g l^{-1} of sulfur the reaction was optimized with respect to the parameters of diesel fraction in water and the fluid hourly space velocity. The diesel fraction was varied between 2.5% (v/v) and 10% (v/v) and the FHSV between 180 h^{-1} and 720 h^{-1}. The temperature was set to 550 °C and the pressure to 250 bar, values which have been identified as best values before.

In general, the yield of higher hydrocarbons is higher than it has been found for low sulfur fuels. Another important trend is that the hydrogen yield decreases with increasing throughput and increasing fuel fraction in the feedstock (cf. Figure 5). With respect to the sulfurous products it was found that hydrogen disulfide is the major sulfur containing product under all conditions. At high throughputs the yield of hydrogen disulfide increases surprisingly with an increased fuel fraction in the feedstock. Also surprisingly the content of thiophene in the product stream is reduced at increased throughput (cf. Figure 6).

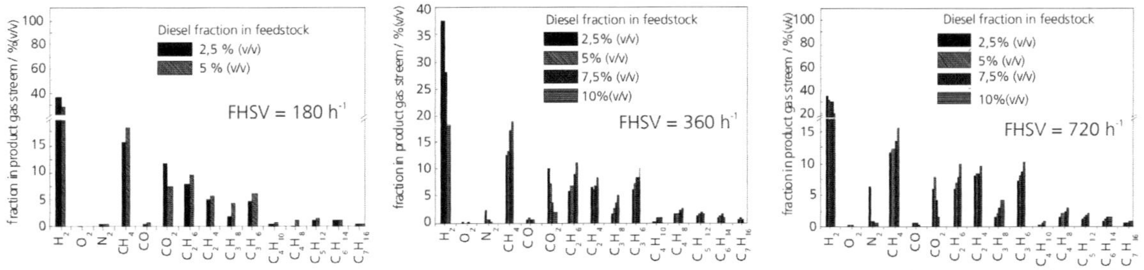

Figure 5: Hydrogen and hydrocarbon product yields for the reforming of sulfur containing diesel at different space velocities and with different fuel concentrations in feedstock.

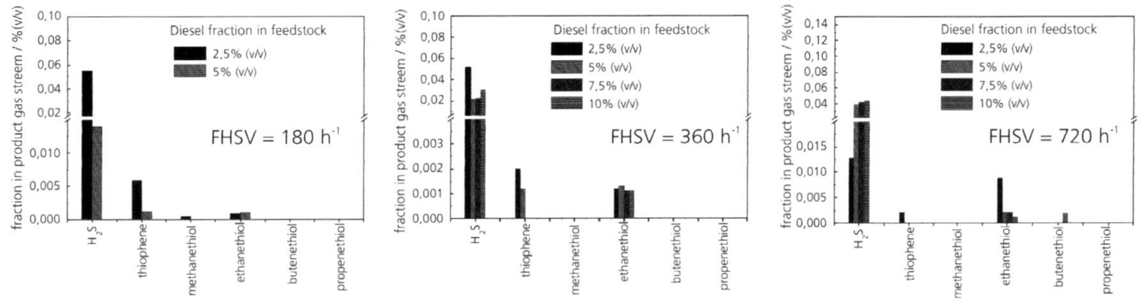

Figure 6: Yields of sulfur containing products from the reforming of sulfur containing diesel at different space velocities and with different fuel concentrations in feedstock

A major drawback is so far the fast poisoning of the catalysts in presence of high amounts of sulfur in the feedstock. Therefore the demonstrator is currently operated with low sulfur content diesel using a catalyst similar to catalyst D from the initial tests.

First tests with the demonstration unit showed, that it is capable to deliver pure hydrogen in the desired quantity.

References

1. Aicher T, Griesser L. Novel process to evaporate liquid fuels and its application to the catalytic partial oxidation of diesel. *Journal of Power Sources.* Feb 2007;165(1):210-216.
2. Pasel J, Latz J, Pors Z, et al. Autothermal Reforming of Jet A-1 and Diesel: General Aspects and Experimental Results. In: Williams M, Krist K, Garland N, Satyapal S, eds. *Fuel Cell Seminar 2007.* Vol 12. Pennington: Electrochemical Society Inc; 2008:589-600.
3. Apfel H, Rzepka M, Tu H, Stimming U. Thermal start-up behaviour and thermal management of SOFC's. *Journal of Power Sources.* Mar 2006;154(2):370-378.
4. Cremers C, Pelz A, Stimming U, et al. Micro-structured methane steam reformer with integrated catalytic combustor. *Fuel Cells.* Apr 2007;7(2):91-98.
5. O'Connell M, Kolb G, Schelhaas KP, Schurer J, Ziogas A, Hessel V. An Investigation into the Transient Behavior of a Microreactor System for Reforming of Diesel Fuel in the kW Range. *Chemical Engineering & Technology.* Nov 2009;32(11):1790-1798.
6. Hu J, Wang Y, VanderWiel D, et al. Fuel processing for portable power applications. *Chemical Engineering Journal.* 2003;93(1):55-60.
7. Franck EU. FLUIDS AT HIGH-PRESSURES AND TEMPERATURES. *Pure and Applied Chemistry.* Jan 1987;59(1):25-&.

8. Tödheide K. Wäßrige Lösungen bei hohen Drucken und Temperaturen. *Berichte der Bunsen-Gesellschaft für Physikalische Chemie.* 1966;70(9/10):1022-1030.
9. Garcia L, French R, Czernik S, Chornet E. Catalytic steam reforming of bio-oils for the production of hydrogen: effects of catalyst composition. *Applied Catalysis a-General.* Jul 2000;201(2):225-239.
10. Dreyer BJ, Lee IC, Krummenacher JJ, Schmidt LD. Autothermal steam reforming of higher hydrocarbons: n-Decane, n-hexadecane, and JP-8. *Applied Catalysis A: General.* 2006;307(2):184-194.
11. Pinkwart K, Bayha T, Lutter W, Krausa M. Gasification of diesel oil in supercritical water for fuel cells. *Journal of Power Sources.* 2004;136(2):211-214.

Catalytic Reforming On Hexaaluminate Based Catalyst Supports

Nicholas E. McGuire[1], Neal Sullivan[1], Robert J. Kee[1] and David T. Wickham[2], (1)Engineering Division, Colorado School of Mines, Golden, CO, (2)Reaction Systems LLC, Golden, CO

Introduction

This work considers dry reforming of methane using rhodium particles supported on a strontium-substituted hexaaluminate. Syngas (CO and H_2) is typically produced via the steam-reforming reaction on a nickel catalyst. Dry reforming of methane to produce syngas offers H_2/CO ratios more suitable for Fischer-Tropsch liquid-fuels synthesis. The Rh-on-hexaaluminate approach offers potential advantages for this process, including resistance to sintering, surface-area loss and carbon formation at the higher temperatures required for the endothermic dry-reforming reaction. Additionally, Rh is a much more active catalyst than is Ni [1] although it is certainly more costly.

The stagnation-flow geometry provides an ideal platform in which catalyst performance can be evaluated. From a fluid-mechanical and chemically reacting-flow perspective, the stagnation-flow geometry has some very important properties. Aside from some small edge effects, the temperature and species profiles within the gas-phase boundary layer below the stagnation surface depend only upon the distance from the surface. That is, the profiles are flat; they are independent of the radial position. The well-known fluid mechanical result is provable mathematically and observable experimentally. The chemical boundary layer that forms above the catalyst surface is amenable to precise, local gas sampling. This enables collection of data that is unattainable when using a conventional foam-based catalyst. These boundary-layer profiles can be used to infer reaction rates in heterogeneous chemical mechanisms, as demonstrated previously [2,3].

In this work we compare various conventional supports to hexaaluminate supports in terms of surface area retained after an accelerated aging process. To study catalytic performance we use microprobe mass spectrometry in a stagnation-flow reactor to measure gas-phase species profiles in the boundary layer normal to the catalyst surface. The experiments are interpreted using a chemically reacting flow model, including a modified version of an elementary heterogeneous chemical reaction mechanism previously published in the literature [4]. The hexaaluminate catalysts are shown to provide high surface area both before and after the accelerated-aging process; additionally, computational models are found to be in good agreement with experimental measurements for predicting the concentrations of most major species.

Experimental

Catalyst Preparation

Hexaaluminate catalyst supports are prepared by coprecipitation of appropriate nitrates. To prepare a calcium hexaaluminate, 3.78 g of $Ca(NO_3)_2$ (Sigma Aldrich) is added to 66 g of aluminum nitrate (Sigma Aldrich) and dissolved in 200 mL of water. The resulting mixture is stirred and 5.0 M ammonium hydroxide is added until a pH between 7 and 8 is measured. This mixture is then dried at 110 °C and calcined at 1300 °C. Similarly, the strontium hexaaluminate is prepared by dissolving 3.14 g of $Sr(NO_3)_2$ with 61.3 g of $Al(NO_3)_3$ in water. Ammonium hydroxide is added until the desired pH is achieved and then dried at 110 °C followed by calcination at 1300 °C. Control supports used are Tosoh-Zirconia YSZ (TZ-8Y) and Praxair gadolinium-doped ceria ($Gd_{0.1}Ce_{0.9}O_2$)

As discussed below, catalyst performance is evaluated in a Stagnation-Flow Reactor (SFR). Thus, additional processing steps are needed to apply the supported catalyst to a cast-ceramic surface that is placed within the reactor. 0.1 g of catalyst support powder is added to 11 ml of aluminum hydroxide, mixed, and then sprayed onto the cast-ceramic substrate until a thickness of approximately 20 μm is achieved. The coated ceramic structure is then calcined at 500 °C, converting the boehmite binder to Al_2O_3. 10 wt-% rhodium (relative to the mass of the hexaaluminate coating) is added to the catalyst layer by incipient wetness followed by heating of the stagnation surface to 800 °C for 1h. Following calcination, the rhodium is in the form of rhodium oxide. Prior to catalyst experiments, the catalyst is reduced in a flow of 3.5 vol % H_2 (balance Ar) at 500 °C for 3h, converting the rhodium oxide to rhodium metal.

Characterization

A number of diagnostics are used to characterize the catalyst and support structures, as well as to show the ruggedness of the hexaaluminate supports. Surface area is measured before and after an accelerated-aging process by single point N_2 adsorption. The aging process consists of exposing the catalyst to a mixture of 50% Ar, 19% H_2 and 31% H_2O for 24h at 900 °C. This process has been shown to effectively 'age' the catalysts to the point where the surface areas and dispersions are similar to catalysts that had been subjected to auto-thermal reforming of gasoline for over 100h [5].

Stagnation-flow Reactor

A detailed description of the Stagnation-Flow Reactor used in these experiments is presented in a previous publication [6]. Briefly, premixed reactive gases enter the stainless steel reactor and impinge upon a heated, downward facing, catalytically active stagnation surface. A steady flow of reactant gases is supplied from compressed-gas cylinders at rates specified via mass-flow controllers. After encountering the stagnation surface, gases continue flowing through an annular space through a controlled downstream butterfly valve that maintains the reactor pressure at 300 torr. Species composition profiles are measured by mass spectrometry via a microprobe that accurately traverses the boundary layer.

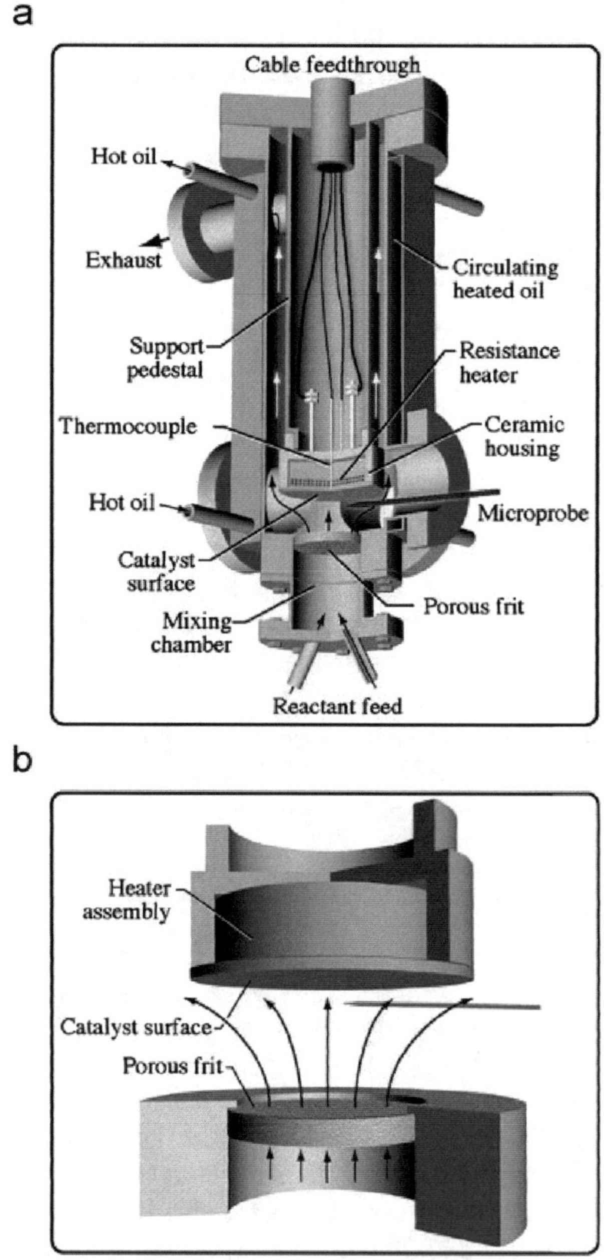

Figure 1: Illustration of the stagnation flow reactor

Boundary-layer measurements

A quartz microprobe is used to sample gas composition within the boundary layer. As gases enter the probe itself, they are expanded rapidly to vacuum conditions which quench any further reaction.

The probe is positioned with a precision stepping motor. Beginning at the stagnation surface, the probe is moved downward through the boundary layer. Upon arriving at each new measurement position, the probe is held stationary. It takes approximately four minutes for the flow field to achieve steady state following probe movement. Once the flow has achieved steady state, the gas-phase composition is measured via the mass spectrometer. A boundary layer profile typically consists of eight measurement points, spanning approximately 5 mm below the stagnation surface.

Once a boundary-layer profile is measured for a certain set of operating parameters, the operating conditions are adjusted (e.g., changing catalyst temperature or reactant composition). Upon changing operating conditions, it takes approximately 30 minutes to establish a new steady state. Once steady state is achieved, a new boundary-layer profile is measured.

The mass spectrometer is calibrated using a premixed gas with known concentrations of CH_4, CO_2, CO, H_2 and Ar. Mass spectrometer signals for the gases of interest (H_2 at 2 amu, CH_4 at 15 amu, CO at 28 amu, O_2 at 32 amu, Ar at 40 amu, and CO_2 at 44 amu) are measured relative to Ar, which is a chemically inert component of the reactant streams. This serves to eliminate signal changes by instrument instabilities. The only cross sensitivity considered is the CO_2 contribution to the CO signal.

Computational Methods

The modeling results in this paper are computed using the SPIN application of the CHEMKIN software package [7]. This software implements kinetic-theory representations of multicomponent gas-phase diffusive transport and elementary, multi-step, chemical-reaction mechanisms for gas-phase and surface chemistry. Details of the molecular-transport and chemical-kinetics formulations are well documented [8].

The catalytic reaction mechanism used here is a modified version of a mechanism published by Schwiedernoch et al [4]. The mechanism is comprised of 42 irreversible elementary reactions involving 13 surface-adsorbed species and 6 gas-phase species. This mechanism was developed and validated using a combination of theoretical and experimental investigations, spanning conditions for steam reforming and partial oxidation. Complete details about the reaction-mechanism development and validation studies are found in papers by the Deutschmann group at the University of Karlsruhe. The effective specific area of the catalyst is taken as a parameter that is adjusted to achieve agreement with the experimentally observed reforming rate. This parameter is constant for all experiments. For methane reforming at temperatures below 900 °C, the gas-phase chemistry can be neglected.

The major objective of the modeling is to assist the understanding and interpretation of the experimental results. Quantitative insight about reaction pathways can be achieved with the use of a validated, multi-step, elementary reaction mechanism. It may be noted that such analysis and interpretation is not available from models that use only global representations of the reforming chemistry.

Results and Discussion

Characterization

Table 1 compares the results of BET surface areas before and after aging for hexaaluminate supports and other commonly used Rh supports. The surface area retained decreases in order of $CaAl_{11}O_{18}$ > 8mol% YSZ > $SrAl_{11}O_{18}$ > $Gd_{0.1}Ce_{0.9}O_2$. Of the supports examined, the $Gd_{0.1}Ce_{0.9}O_2$ appears to have lost the most surface area at around 35%, while the other three supports lost only 5-10% by the aging process. The Sr-based hexaaluminate has the highest initial and final surface areas of these three and thus is chosen as the support for catalytic reforming tests in the stagnation flow reactor.

Table 1: BET Surface Area measurements

Material	Area
$CaAl_{11}O_{18}$	7.0
$CaAl_{11}O_{18}$, aged	6.6
$SrAl_{11}O_{18}$	14.3
$SrAl_{11}O_{18}$, aged	12.9
8 mol% YSZ	12.8
8 mol% YSZ, aged	11.8
$Gd_{0.1}Ce_{0.9}O_2$	9.3
$Gd_{0.1}Ce_{0.9}O_2$, aged	6.0

Dry Reforming

The catalyst formulation on the stagnation surface is 10% Rh on Sr-hexaaluminate. The inlet composition is 10 % CH_4 and 12 % CO_2 (balance Ar). The total inlet flow rate is 19.3 slm resulting in an inlet velocity of 123 cm/s. For the initial baseline case, surface temperature is maintained at 700 °C and reactor pressure is maintained at 300 torr.

Figure 2a shows measured (points) and predicted (lines) species profiles within the stagnation-flow boundary layer. The catalyst surface is at the left edge of the graph (x=0 mm), with gases flowing from right to left. It is clear from Fig 2a that there is little reforming activity at these conditions. The result is consistent with the model. The lack of reforming is due to formation of carbon on the catalyst surface, as the catalyst had to be regenerated to acquire reforming on all subsequent tests (below). Regenerating the catalyst requires exposure of the catalyst at temperature to oxygen followed by a re-reduction.

Figure 2b shows mole-fraction profiles for dry reforming methane at a catalyst surface temperature that is maintained at 800 °C. Figure 2b shows that there is reforming activity at the catalyst surface. The measurements capture the entire chemical boundary layer. In other words, all species mole fractions

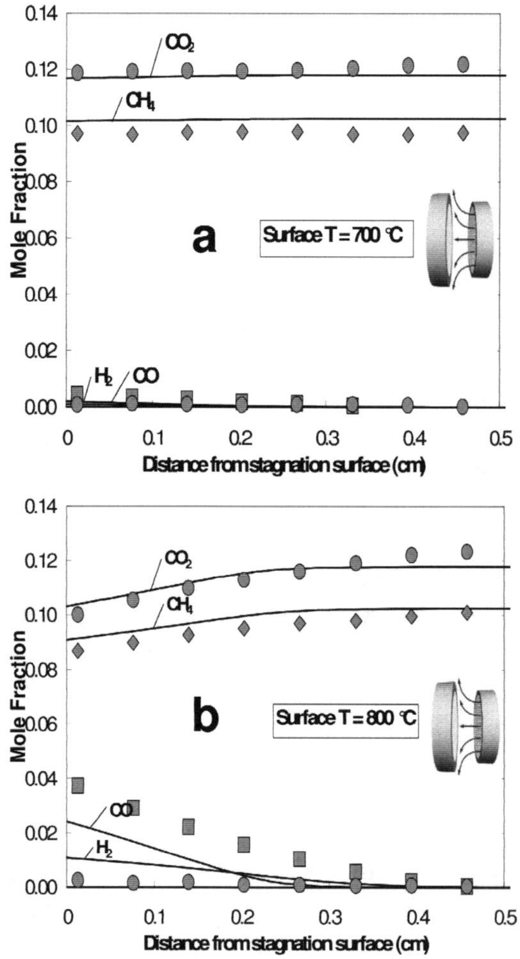

Figure 2: Boundary-layer mole-fraction profiles for CH_4/CO_2 = 0.1/0.12 and surface temperatures of (a) 600 °C (b) (a) 700 °C.

asymptotically approach the inlet values at a distance of about 4 mm from the stagnation surface. Approximately 14% of the methane supplied in the reactant stream is reformed and 18 %of the CO_2 is reformed.

To use the reaction mechanism, the effective surface area of the porous catalyst coating and catalytically active site density must be known. However, the effective catalyst area is not measured after the catalyst is applied to the stagnation surface. The effective catalyst area is taken to be a single parameter in the model that can be adjusted to match the overall measured reforming rate. The effective area is found to be a factor of 40 greater than the geometric surface area. The effective area, which is assumed to be a property of a particular catalyst-support configuration, is held fixed for all subsequent simulations. The model predictions agree reasonably with experimentally measured results.

Dry Reforming with O_2 addition

In a recent work, Donazzi et al [9] showed that a small addition of oxygen to the dry-reforming system increases the amount of CH_4 conversion. Similarly, we tried to perturb the system with more oxygen coverage by adding a small amount of oxygen to the feed stream. The catalyst formulation on the stagnation surface is the same as that of the previous section. The inlet composition is 10 % CH_4, 12 % CO_2, 1% O_2 (balance Ar). The total inlet flow rate is 19.3 slm resulting in an inlet velocity of 123 cm/s. Surface temperature is maintained at 600 °C for the baseline case and reactor pressure is maintained at 300 torr.

Figure 3a shows mole fraction profiles for the dry reforming case with oxygen addition. In this lower temperature condition it appears that there is no conversion of CO_2 and any conversion of methane is due to the presence of oxygen (catalytic partial oxidation). Approximately 11% of the methane is consumed at the surface and 100% of the oxygen is converted. The model predictions for this case agree reasonably well with the measured profiles.

Figure 3b shows mole fraction profiles for dry reforming with O_2 addition at a surface temperature of 700 °C. Approximately 14 % of the methane is converted and 2 % of the carbon dioxide is converted.

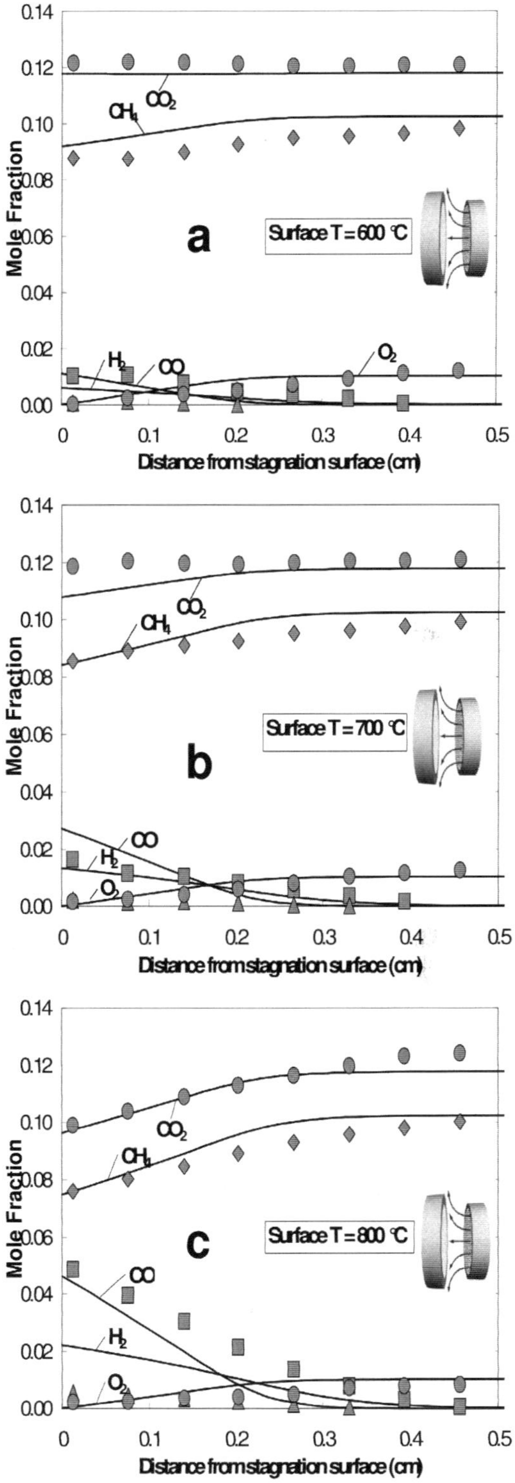

Figure 3: Boundary-layer mole-fraction profiles for $CH_4/CO_2/O_2 = 0.1/0.12/0.01$ and surface temperatures of (a) 600 °C, (b) 700 °C and (c) 800 °C

In this case it appears once again that the carbon dioxide, as experimentally measured, does not play a role in the conversion of methane. This is in contrast to the predictions of the SPIN model that show some CO_2 consumption.

Figure 3c shows mole fraction profiles for dry reforming with O_2 addition at a surface temperature of 800 °C. Approximately 25 percent of the methane is converted and 20 percent of the carbon dioxide is converted. Here, the model predictions agree with measured profiles with the exception of H_2 which was measured to be non-existent at the catalyst surface.

Conclusion

Rhodium supported on Sr-substituted hexaaluminate is found to perform well for the dry reforming of methane. BET surface-area measurements show that although most supports tested retained much of their initial surface area, it was found that the Sr-substituted hexaaluminate has the highest initial surface area and thus is chosen as the support for catalytic reforming tests.

The reforming activity is studied using a stagnation-flow reactor. Microprobe mass spectrometry is used to measure species profiles in the boundary layer adjacent to the catalyst surface. Indeed, the stagnation-flow configuration provides diagnostics access that would be extraordinarily difficult, if not impossible, in technologically viable catalyst configurations such as porous foams or channel monoliths.

The stagnation-flow geometry also facilitates computational modeling of the surface chemistry coupled with the gas-phase fluid flow and molecular transport. This paper reports comparisons between predicted and measured gas-phase boundary-layer profiles. A modified version of a previously developed detailed reaction mechanism is used. Despite the fact that the mechanism was developed for a Rh-Al_2O_3 catalyst-support system, it performs reasonably well for the current hexaaluminate-supported system.

The concerted experimental-modeling effort provides new fundamental data and insight concerning the structure and performance of hexaaluminate-supported catalysts. These results can be used in the design and development of catalytic reforming technologies in other geometric configurations.

Acknowledgement

This work has been funded in part by the Department of Energy, Office of Energy Efficiency and Renewable Energy, Contract Number DE-FG36-08GO88100.

References:

[1] J. Rostrup-Nielsen, J.-H.B. Hansen, "CO_2 reforming of methane over transition metals," *Journal of Catalysis*, **144** 38-49 (1993).

[2] J.D. Taylor, M.D. Allendorf, A.H. McDaniel, S.F. Rice, "In Situ Diagnostics and Modeling of Methane Catalytic Partial Oxidation on PT in a Stagnation-Flow Reactor," *Industrial & Engineering Chemical Research*, **42** 6559-6566 (2003).

[3] R.W. Sidwell, H. Zhu, R.J. Kee, D.T. Wickham, "Catalytic combustion of premixed methane-in-air on a high temperature hexaaluminate stagnation surface," *Combustion and Flame*, **134** 55-66 (2003).

[4] R. Schwiedernoch, S. Tischer, C. Correa, O. Deutschmann, "Experimental and numerical study on the transient behavior of partial oxidation of methane in a catalytic monolith," *Chemical Engineering Science*, **58** 633-642 (2003).
[5] M. Ferrandon and T. Krause, "Role of the oxide support on the performance of Rh catalysts for the autothermal reforming of gasoline and gasoline surrogates," *Applied Catalysis A:General*, **311** 135-145 (2006).
[6] N.E. McGuire, N.P. Sullivan, R.J. Kee, H. Zhu, J.A. Nabity, J.R. Engel, D.T. Wickham, M.J. Kaufman, "Catalytic steam reforming of methane using Rh supported on Sr-substituted hexaaluminate," *Chemical Engineering Science*, **64** 5231-5239 (2009).
[7] M. Coltrin, R.J. Kee, G. Evans, E. Meeks, F. Rupley, J. Grcar, *SPIN: a FORTRAN program for modeling one-dimensional rotating-disk/stagnation-flow chemical vapor deposition reactors.* Technical Report SAND91-8003, Sandia National Laboratories (1991).
[8] R.J. Kee, M.E. Coltrin, P. Glarborg, *Chemically Reacting Flow: Theory and Practice*, John Wiley, Hoboken, NJ, 2003.
[9] A. Donazzi, A. Beretta, G. Groppi, P. Forzatti, "Catalytic partial oxidation of methane over a 4% $Rh/\alpha\text{-}Al_2O_3$ catalyst Part II: Role of CO2 reforming," *Journal of Catalysis*, **255** 259-268 (2008).

(113d) Ni-Substituted Oxide Catalysts with Oxygen Ion Conductivity for Methane Steam Reforming

Mark Smith, REM Engineering Services, U.S. DOE NETL, Morgantown, WV

David A. Berry, Fuel Processing R&D, National Energy Technology Laboratory, Morgantown, WV

Dushyant Shekhawat, National Energy Technology Laboratory, Morgantown, WV

Daniel J. Haynes, Parsons, Inc., Morgantown, WV

James J. Spivey, Chemical Eng, Louisiana State University, Baton Rouge, LA

Nickel-substituted pyrochlore and hexaaluminate catalysts were studied for hydrogen production from methane steam reforming. The reaction conditions were as follows: T = 750°C, GHSV = 25,000 scc/gcat•hr, S/C = 0.9. Each reaction test was run for 5 hrs. The hexaaluminate catalyst, BaNixAl12-xO19-y, contained 6wt% nickel and was used as the base case. The pyrochlore catalysts, A2NixZr2-xO7-y (A = Mg, Ca, Sr), contained 3, 6, and 8wt% nickel. Further, the pyrochlore catalysts were calcined at three different temperatures (700, 800, and 900°C) to examine the effect, if any, on steam reforming performance. The use of an oxygen-conducting support, gadolinium-doped ceria, was also investigated. The catalyst that provided the highest, and most stable level of hydrogen production was the pyrochlore with A = Sr, a nickel loading of 6wt%, and a calcination temperature of 800°C. Under the reaction conditions given this catalyst provided a product stream with an initial hydrogen concentration of ~54% and a final concentration of 51%. Substitution of nickel into an oxide structure, like the pyrochlore, is believed to provide additional stability under methane steam reforming conditions. Further, the presence of oxygen-conductivity is believed to prevent the formation of deactivating carbon species on the active sites.

See more of this Session: Fuel Processing for Hydrogen Production I
See more of this Group/Topical: Topical 2: Advanced Fossil Energy Utilization

HYDROGEN AND CARBON NANOTUBES PRODUCTION OVER Co-Mg-Al CATALYSTS BY CATALYTIC DECOMPOSITION OF METHANE

N. Latorre, C. Royo, E. Romeo and A. Monzón
Institute of Nanoscience of Aragon (INA). Department of Chemical and Environmental Engineering. University of Zaragoza. Zaragoza. SPAIN.

Abstract

The catalytic decomposition of hydrocarbons, mainly methane, over catalysts based on transition metals is receiving an increasing attention as an alternative route to the production of hydrogen, can be considered as a long-term replacement for natural gas and with great potential of the new uses, and carbon nanotubes, CNTs, which have gained a huge interest due to their potential for unique applications.

In the present paper, we report the results of characterization and catalytic behavior of a Co-Mg-Al catalyst prepared by coprecipitation. The influence of the operating temperature and feed composition on nanocarbonacous materials content, hydrogen formation rate and carbon morphology has been studied. The obtained results indicate that the catalyst shows the highest activity and stability at low reaction and calcination temperatures (625 and 400ºC respectively). The change of the calcination temperature strongly modifies the interactions developed between the metallic nanoparticles and the support, and, in consequence, its reducibility and reactivity. Furthermore, it was found that the addition of low H_2 concentrations introduced in the reaction feed causes an increment of the catalyst activity and a diminution of the deactivation rate, due to hydrogen prevents the formation of encapsulating carbon at low concentrations. TEM observations indicate that the carbonaceous products are carbon nanotubes (diameters ~8-10 nm) and encapsulating carbon, which causes the catalyst deactivation. The amount of each type of carbonaceous material depends on the operating conditions used.

Introduction

It is considered that hydrogen will become an efficient renewable and sustainable energy carrier for the future [1]. Hydrogen is a clean fuel that emits no CO_2 when burned or used in H_2-O_2 fuel cells, can be stored as a liquid or gas, is distributed via pipelines, and has been described as a long-term replacement for natural gas [1]. Hydrogen is usually produced by steam reforming, partial oxidation or autothermal reforming of hydrocarbons and alcohols. These processes are based on mature technologies, but are quite complex involving many steps. In addition, a common problem of these processes is the formation of carbon monoxide and carbon dioxide, which are difficult to separate from hydrogen. Furthermore, for certain applications such as in polymer electrolyte membrane (PEM) fuel cells, carbon monoxide has to be removed from hydrogen in order to avoid poisoning of the electrocatalyst, increasing the costs of the traditional production routes for hydrogen. The catalytic decomposition of methane (CDM) has been proposed in recent years as an interesting alternative route for the production of hydrogen. This process is less energy consuming than steam reforming of methane, and does not directly produce CO and CO_2 [2]. Consequently, additional steps such as water gas shift, WGS, and preferential oxidation of CO are not necessary, and this considerably simplifies the process and decreases the costs of production. Moreover, during the CMD, nanocarbonaceous materials (NCMs) are produced such as carbon nanotubes (CNTs) and carbon nanofibres (CNFs). Since the discovery of carbon nanotubes [3], they have received considerable attention due to their excellent physical and chemical properties which allows their use in a large number of potential industrial applications. As regards of the catalysts used for the CDM process, Fe, Co, Ni and their alloys are the active metals usually used in the hydrocarbon decomposition [4]. The role of the catalyst is crucial for obtaining high activity and, in particular, selectivity towards the desired NCMs [5]. Thus, in the majority of the cases studied, the catalysts used in the hydrogen production by CMD are based on Ni, while Fe and Co based catalysts are specifically used in the production of CNTs.

Our research group has worked on the development of Ni based catalysts, using hydrotalcite-like compounds as precursors, for the application in the CDM process [2,5,6]. We have found that the modification of catalyst composition, by insertion of MgO to obtain Ni-Mg-Al catalysts, notably increases their productivity and stability during the CDM [6]. However, the morphology of the carbon formed is mainly CNFs. In this case, we study the use of the Co, in a Co-Mg-Al catalyst, as the active phase with the aim of improving the selectivity to CNTs. The influence of the activation, including calcination and reduction temperatures, and operating conditions, on hydrogen production and carbon formation has been studied.

Experimental

The catalyst, with nominal composition $(CoO)(MgO)(Al_2O_3)$, was prepared by coprecipitation of the metallic nitrate mixture (Co, Mg and Al) with $Na_2CO_3/NaOH$, at constant pH (10.2 ± 0.2) and temperature (60° C). The corresponding mixed oxide was obtained by calcination of the dried hydrated precursor in N_2 at different calcination temperatures (from 400 to 800ºC) for 11 h. The catalyst was characterized by AA, XRD, TPR and measurement of S_{BET}. Methane catalytic decomposition reaction was carried out in gas phase using a thermobalance (CI Electronics Ltd., UK, model MK2) equipped with mass flow and temperature controllers and operated as a differential reactor (CH_4 conversion < 8%). This experimental system allows continuous recording of the sample weight and temperature during reaction. After reaction, the catalyst and the carbon obtained was characterized by TEM, XRD and TPO.

Results

Influence of the calcination temperature

Figure 1 shows the XRD patters of the catalyst after calcination at different temperatures ranged between 400 and 800 ºC. As it can be deduced from the sharpening of the peaks corresponding to the detected phases, the lower is the calcination temperature; the lower is the crystallinity of the solid. Simultaneously, the specific surface area of the solid, which will ensure an adequate dispersion of the active metal, and therefore a good activity, augments, from 83 m^2/g to 264 m^2/g, as the calcination temperature decreases from 800 ºC at 400 ºC respectively.

After calcination at low temperature, the solid is formed for a mixture of metallic oxides (CoO, MgO). At higher temperatures, the XRD patterns show the presence of peaks corresponding to a spinel-like phase ($CoAl_2O_4$ /$MgAl_2O_4$) in addition to the oxides. In this point, it is noteworthy to remember that a that a partial change, from octahedral to tetrahedral, in the coordination of bulk M^{+3} cations can occur depending on the activation conditions [7,8]. This transformation converts the so-called "normal spinel" to a partially "inverted spinel". At low calcination temperatures, M^{+2} cations engage the surface octahedral hollows of the spinel, which have larger coordination index, better accessibility to the reactants, and therefore will be more active [7, 8].

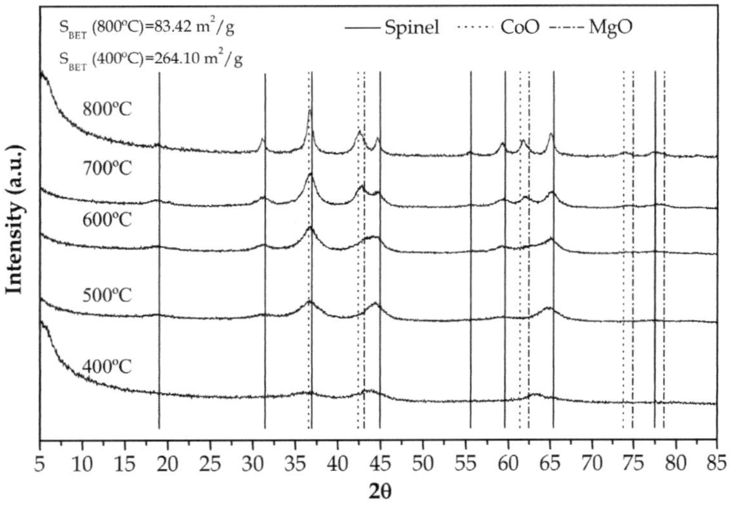

Figure 1. XRD difractograms and BET areas of Co-Mg-Al catalyst calcined at different temperatures.

Figure 2 shows the effect of calcination temperature on hydrogen formation rate and on the amount of CNTs accumulated over the catalyst, that is a direct measurement of the H_2 produced.

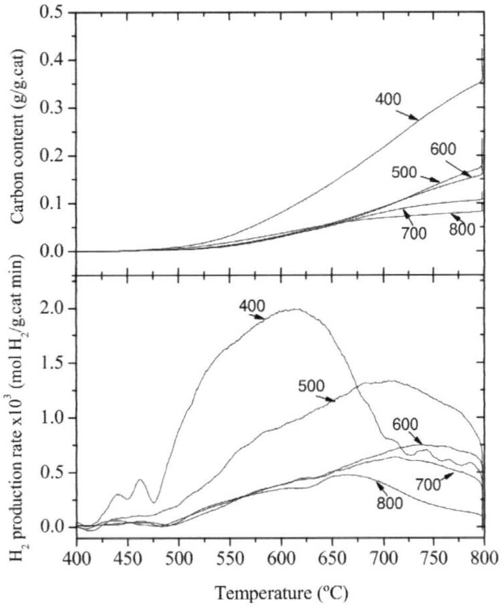

Figure 2. Evolution with time and temperature of the carbon content and H_2 formation rate. Influence of calcination temperature. P_{CH4}=5%, P_{H2}=0%.

These experiments have been carried under dynamic conditions 5° C/min from 400 to 800 °C in order to discriminate simultaneously the effects of the reaction temperature and of the operating time. The catalyst was previously reduced "in-situ" at 750°C for 2 hours.

As Figure 2 indicate, the quantity of CNTs concentration, and the range of temperatures where the catalyst is active, diminishes as the calcination temperature increases. Thus, a lower amount of reduced metallic Co is obtained after the reduction, when the catalyst has been calcined at high temperatures. This fact is due to the high stability and low metallic dispersion of the spinel structure. Besides, at high calcination temperatures a large incorporation of the Co inside the spinel network is produced, due to the migration of the Co from octahedral to tetrahedral positions, which are much less active [7,8]. In addition, the formation of a major proportion of deactivating encapsulating carbon is favored by large sized metallic Co particles, due to the difficulty in the diffusion through big particles [9]. In consequence, the deactivation rate of the catalyst calcined at elevated temperatures is higher.

Influence of the reaction temperature

The effect of the reaction temperature on the carbon content and hydrogen production rate is shown in the Figure 3. The hydrogen formation rate curves have been calculated from the numerical derivative of the experimental carbon content versus time curves and taking into account the stoichiometry of the reaction, which has previously been proved. The operating conditions used for these experiments were: calcination temperature: 400ºC; feed composition: 5%CH_4/95%N_2.

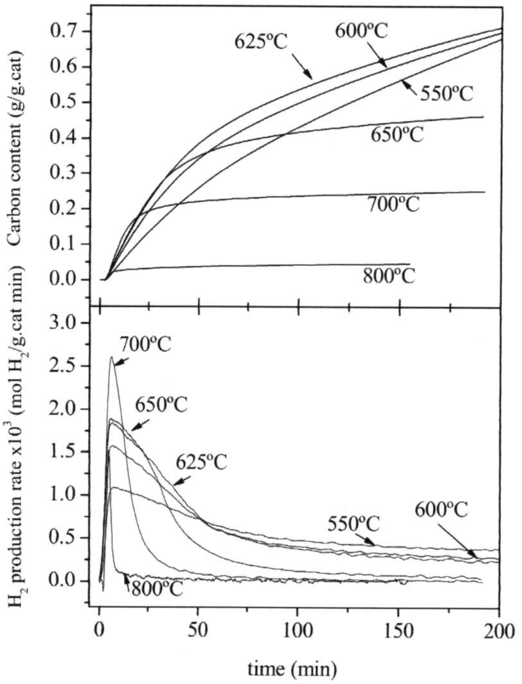

Figure 3. Evolution with time of the carbon content and H_2 formation rate. Influence of operating temperature. P_{CH4}=5%, P_{H2}=0%.

In all cases, the catalyst shows a fast activation period until a maximum rate is attained after which the catalyst is deactivated. However, it is worth pointing out that at temperatures below 625ºC the deactivation rate is low, while above 650ºC it increases dramatically. Furthermore, it can be observed that an augment in the operating temperature produces the increment of the rate at the initial moments of the reaction (i.e. activation period). That is, the higher is the operating temperature the greater are the activation (i.e. methane decomposition) and deactivation reaction rates [6]. In consequence, the catalyst shows a maximum yield at 625ºC. When the operating temperature is lower, the carbon content is smaller but deactivation is less severe and deactivation rate is faster at high reaction temperatures and, as a result, a reduction in the carbon content is observed.

Figure 4 shows the evolution with time of the carbon content and the hydrogen formation rate at 625ºC for the catalyst calcined at 800 and 400ºC. An improvement of catalyst activity and stability can be observed when the calcination temperature is 400ºC. As it has been explained, at this temperature higher amount of reduced metallic Co is obtained after the reduction and an appropriate interaction is developed between the active metal and the support, which prevents the formation of large metallic particles that lead to the appearance of a larger proportion of deactivating carbon.

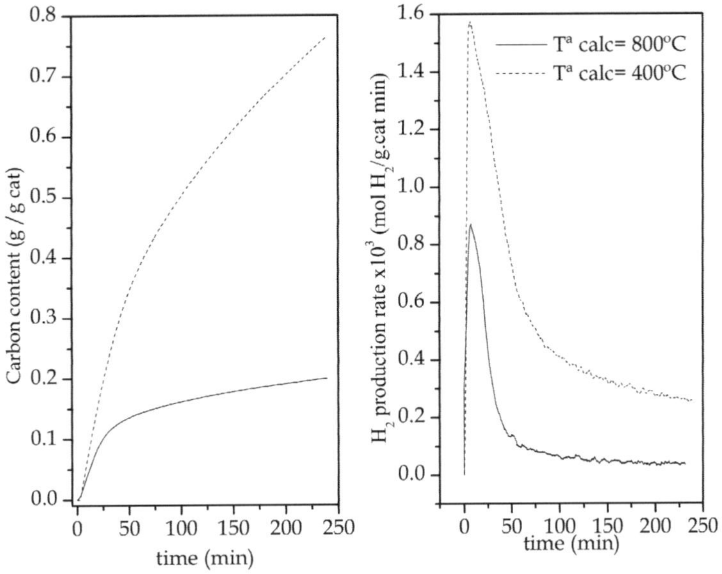

Figure 4. Evolution with time of the carbon content and H_2 formation rate. Temp. reaction= 625ºC; P_{CH4}=5%, P_{H2}=0%.

Influence of the partial pressure of hydrogen

The effect of the partial pressure of hydrogen on the carbon content and H_2 production rate is shown in Figure 5. The study has been developed using a calcination temperature of 400 ºC and a reaction temperature of 625 ºC. P_{CH4}=5%. A small amount of hydrogen in the feed provokes an increase in the catalyst activity and diminishes the deactivation. When the partial pressure of hydrogen goes on increasing, the amount of deposited carbon on the catalyst diminishes and the deactivation disappears.

These results are explained taking into account that hydrogen causes two opposite effects: on the one hand, hydrogen prevents the formation of encapsulating carbon, and consequently the formation of carbon nanotubes is not hindered by this carbon [6]. This fact explains the increase in the amount of carbon deposited when the partial pressure of H_2 is 0.02 atm. On the other hand, the competence between the methane and the hydrogen for the metallic surface sites causes a diminution in the formation rates of both, the carbon nanotubes (negative effect) and the encapsulating carbon species that deactivate the catalyst (positive effect) [6]. Therefore, hydrogen concentrations that are too high, in this case $P_{H2} \geq 0.07$ atm, inhibit the formation of carbon nanotubes but also prevent the deactivation of the catalyst.

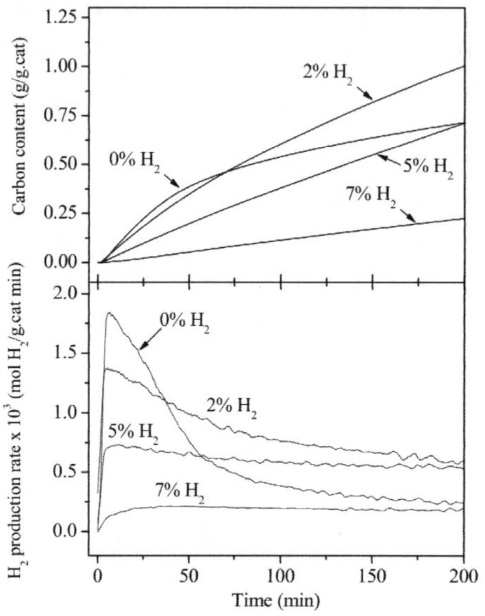

Figure 5. Evolution with time of the carbon content and H_2 formation rate. Influence of hydrogen partial pressure.

Carbon characterization after reaction

The quality of the materials obtained with the Co-Mg-Al catalyst was determined by Transmission Electron Microscopy. Figure 6 shows several TEM micrographs corresponding to the carbon deposited on the catalyst at different operating conditions.

As the kinetic experiments show (Figures 2 to 5), in most cases it is observed that some degree of catalyst deactivation takes place, that is attributed to the formation of a carbonaceous layer of less graphitized material that encapsulates the metallic particles hindering the decomposition of the methane, and the subsequent carbon diffusion to form the carbon nanotubes [6,9]. In consequence, two types of carbonaceous material are always present in the samples: carbon nanotubes and encapsulating carbon. However, for a given catalyst, the relative quantities of each type depend on the different operating conditions. The lower is the deactivation observed, the lesser is the formation of encapsulating carbon. Thus, when the calcination temperature is 800 (Figure 6a), the morphology of the formed carbon is mainly encapsulating. In consequence, the catalyst activity in these conditions is very low (Figure 4). A small proportion of multiwalled nanotubes (MWNT) and metal agglomerations owing to the sinterization caused by the high calcination temperatures can also be observed. Figure 6b shows the sample obtained after calcination at 400ºC and reaction at 625ºC. It also contains two types of carbon: MWNT (diameter ~10nm) and encapsulating carbon. However, the proportion of encapsulationg carbon is few and, in consequence, the deactivation observed in the experiment is also small (Figure 3 and 4). Furthermore, sinterization is less severe at this calcination temperature, not appearing big metal agglomerations.

As it has been explained, the introduction of a small quantity of H_2 in the feed produces the cleaning of the catalytic surface and prevents the formation of encapsulating carbon. In consequence, the sample (Figure 6c) contains a lower amount of amorphous carbon, according to the obtained results of activity (Figures 5). The MWNT obtained show similar diameters that in the case without hydrogen, ranged between 8 and 10 nm and length of several microns.

a

Temp. Calc.= 800ºC

Temp. React.= 625ºC

Feed composition:

5% CH_4 / 95% N_2

b

Temp. Calc.= 400ºC

Temp. React.= 625ºC

Feed composition:

5% CH_4 / 95% N_2

c

Temp. Calc.= 400ºC

Temp. React.= 625ºC

Feed composition:

5%CH_4 /2% H_2/93% N_2

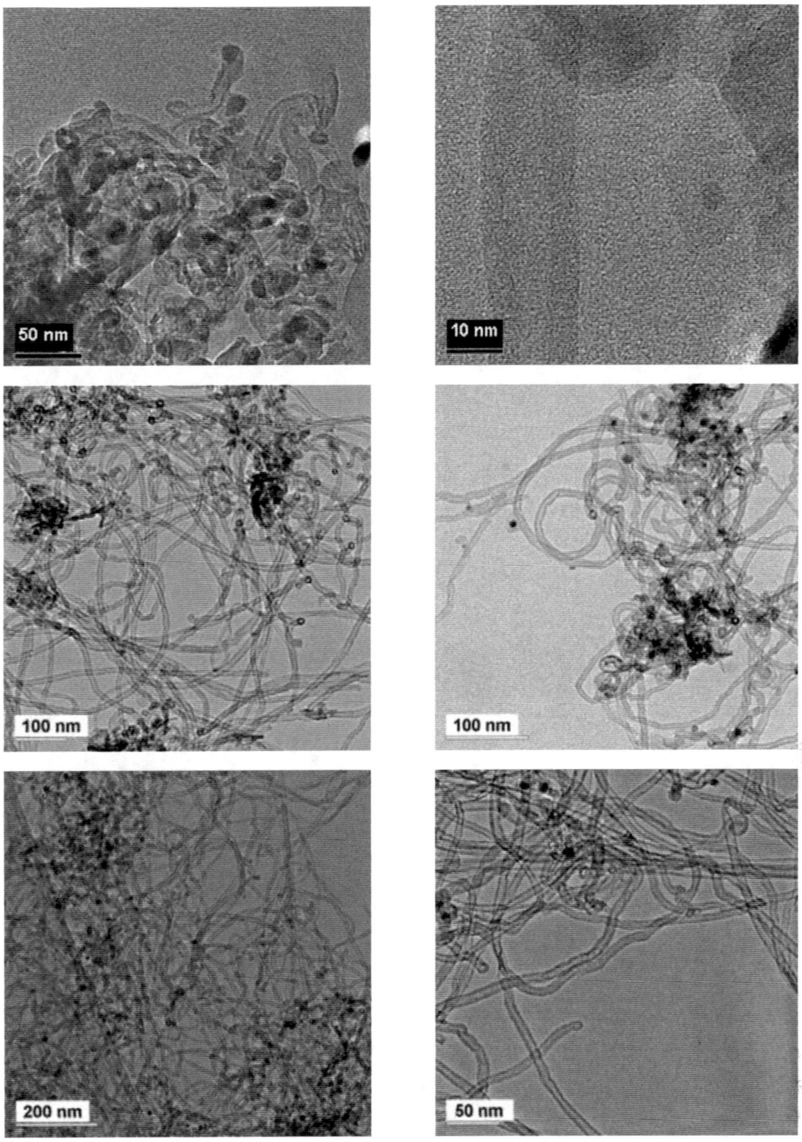

Figure 6. TEM images of the samples after reaction at different operating conditions.

Conclusions

The direct catalytic decomposition of methane catalysts is a potential alternative route for the production of hydrogen as well as of nanocarbonaceous materials from natural gas. In particular, the $(CoO)(MgO)(Al_2O_3)$ catalyst is active and selective for the synthesis of carbon nanotubos of good quality.

This catalyst shows a maximum activity at low calcination and reaction temperatures, being the best results obtained for 400 and 625 ºC, respectively. The effect of the calcination temperature strongly modifies the interactions developed between the metallic nanoparticles and the support, and, in consequence, its reducibility and reactivity. At low calcination temperatures higher amount of reduced metallic Co is obtained after the reduction and an appropriate interaction exists between the active metal and the support, which prevents the formation of large metallic particles that lead to the appearance of a larger proportion of deactivating carbon. On the other hand, the optimum operating temperature offers the best equilibrium between the augment produced for the increment of this variable in the activation (i.e. methane decomposition) and deactivation reaction rates. The presence of little amounts of hydrogen in the feed favours the synthesis of nanotubes and increases the catalyst stability due to the regenerating effect of hydrogen, which inhibits the encapsulating carbon deposition that deactivates the catalyst.

The transmission electron microscopy results indicate the presence of two types of carbonaceous material in the samples: MWNT (diameters~8-10 nm) and encapsulating carbon, responsible for the catalyst deactivation.

In conclusion, the activity, selectivity and resistance against deactivation of Co-Mg-Al catalysts prepared by coprecipitation strongly depend on the activation conditions, in particular on the calcination temperature. This result opens a route to optimize the catalyst performance in the simultaneous production of H_2 and CNTs by CMD.

Acknowledgements

The authors acknowledge financial support from MICINN (Spain)-FEDER, Project CTQ 2007-62545/PPQ, and the Regional Government of Aragón, Departamento de Ciencia, Tecnología y Universidad, Project CTP P02/08.

References

[1] J.N. Armor, *Appl. Catal. A: General* 176 (1999) 159.

[2] J.I. Villacampa, C. Royo, E. Romeo, J.A. Montoya, P. Del Angel and A. Monzón, *Appl. Catal. A: General* 252 (2003) 363.

[3] S. Ijima, *Nature* 354 (1991) 56.

[4] S.B. Sinnott, R. Andrews, D. Qian, A.M. Rao, Z. Mao, E.C. Dickey and F. Derbyshire, *Chem. Phys. Lett.* 315 (1999) 25.

[5] M. Monthioux, L. Noé, L. Dussault, J.C. Dupin, N. Latorre, T. Ubieto, E. Romeo, C. Royo, A. Monzón and C. Guimon, *J. Mater. Chem.* 17 (2007) 4611.

[6] N. Latorre, J.I. Villacampa, T. Ubieto, E.Romeo, C. Royo, A. Borgna and A. Monzón, *Topics Catal.,* 51 (2008) 158.

[7] Trifirò, F.: Vaccari, A.*; Stud. Surf. Sci. and Catal.,* 67 (1991) 157.

[8] Jacobs, J.P.; A. Maltha, A.; Reintjes, J.G.H. ; Drimal, J.; Ponec V.; and Brongersma, H.H., *J. Catal.,* 147 (1994) 294.

[9] De Chen, Kjersti O. Christensen, E. Ochoa-Fernández, Z. Yu, B, Tøtdal, N. Latorre, A. Monzón and A. Holmen, *J. Catal.,* 229 (2005) 82.

NON-CATALYTIC MULTI-FUEL REFORMER FOR SYNGAS PRODUCTION FOR FUEL CELL APPLICATIONS

A. Pastore and E. Mastorakos
Department of Engineering, University of Cambgirdge, Cambridge, UK

Abstract

This paper investigates rich combustion of *n*-heptane, Diesel oil, Jet A-1 Kerosene and Bio-diesel (Rapeseed-oil Methyl Ester) for the purpose of producing hydrogen enriched gas, ready for the clean-up stages for fuel cell applications. Rich flames have been stabilised in a two-layer porous combustor and a range of rich equivalence ratios and porous material configurations have been examined. *n*-heptane was successfully reformed up to an equivalence ratio of 3, reaching a conversion efficiency up to 83% for a packed bed of alumina bead burner. Similarly, Diesel, Kerosene and Bio-diesel were reformed to syngas in a Zirconia foam burner with conversion efficiency up to 65%. Soot emission has been also assessed, demonstrating a high particle formation for Diesel oil above φ=2 whereas Bio-diesel showed the lowest values among all the fuels tested.

Introduction

Recently, a two-layer porous burner has demonstrated the capability of producing hydrogen-rich gas from gaseous and light liquid fuels in a steady rich flame [1,2]. In this paper, heavy liquid fuel reforming to syngas is presented for a similar configuration. Previous works have also been published by several authors about the thermal (non-catalytic) partial oxidation of fossil fuels by using filtration wave, parallel channel, spouted bed combustor and two-layer porous burner concepts [3-6].

This work is based on the principle of superadiabatic combustion. The superadiabatic effect consists in the local occurrence of higher flame temperatures compared to the adiabatic temperature of a free flame at the same equivalence ratio. Specifically to this experimental investigation, this local effect is obtained by inserting a porous ceramic material in the flame zone. The recirculation of the sensible enthalpy of the flame products occurs through convection from the hot gas to the inert matrix, which, in turn, transfers the heat through conduction and radiation to the incoming unburned mixture. The higher thermal conductivity of the solid matrix compared to the conductivity of the gas enhances the effective heat transfer to the reactants from the products. Therefore, the global effect is a substantial preheat of the fresh mixture. The subsequent faster kinetics leads to higher laminar flame speed and to the extension of the flammability limit of the fuel/oxidiser mixture [7].

In this work, the stabilization of rich premixed flames has been achieved by using a two-layer porous burner [8-10]. Flashback would arise in a single piece of foam and hence a means to stabilize the flame is necessary. The material property discontinuity between the first (flame holder) layer and the second (diffuser) layer can fulfil this requirement. Trimis et al. [11] summarized this effect by defining a critical Peclet number limiting the propagation of a combustion wave in porous media. In this project, the flame support layer is a thin layer with small pores, whereas the diffuser layer is a thick layer with larger pores.

The extended rich flammability limit allows partial oxidation conditions to be established, thus allowing the stripping out of the hydrogen from the original fuel molecule. Consequently, the main products of the reactions are molecular hydrogen and carbon monoxide (syngas), along with nitrogen

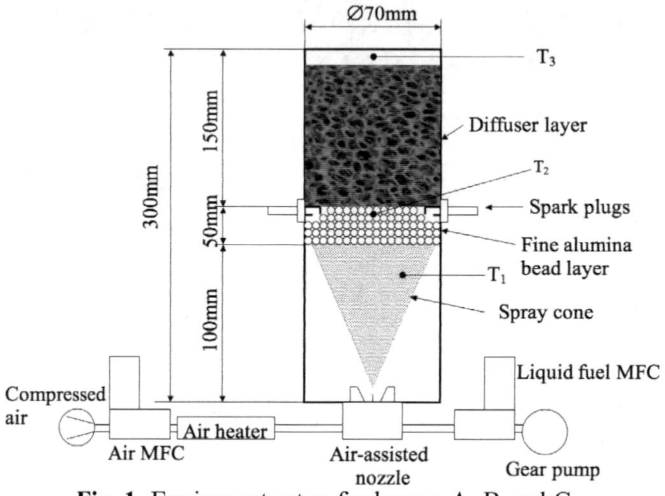

Fig. 1. Equipment setup for burner A, B and C.

(air provides the oxidiser) and some other species in lower concentration (carbon dioxide and light hydrocarbons). The absence of catalyst makes the process robust and fuel flexible, as this paper demonstrates by discussing the reforming of *n*-heptane, kerosene, diesel and bio-diesel.

Experimental Methods

Apparatus

The burner (Fig. 1) has a two-layer porous structure, the first of which assists evaporation and prevents the flame to propagate upstream in most of the conditions investigated, while the second (downstream) layer is where the reforming occurs. The first layer is a flame holder and it consists of a 50 mm long packed bed of 3-4 mm Al_2O_3 beads. On top of this first layer, different porous materials were tested. The following sections deal with the different configurations adopted. In the optimum condition, the flame sits at the interface of the two layers and the reaction zone extends into the reforming layer.

The liquid fuel is delivered through an air-assisted commercial nozzle (Delavan AL-15 external mix nozzle) modified to withstand the high temperature of the air flow. The fuel is delivered by a magnetic driven gear pump controlled by a MINIcori Bronkhorst Mass Flow Controller. The air flow is supplied by the laboratory compressed air line and it is controlled by a Bronkhorst Mass Flow Controller. A 750 W heater heats the air flow up to 300 °C. This air ensures atomization, subsequent evaporation because it is heated, but also provides all the oxidizer. The flow controller manufacturer gives an accuracy of ±0.2% of the reading plus zero stability (better than ±0.2 g/h) for the liquid mass flow controller and of 0.5% of the reading plus ±0.1% of the full scale (120 ln/min) for the air mass flow controller. The spray flows in a conical section 100 mm long and the evaporated fuel-air mixture reaches the first porous layer. A thermocouple placed just before the upstream side of the first porous section (i.e. at the end of the atomiser-vapouriser conical section) gives the voltage feedback to the air heater through a temperature controller (T_1). Two R-type thermocouples detect the temperature at the layer interface (T_2) and the gas temperature on the top of the reformer (T_3) respectively.

A water-cooled probe allows the collection of a sample of the exhaust gas at the burner's exit. Subsequently, the sample is passed through a calcium chloride cartridge for the water content removal and then injected into a gas chromatograph (GC). The GC is equipped with a Thermal Conductivity Detector (TCD) and a Flame Ionization Detector (FID). The Molecular sieve and Haysep N porous polymer columns embedded in the GC oven can separate permanent gases such as H_2, O_2, N_2, CO and

CO_2 and hydrocarbons up to C_2 (methane, acetylene, ethylene and ethane). The machine is calibrated by a gas mixture with species concentrations similar to the one expected in the reformate. The uncertainty associated with the GC is at least ±2% which is the uncertainty assessed for the calibration gas by the supplier company (Scientific & Technical Gases Ltd). Furthermore, the sample line is periodically checked for leaks and the samples are injected at the same temperature and pressure.

Soot emission was also investigated by means of a filter-paper-method machine (AVL 415S). A sample of the off-gas is taken from the exit of the burner with a probe and sucked through a filter paper. The blackening of the filter paper is measured with a reflectometer and indicates the soot content in the exhaust gas. The output value (*FSN*) is the Filter Smoke Number and it corresponds to the Paper Blackening (*PB*), as defined by Eq. 1, for a sampling length of 405 mm at 298 K and 1 bar:

$$PB = \frac{100 - R_R}{10} \tag{1}$$

with $R_R = \frac{R_P}{R_F} \cdot 100\%$ and R_P the reflectometer value of sample, R_F the reflectometer value of the unblackened paper and R_R the relative brightness of the sample.

Procedure

The test procedure starts from the warming up of the reactor by a flow of only the oxidiser (i.e. the air) through the reactor to bring the temperatures up to the appropriate value corresponding to the fuel characteristics to avoid any condensation of the liquid fuel once injected. This first operation takes approximately from 10 to 20 minutes. At this point, the fuel is injected. An air/fuel ratio close to stoichiometry and a thermal power of 2 kW is used first to achieve a prompt ignition. Four automotive spark plugs are fixed on the burner. The development of four flame kernels by the spark plugs reduce the time the flame needs to spread over the entire cross section of the burner, avoiding the emission of any unreacted mixture. After the ignition, the reaction zone quickly extends inside the porous matrix and the equivalence ratio is switched to rich values. When the thermocouple on the top of the burner gives a constant reading, the system has reached thermal equilibrium and samples can be extracted through the sampling line. From cold to fully stabilised operation, the system takes about 20-30 minutes.

Fuels

The fuels investigated in this work are *n*-heptane, Jet A-1 Kerosene, commercial Diesel and RME (Rapeseed-Oil Methyl Ester) Bio-diesel. Laboratory microanalyses were performed for Kerosene, Diesel and Bio-diesel and the fuel specifications are summarised in Table 1.

Table 1. Fuel specifications.

	***n*-heptane**	**Jet A-1 Kerosene**	**Diesel**	**RME Bio-diesel**
Chemical formula	C_7H_{16}	$C_{12}H_{23}$	$C_{12}H_{22}$	$C_{19}H_{36}O_2$
LHV [MJ/kg]	44.926	42.8	43	37.6
Supplier	Acros Organics	CPS fuels	Shell	ADM

Different evaporation temperatures (T_1) have been set for ensuring full evaporation for each fuel. *n*-heptane has been vaporised in a 150 °C environment, whereas Diesel at 300 °C and Jet A-1 and Bio-diesel at 250 °C. The choice of the lower temperature for Bio-diesel/air mixture preparation compared to Diesel oil has been suggested by the risk of cracking at high temperatures, due to the thermal instability of this fuel due to the presence of oxygen atoms in the fuel molecule.

Burners

Different configurations have been tested in order to assess the sensitivity of the reformer to parameters such as porosity, type of material and thickness of the porous matrix. The setups investigated are summarized in Table 2.

Table 2. Burner configurations.

Burner	**A**	**B**	**C**	**D**	**E**	**F**
ID [mm]	70	70	70	70	100	100
Material	ZrO_2 foam	ZrO_2 foam	SiC, Al_2O_3, SiO_2 compound foam	5-6mm diameterAl_2O_3 beads	ZrO_2 foam	ZrO_2 foam
Porosity	30 ppi	10 ppi	30 ppi	40%	30 ppi	30 ppi
Length [mm]	100	100	100	100	50	25

Results and discussion

n-heptane

The choice of reforming *n*-heptane has been dictated by the possibility of comparing experimental data with numerical simulations and computational models, being the mechanisms and the chemical kinetics modelled in the past [12] for such a reference fuel. In a previous paper [2], the air/fuel mixture uniformity across the pre-vapouriser and the operational points of an equivalent burner to configuration C were investigated. In this work, further parameters have been explored.

The first task consists in an analysis of the reformate composition over a range of equivalence ratios from φ=1.1 to φ=3 with 0.1 steps for a thermal power of 7 kW for burner A. The results are shown in Fig. 2. The general trend is a monotonic increase of the hydrogen and carbon monoxide molar fraction and decrease of carbon dioxide molar fraction with increasing equivalence ratio, following the trends of thermodynamic equilibrium calculations. Furthermore, the hydrocarbon (HC) presence increases dramatically with φ. The disparity of the hydrogen and carbon monoxide values between the experimental and equilibrium data is due to the formation of light hydrocarbons at high equivalence ratios. Fig. 3a depicts the conversion efficiency defined by Eq. 2:

$$\eta_{LHV} = \frac{\dot{m}_{H_2} LHV_{H_2} + \dot{m}_{CO} LHV_{CO}}{\dot{m}_{C_7H_{16}} LHV_{C_7H_{16}}} \qquad (2)$$

and Fig. 3b shows the H_2 distribution in molecular hydrogen, water and hydrocarbons (this was calculated by further processing the measurements by using the atomic ratios in the reactants and assuming that all carbon is contained in the HCs measured). As the equivalence ratio increases, the hydrogen content in the HCs increases. The conversion efficiency peaks for equivalence ratios between 1.8 and 2.1, with values over 65%. The operating point at φ=2 has demonstrated better performances in terms of high stability and repeatability, high conversion efficiency and low soot formation. This aspect is underlined by Fig. 3b, where the H_2 content is maximised at φ=2. High water content is exhibited at lower φ and high HC content is observed at higher φ. Both water and light hydrocarbons act as hydrogen sinks.

The second task focused on different porous materials for the reforming stage. Fig. 4 compares the experimental results obtained from a similar investigation as before (φ ranging from 1.1 to 3 at P=7 kW), but carried out in burners A, B, C and D. The only sensible difference in the reformer

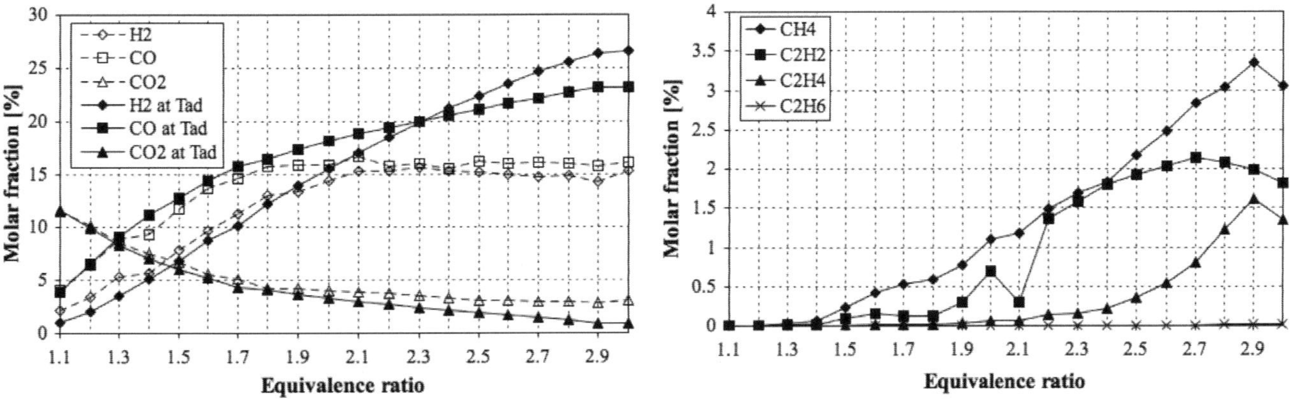

Fig. 2. Calculated equilibrium at the adiabatic flame temperature (1663 K), and the experimental mole fractions (on a dry basis) for a inlet air/fuel temperature of 423 K at various equivalence ratios for *n*-heptane reforming at P = 7 kW for burner A. (a) Hydrogen, carbon monoxide and carbon dioxide. (b) Methane, acetylene, ethylene and ethane. The thermodynamic equilibrium calculations predict negligible concentrations of these hydrocarbons.

performances has been demonstrated by burner D, which is characterised by a packed bed of alumina beads. This enhancement is especially relevant at equivalence ratios higher than φ=2.2. The temperature profile graph in Fig. 5 may aid the understanding of this different trend; it appears that the beads are able to maintain a higher reforming temperature (by 100 K) compared to the foam matrices. As the equilibrium calculations demonstrate, higher flame temperatures lead to higher conversion efficiency. Therefore, as shown by the experimental results, higher reforming temperatures lead to a lower presence of HCs in the reformate, resulting in higher H_2 and CO molar fraction. Consequently, the conversion efficiency peaks at 83% at φ=2.9. The drawback of the utilization of beads in place of foams is the higher thermal inertia and hence the lower response time of the configuration D. The other parameters such as porosity and foam material (burner A, B and C) seem to not affect appreciably the reformate composition.

The last task conducted with *n*-heptane concerns the length of the second porous medium (reforming stage). Burners E and F were set for this purpose. The condition investigated is thermal input of 7 kW and equivalence ratios ranging from 1.4 to 2.5. Results are displayed in Fig. 6. The graphs demonstrate a slightly higher performance of burner E compared to burner F. This may be due to the larger amount of heat "captured" and therefore recirculated by a thicker matrix, but also to the higher residence time in a hot environment for the partial oxidation products.

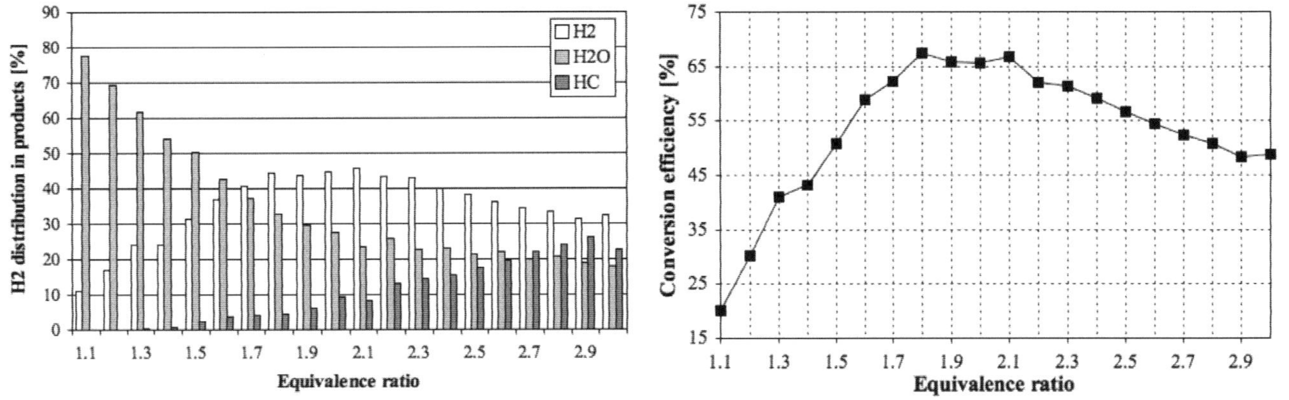

Fig. 3. (a) H_2 distribution in hydrogen, water and hydrocarbons present in *n*-heptane reformate as a function of equivalence ratio (conditions as in Fig. 1). (b) Conversion efficiency defined by Eq. (2) against the equivalence ratio for *n*-heptane (conditions as in Fig. 1).

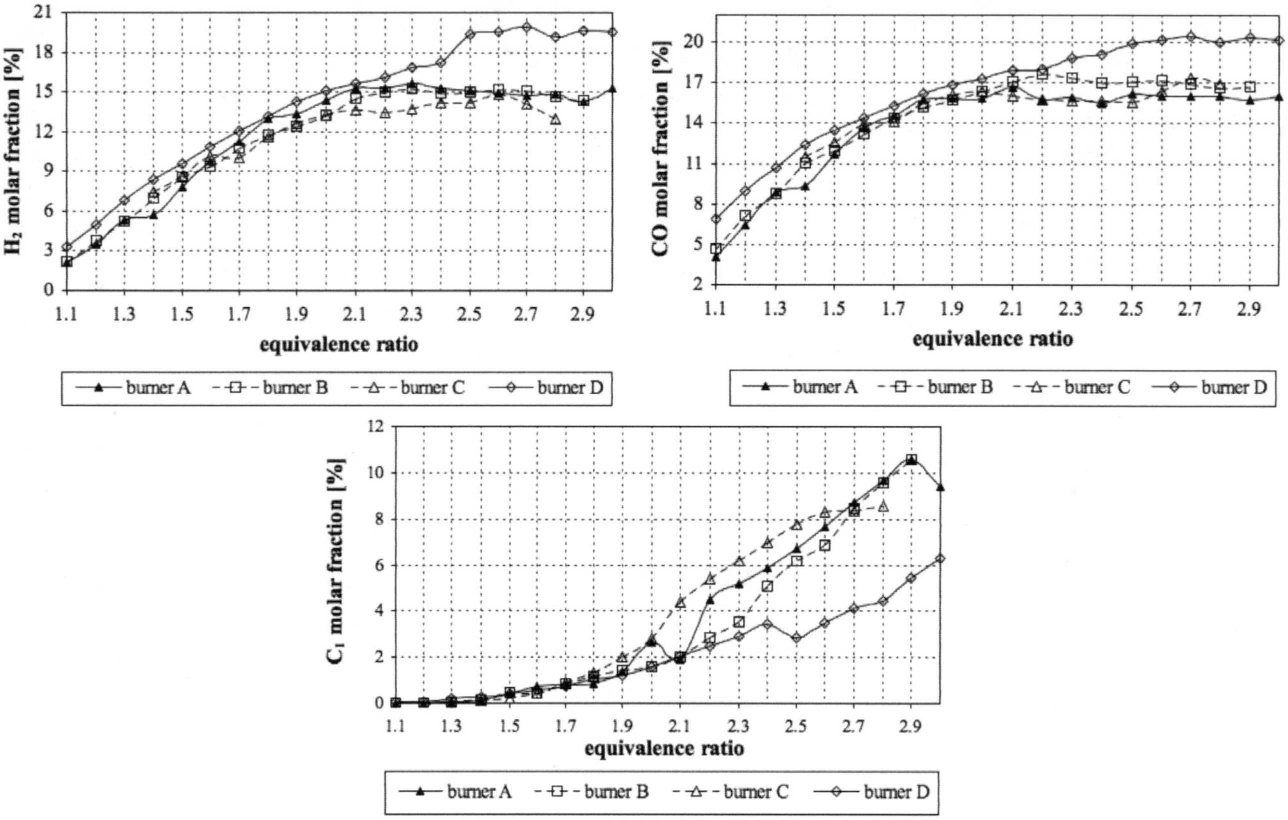

Fig. 4. Experimental mole fractions (on a dry basis) of (a) H_2, (b) CO and (c) C_1 (contributes of CH_4, C_2H_2, C_2H_4, C_2H_6) for a inlet air/fuel temperature of 423 K at various equivalence ratios for *n*-heptane reforming at P = 7 kW for burner A, B, C and D at P=7 kW.

Numerical simulations [13, 14] suggest that the fuel goes through combination of partial and complete oxidation reactions. The water produced downstream drives steam reforming and produces H_2. Being an endothermic process, the presence of a hot material may enhance this second stage of the reforming. However, the limited enhancement of the performances observed, suggested that any further tests with longer pieces of foam would not be very productive.

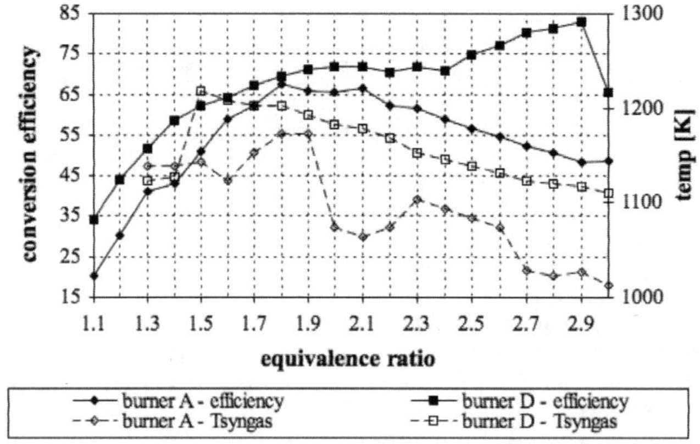

Fig. 5. Temperature profile and conversion efficiency, as defined by Eq. 2, for a inlet air/fuel temperature of 423 K at various equivalence ratios for *n*-heptane reforming at P = 7 kW for burner A and D.

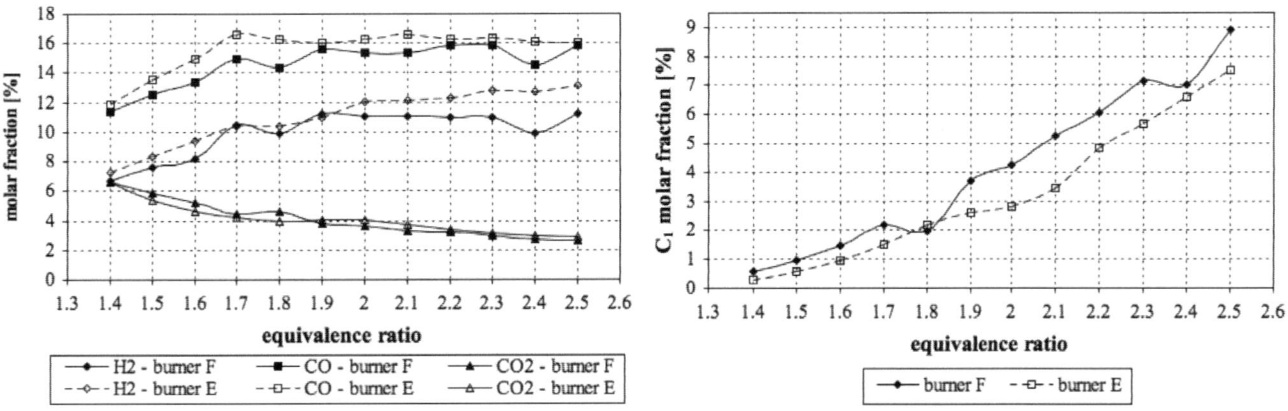

Fig. 6. Experimental mole fractions (on a dry basis) for a inlet air/fuel temperature of 423 K at various equivalence ratios for *n*-heptane reforming at P = 7 kW for burner E and F. (a) Hydrogen, carbon monoxide and carbon dioxide. (b) C_1 as contributes of methane, acetylene, ethylene and ethane.

Jet A-1, Diesel and Bio-diesel

After having explored different parameters for *n*-heptane reforming, realistic oil-based liquid fuels have been processed. The main difference between these fuels and *n*-heptane is the complexity of the molecule. The presence of sulphur and/or aromatics make the chemical kinetics much complex, especially at rich conditions.

Results are shown in Fig. 7, 8 and 9 for Jet A-1, Diesel and Bio-diesel respectively reformed in burner A at P=7 kW. The data obtained show a similar trend compared to *n*-heptane reforming up to φ=2.5. For fuel-richer conditions, the syngas quality (i.e. CO and H_2 against HC molar fraction) decreases dramatically. This effect is possibly due to the quenching of the reaction. Another difference with the *n*-heptane data is the higher presence of ethylene compared to acetylene for Jet A-1 and especially Diesel and Bio-diesel. This suggests different kinetic pathways compared to *n*-heptane for the HCs chain breaking and recombination.

Overall, the composition of the reformate for the three fuels is very similar. Bio-diesel has been easily reformed up to φ=3, whereas Jet A-1 and Diesel presented high deposit formation above φ=2.5 which led to a reddish flame on the top of the foam associated with a large particulate emission. Fig. 10 compares the conversion efficiency for these 3 fuels. Bio-diesel and Diesel present very similar results, whereas Kerosene has a slightly higher efficiency, perhaps due to the higher H/C ratio compared to the former fuels.

Soot formation

Fig. 11 gives an indication of the soot content in the off-gas for the four different fuels described in this paper at P=7 kW for burner A at different equivalence ratios. The graph represents a qualitative comparison between the fuels and the soot emission trend, rather than a quantitative evaluation of the particulate properties (soot number and size) at the different conditions considered. The main considerations drawn from the graph are that soot start to turn up in the off-gas above φ=2-2.1. Secondly, diesel presents the highest *FSN* value and hence the largest particulate emission, as expected. The last important information given by the measurement consists in the low soot generation observed in Bio-diesel reforming. This trend is due to the presence of oxygen atoms in the fuel molecule, which lowers the rate of particulate formation.

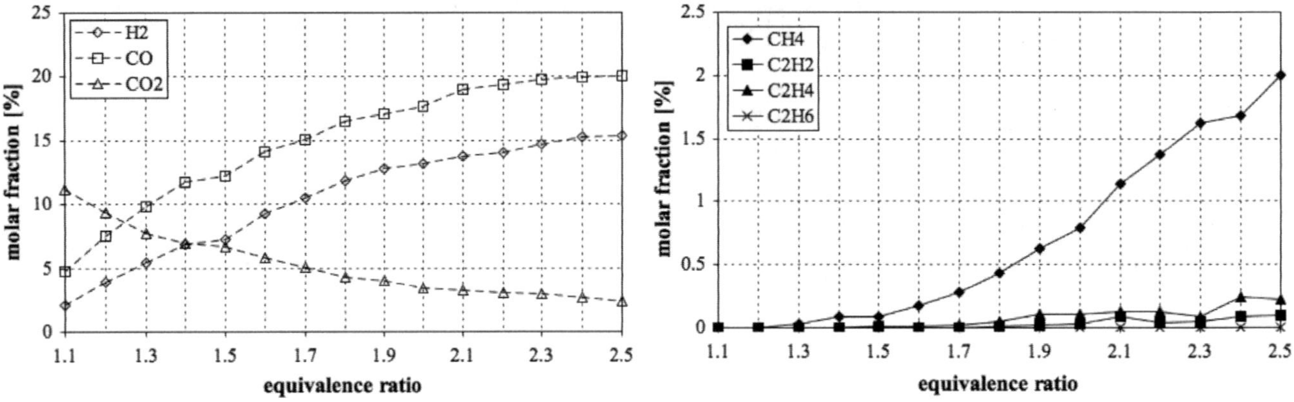

Fig. 7. Experimental mole fractions (on a dry basis) for an inlet air/fuel temperature of 523 K at various equivalence ratios for Jet A-1 reforming at P = 7 kW for burner A. (a) Hydrogen, carbon monoxide and carbon dioxide. (b) Methane, acetylene, ethylene and ethane.

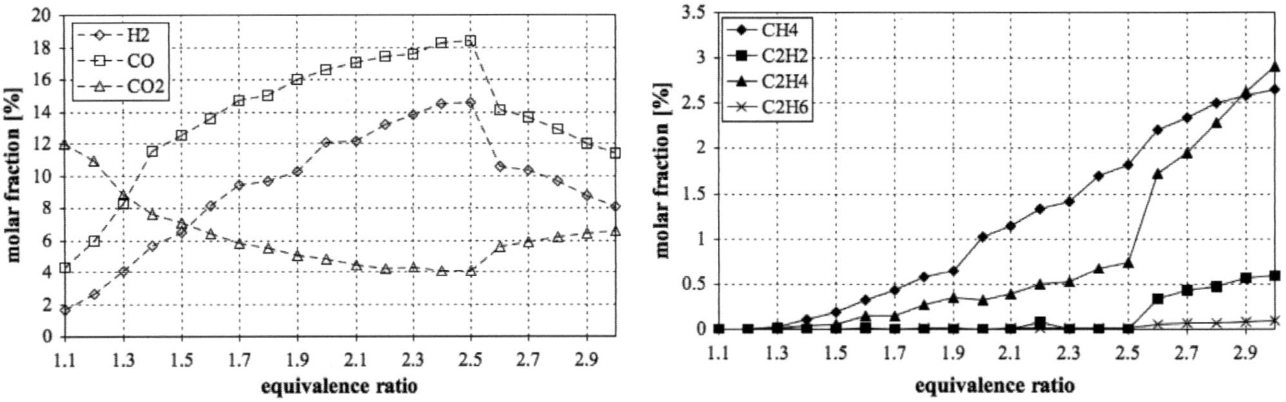

Fig. 8. Experimental mole fractions (on a dry basis) for a inlet air/fuel temperature of 573 K at various equivalence ratios for Diesel oil reforming at P = 7 kW for burner A. (a) Hydrogen, carbon monoxide and carbon dioxide. (b) Methane, acetylene, ethylene and ethane.

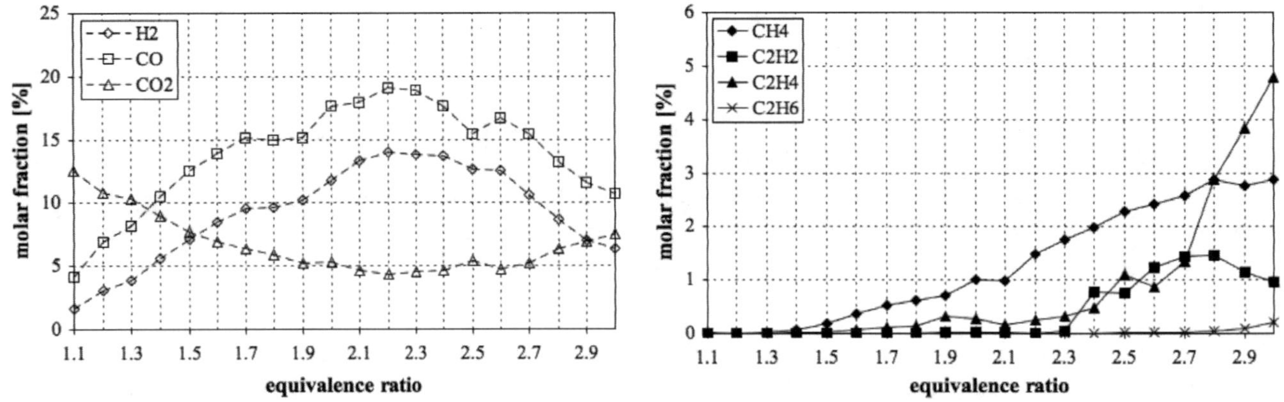

Fig. 9. Experimental mole fractions (on a dry basis) for a inlet air/fuel temperature of 573 K at various equivalence ratios for RME Bio-diesel reforming at P = 7 kW for burner A. (a) Hydrogen, carbon monoxide and carbon dioxide. (b) Methane, acetylene, ethylene and ethane.

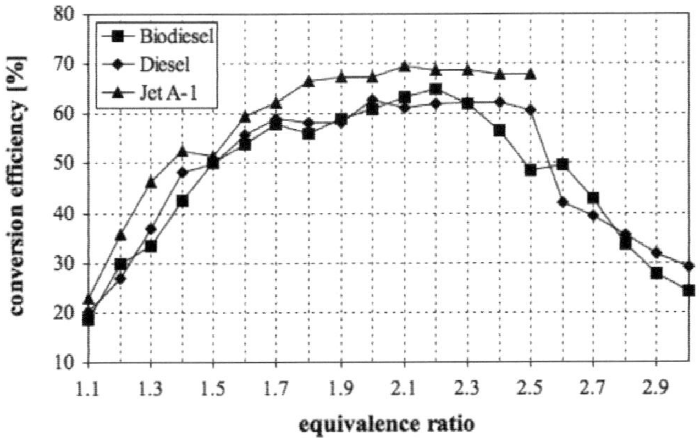

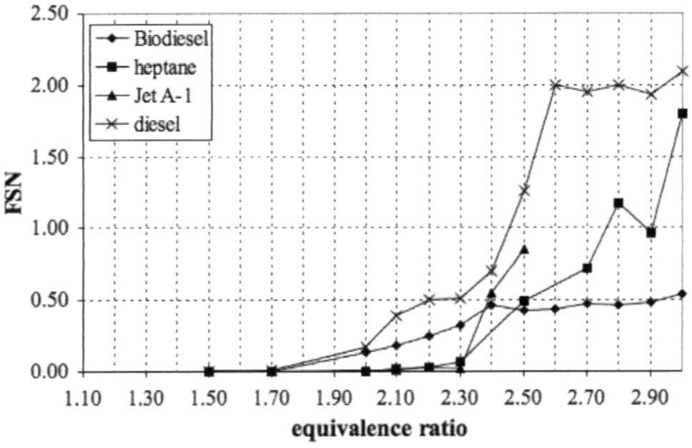

Fig. 10. Comparison of the conversion efficiency defined by Eq. 2 of Jet A-1, Diesel oil and Bio-diesel at P=7 kW for burner A.

Fig. 11. Soot emission trends for *n*-heptane, Jet A-1, Diesel and Bio-diesel fuel at P=7 kW for burner A.

Conclusions

A multi-fuel non-catalytic reformer has been designed and developed for syngas production. Rich stationary *n*-heptane, diesel, kerosene and bio-diesel flames have been studied experimentally over a range of equivalence ratios for different burner configurations at a thermal load of P=7 kW. The experimental data for *n*-heptane have been compared to the calculated equilibrium results showing the consistency of the reforming process. Similar reforming performances have been observed for real fuel such as diesel, kerosene and bio-diesel up to φ=2.5. Above this value soot formation and quenching deteriorate considerably the syngas quality. A working condition of φ=2 and P=7 kW has demonstrated high stability, low HC formation, high conversion efficiency and low soot emission. A common syngas composition around this condition is 15% and 13% H_2, 15% and 17% CO and 4% and 4.5% CO_2 for *n*-heptane and Diesel, Jet A-1 and Bio-diesel respectively for burner A. The conversion efficiency peaked at around 66% for all the fuels tested.

References

1. Pedersen-Mjaanes H., Chan L. and Mastorakos E. (2005), "Hydrogen production from rich combustion in porous media", *International Journal of Hydrogen Energy*, 30, 579-92.
2. Pastore A. and Mastorakos E. (2009), "Rich *n*-heptane and diesel combustion on porous media", *Experimental Thermal and Fluid Science*, Article in press.
3. Drayton MK., Saveliev AV., Kennedy LA., Fridman AA. and Li Y., (1998) "Syngas production using superadiabatic combustion of ultra-rich methane-air mixtures", *Proceeding of the Combustion Institute*, 1361–7.
4. Schoegl I., Newcomb SR. and Ellzey JL., (2009) "Ultra-rich combustion in parallel channels to produce hydrogen-rich syngas from propane", *International Journal of Hydrogen Energy*, 34, 5152-63.
5. Al-Hamamrea Z., Voß S. and Trimis D., (2009) "Hydrogen production by thermal partial oxidation of hydrocarbon fuels in porous media based reformer", *International Journal of Hydrogen Energy*, 34, 827-32.
6. Weinberg FJ., Bartleet TG., Carleton FB. and Rimbotti P., (1988) "Partial oxidation of fuel-rich mixtures in a spouted bed combustor". *Combustion and Flame*, 235–9.
7. Diamantis DJ., Mastorakos E. and Goussis DA. (2002), "Simulations of premixed combustion in porous media", *Combustion Theory and Modelling*, 6, 383-411.

8. Hsu P., Evans W. and Howell J., (1993) "Experimental and numerical study of premixed combustion within nonhomogenous porous ceramics", *Combustion Science and Technology*, 90, 149–172.
9. Kaplan M. and Hall MJ., (1995) "The combustion of liquid fuels within a porous media radiant burner", *Experimental Thermal and Fluid Science*, 11, 13–20.
10. Vijaykant S. and Agrawal AK. (2007), "Liquid fuel combustion within silicon-carbide coated carbon foam", *Experimental Thermal and Fluid Science*. 32, 117–125.
11. Trimis D. and Durst F., (1996) "Combustion in a Porous Medium-Advances and Applications", *Combustion Science and Technology*, 121, 153-68.
12. Chaos M., Kazakov A., Zhao Z.W., and Dryer F.L. (2007), "A high-temperature chemical kinetic model for primary reference fuels", *International Journal of Chemical Kinetics*, 39 (7), 399-414.
13. Mhadeshwar AB. and Vlachos DG. (2005), "Hierarchical multiscale mechanism development for methane partial oxidation and reforming and for thermal decomposition of oxygenates on Rh", *Journal of Physical Chemistry B*, 109 (35), 16819-35.
14. Dobrego K.V., Gnesdilova N.N., Leeb S.H. and Choib H.K. (2008), "Overall chemical kinetics model for partial oxidation of methane in inert porous media", *Chemical Engineering Journal*, 144 (1), 79-87.

2010 AIChE Spring National Meeting (San Antonio, TX)

Integrated hydrogen production, purification & compression demonstration project

Authors
Tony Boyd, Ali Gulamhusein, Anwu Li (Membrane Reactor Technologies Ltd.)
Satish Tamhankar (Linde North America Inc.)
David DaCosta, Mark Golben (Ergenics Corp.)

1. Introduction

As part of an ongoing US Department of Energy sponsored project, a novel hydrogen production demonstration unit using a natural gas feed was built and tested by the project partners (Linde North America Inc., Membrane Reactor Technologies Ltd. ("MRT") and Ergenics Corp.). The technical performance targets of this phase of the DOE project are:

- Production capacity of 15 Nm^3/hr
- Hydrogen output pressure of 100 bar
- Hydrogen purity of 99.99%
- Production unit efficiency of > 70% and compression efficiency of > 70%
- Unattended operation

The system is comprised of an autothermal fluidized bed membrane reformer coupled with a non-mechanical, metal hydride compressor ("MHC"). A schematic of the traditional reforming process to produce hydrogen compared to our approach is presented in Figure 1. The new scheme reduces the number of process components and system complexity.

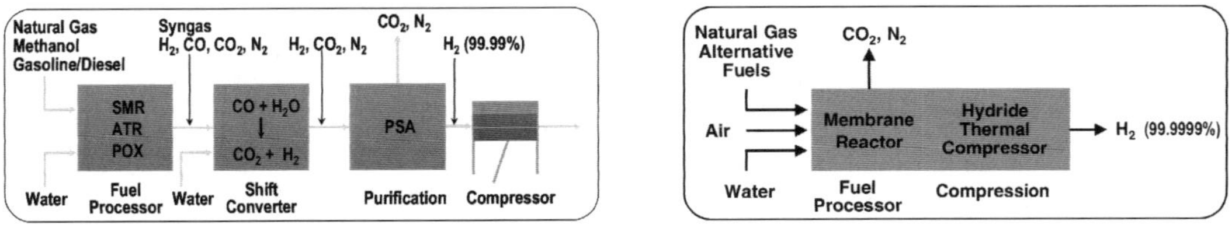

Figure 1: Conventional reforming vs. membrane reforming / hydride compression

The novel reformer provided by MRT includes 25 palladium-alloy membranes (0.15 x 0.28 x 0.006 m, double-sided, 25 μm thick foil panels) to provide in-situ separation of hydrogen from the reforming zone. Figure 2 illustrates how the membranes are assembled and inserted into the reformer. Removal of hydrogen from the reactor shifts the thermodynamic equilibrium forward, so that high conversion rates of methane are achieved at relatively mild operating conditions (550°C, 25 barg). Palladium membranes have been used to enhance reformer performance by a number of researchers (see reviews in Ref. 1, 2, 3), but their use in a fluidized bed environment is less common (Ref. 4).

Figure 2: Reformer assembly and single membrane panel

Air is directly added to the fluidized reforming bed to provide the endothermic reaction heat, thereby eliminating the need for costly indirect heat transfer. As a result, the reformer is made from lined carbon steel. The fluidized bed environment facilitates an isothermal reactor, which is important for membrane performance. Heat is transferred from the heating (air addition) zones to the reforming (membrane) zones through solids movement. The membrane retentate (reactor off-gas), which is essentially depleted syngas with nitrogen, exits the reactor via an internal filter. The residual fuel value in this stream is recovered in two burners, where it is used to raise the reformer steam and provide a portion of the metal hydride compression duty.

An autothermal reforming catalyst for fluidized bed service is not commercially available. In conjunction with a catalyst supplier, MRT prepared a non-Ni catalyst on an alumina powder (~100 μm diameter). The minimum fluidization velocity of this material was approximately 0.01 m/s at operating conditions.

The MHC is a 3-stage system designed and fabricated by Ergenics. Hydrogen from the reformer is absorbed in cool metal hydride beds, and then desorbed at higher pressure after heating (Figure 3). The system has two parallel compression paths (6 beds in total) to provide a near-continuous inlet flow of hydrogen. The system compresses hydrogen permeate from sub-atmospheric to 100 bar in a single system by engineering the composition of the metal hydrides in each compression stage. The first-stage hydride bed is cooled, and then hydrogen is absorbed at low pressure. The bed is then heated to desorb the hydrogen at higher pressure, which is then absorbed in the second-stage (cool) bed. Three compression stages boost the hydrogen pressure from sub-atmospheric (0.5 bara) to over 100 barg. Compression work is provided by thermal energy, which in the current demonstration system is a hot glycol solution heated by a small natural gas burner.

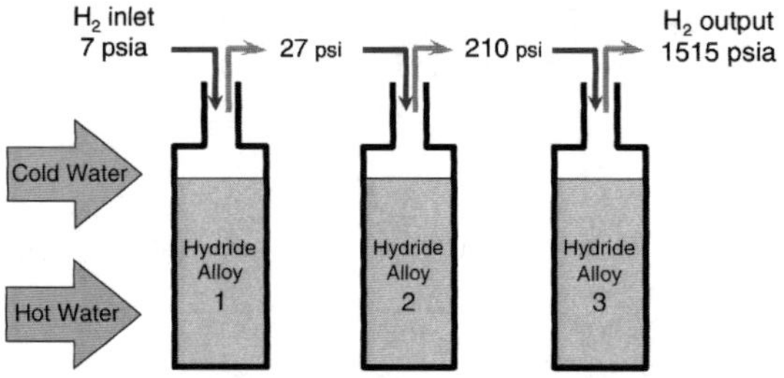

Figure 3: Three stage metal hydride compression

One of the engineering challenges for the MHC is maintaining a high heat transfer rate to and from the hydride beds in order to minimize the volume of hydride powder. In the Ergenics system, a proprietary design incorporates numerous small diameter tubes packed with hydride powder assembled into a ring manifold. Hydrogen then flows to and from the hydride tubes through a central distribution pipe.

Figure 4: Metal hydride bed assembly

Figure 5: Membrane reformer and metal hydride compressor demonstration unit installed at NRC's Institute for Fuel Cell Innovation

Coupling the membrane reformer and MHC provides thermal efficiency and capital cost benefits:

- The MHC can provide sub-atmospheric pressure (suction) on the hydrogen membranes. This increases the hydrogen permeation driving force, thereby reducing the membrane surface area in the reformer.

- Excess energy from the reformer can provide a portion of the compression work.
- Common heat transfer systems (e.g. burner, gas cooling).
- Potentially lower capital cost compared to conventional fuel processors by reducing component count and sub-system complexity.

2. Test Highlights

The reformer and compressor were initially operated separately to characterize performance of each system.

2.1 Fluidized Bed Membrane Reformer

The reformer design conditions were as follows:

- 25 barg reactor operating pressure, 0.5 bara hydrogen permeate pressure
- 550°C reactor temperature
- Steam to carbon (natural gas) molar ratio of 3.1:1
- Air to carbon (natural gas) molar ratio of 1.7:1

A number of separate test runs were performed, some without a full complement of membranes. The goal of these tests was to characterize the reformer performance, as well as to test out the balance of the skid equipment. Some of the key outcomes of this testing:

- Operations:
    - Unattended operation proved (7 days).
    - Autothermal operation achieved through air addition, though system start-up was hampered by under-performing start-up heaters and pilot-scale heat losses.
    - Reforming (membrane) zone in the reactor was very isothermal (typically < 10°C variation). This is an important factor for membrane performance.
    - Reactor operation was very stable.
- Hydrogen production: The maximum hydrogen production achieved through the membranes without coupling to the compressor was 9 Nm^3/h. Models predict that the design production rate of 15 Nm^3/h would be achieved with the sub-atmospheric (0.5 bara) suction provided by the MHC.
- Hydrogen purity: For the longest run (17 days), the hydrogen permeate purity was greater than 99.9%. No performance decline due to catalyst abrasion was noted. Purity above 99.99% was achieved on other trials. During initial testing, it was found that some membranes exhibited lower hydrogen purity after the system shutdown due to development of a small number of pinholes. This was addressed by modifying the reactor shutdown procedure.

2.2 Membranes

The reformer was designed to accommodate 25 double-sided membranes, for a total effective membrane area of 1.0 m^2. The palladium alloy foil used in the reformer runs was 25 μm thick. Testing of thinner (15 μm) foil membrane modules was also conducted. As shown in the figure below, the hydrogen flux is inversely proportional to the foil thickness. Using thinner foils in the membrane reformer is very desirable as it reduces the membrane area and reactor size for a given production rate. However, this needs to be balanced with hydrogen purity requirements.

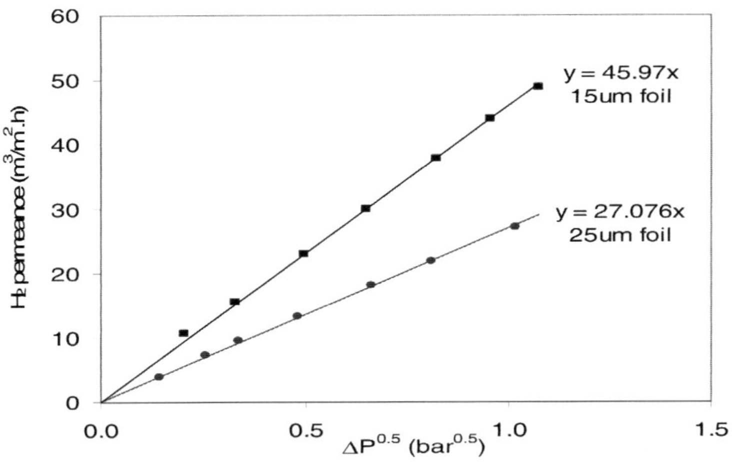

Figure 6: Measured hydrogen flux of 15 and 25 μm thick membrane modules (550°C)

2.3 Metal Hydride Compressor

The MHC system has two parallel hydride beds, each of which was designed to contain three compression stages. Due to time restrictions and a manufacturing issue, one of the beds had only two stages, which somewhat restricted the system performance. However, the MHC was successfully operated and proven to be able to compress hydrogen from 0.5 bara to 105 bara (at slightly below the design target) and meet the design flow target (at lower discharge pressure).

The figure below shows the results from one test using two compression stages. 310 L/min (purple line, 18.6 Nm^3/h) of hydrogen is compressed from sub-atmospheric pressure (green line, 6-9 psia) to over 200 psia (orange line). The MHC is a batch system, and the heating / cooling cycle (50 seconds) is also shown on the graph (gray line). Hydrogen is released from last compression stage once the pressure exceeds the setpoint of the discharge pressure regulator, thus the periodic spikes in the outlet pressure (orange line). Despite some variation in the inlet (suction) pressure, the inlet hydrogen flow is relatively stable.

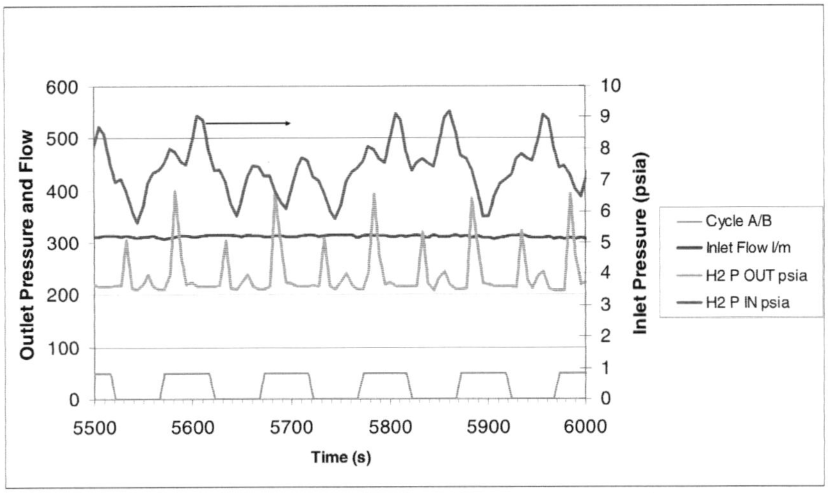

Figure 7: Example of MHC performance test results

2.4 Combined Reformer / MHC Operation

The run time for the combined reformer / MHC was significantly curtailed due to an operational issue (deactivation of catalyst due to trace sulfur breakthrough in the natural gas treatment system). However, the following was observed:

- Membrane flux increased as expected when the MHC was connected.
- The cyclic nature of the MHC suction did not noticeably affect the steady operating performance of the reformer.

3. Cost Modeling

Using the DOE's "H2A" analysis program (Ref. 5), costs for the integrated membrane reactor / MHC dispensing system were calculated at $7.98 and $3.30/kg for a 100 and 1,500 kg/day hydrogen production facility respectively (430 barg). The cost for the 1,500 kg/day system compared favorably to the H2A standard of $3.50/kg. Overall thermal efficiency for production of compressed hydrogen was estimated to be 50.4% (LHV).

The key component items that impact the cost estimate were identified to be, in decreasing importance:

- Membrane replacement period
- Membrane thickness
- MHC thermal efficiency

4. Summary

This project couples two novel technologies in an attempt to develop a more cost-effective route of converting natural gas to hydrogen at smaller scale. Compared to the conventional process (small-scale SMR / mechanical compression), the main attractions to the new process are:

- Low temperature, autothermal reformer eliminates indirect heat transfer (costly and high-maintenance materials of construction in SMR).
- In-situ hydrogen separation eliminates post-reformer water gas shift and PSA, lowering the system component count.
- The MHC can achieve very high discharge pressures, simplifying the hydrogen compression train.
- High thermal efficiencies can be achieved (no steam export, waste heat can be used effectively in the MHC)

The membrane reformer and MHC were successfully commissioned and shown to operate well together in a demonstration system, though operational issues prevented the system operating at its full design rate. There are some very attractive technical features and potential cost benefits of this novel system. However, there are several technical issues that must be addressed before this system could be considered commercial-ready:

Hydrogen purity: MRT is working with its palladium supplier and developing improved membrane substrate fabrication techniques to increase membrane robustness.

MHC mechanical integrity: Ergenics is addressing the mechanical reliability of its MHC manifolds.

Integration: The reformer and MHC must be more closely integrated to improve thermal efficiency, reduce skid size and lower capital costs.

Acknowledgements: We gratefully acknowledge the support of the US Department of Energy (project DE-FG36-05GO15017). Additional engineering and project support was provided by NORAM Engineering and Constructors Ltd. (Vancouver, Canada).

References

1. Shu, J., Grandjean, B.P.A., Van Neste, A., Kaliaguine, S., "Catalytic Palladium-based Membrane Reactors: A Review", Can. J. Chem. Eng., Vol. 69, 1036-1060 (1991).

2. Paglieri, S.N., Way, J.D., "Innovations in palladium membrane research", Separation and Purification Methods, Vol. 31, No. 1, 1-169 (2002).

3. Dixon, A.G., "Recent research in catalytic inorganic membrane reactors", International Journal of Chemical Reactor Engineering, 1, Review R6 (2003).

4. Deshmukh, S.A.R.K, Heinrich, S., Mörl L., van Sint Annaland, M., Kuipers, M., "Membrane assisted fluidized bed reactors: Potentials and hurdles", Ind. Eng. Chem. Res, 44, 5966-5976 (2005).

5. www.hydrogen.energy.gov/h2a_analysis.html

PEROVSKITE SORBENTS FOR REMOVAL OF SULFUR, ARSENIC AND PHOSPHORUS

Sara L. Rolfe, David A. Gribble, Jr., David H. Anderson, Douglas S. Jack, and Michael V. Mundschau, Eltron Research & Development Inc., 4600 Nautilus Court South, Boulder, CO

Solid sorbents are being developed to reduce concentrations of sulfur, arsenic and phosphorus in synthesis gas and water-gas shift mixtures from parts-per-million by volume (ppmv) concentrations to no more than 1–2 parts-per-billion by volume (ppbv) in the presence of steam. To prevent poisoning of nickel catalysts used in solid oxide fuel cells (SOFC), sulfur concentrations must be reduced to below 60 ppbv [1]. Arsenic and phosphorus concentrations must also be reduced to below 5 and 1–2 ppbv, respectively. The platinum catalysts of proton exchange membrane (PEM) fuel cells are even more susceptible to poisoning, unable to withstand over 4 ppbv of sulfur [1]. PEM fuel cells will require hydrogen of the highest purity—likely requiring hydrogen purification membranes—which in turn require protection from sulfur, arsenic and phosphorus. Commercial palladium-silver based membranes for hydrogen purification will not tolerate more than 4 ppbv H_2S, depending upon operating conditions, due to formation of a bulk Pd_4S. The sorbents under development bind sulfur more tightly than any other materials, thus greatly reducing the pressures of H_2S and COS relative to conventional sorbents based upon zinc or iron. These polishing sorbents are designed to operate at 300°C, which complements bulk sulfur sorption, water-gas shift and palladium membrane technologies. In addition, these sorbents form phosphates and arsenates with high thermal stabilities, thereby reducing the concentrations of arsenic and phosphorus species, including As_4, AsH_3, P_4, PH_3, HPO_2 and HPO_3. Calculations indicate that strongly binding sulfur will reduce the equilibrium partial pressure of H_2S even in the presence of high-pressure steam to immeasurably low concentrations (≤ 1 pptv). Thermogravimetric analysis (TGA) shows that preferred sorbents capture over 7 wt% sulfur at 300°C. In addition, TGA has been used to study regeneration by temperature-swing desorption at 500°C and above—showing over 15 wt% cumulative sulfur capture and regenerability through 20 cycles. Breakthrough tests under simulated sulfur-contaminated water-gas shift conditions (40% steam by volume) indicate total sulfur levels well below the current limits of detection (< 70 ppbv). Control experiments using commercial Cu/ZnO sorbents under identical breakthrough conditions show total sulfur concentrations at or above 117 ppbv throughout the breakthrough test—indicating that the Eltron sorbents are superior to zinc-based sorbents under steam. In addition to high sulfur capacity, Eltron sorbents show arsenic and phosphorus capture of nearly 5 wt% during 3-hour exposure tests.

Coal Derived Hydrogen for Electricity and Chemicals

The United States consumes 1 billion metric tons of coal annually [2]. Of this, approximately 900 million metric tons is utilized for generation of electric power [2]. Integrated Gasification Combined Cycle (IGCC) electric power plants utilizing hydrogen turbines or SOFCs have the potential for improving overall efficiency from about 33% (achieved by conventional fossil fuel and nuclear electric power plants) to 45–50% [3]. Improved efficiency across the entire fleet of coal-burning power plants could save up to 300 million tons of coal annually while simultaneously cutting CO_2 production at U.S. power plants by one third. In addition, new IGCC power plants have the potential to essentially eliminate all CO_2 emissions. Use of dense hydrogen transport membranes to separate hydrogen from CO_2 downstream of water-gas shift reactors would provide non-polluting hydrogen to turbines and fuel cells while retaining CO_2 at the high concentration and pressure necessary to further reduce the costs associated with CO_2 sequestration (i.e., compression and pipeline transportation) [4].

However, impurities originating from coal will need to be reduced to extremely low levels to avoid damaging high-speed, hydrogen-powered turbines or avoid poisoning the nickel catalysts used in the anodes of SOFCs [5,6] envisioned by the U.S. Department of Energy (DOE) for Integrated Gasification Combined Cycle Fuel Cell (IGCCFC) power plants. SOFCs will be permanently damaged if bulk nickel sulfides form and melt (m.p. NiS = 797°C and Ni_3S_2 = 790°C), destroying the physical and mechanical integrity of the anodes. Table 1 lists some stated target concentrations of select impurities as recommended by the DOE [1]. Note that to protect nickel catalysts in SOFCs for IGCCFC power plants, total sulfur will need to be lowered to 60 ppbv and arsenic to 5 ppbv.

Table 1. DOE Polishing Filter Technical Targets [1]

Contaminant	Gas Turbine	SOFC	PEM Fuel Cells
Total Sulfur (H_2S, COS, CS_2, etc.)	20 ppmv	60 ppbv	4 ppbv
Metals	20 ppbw Pb 10 ppbw V 40 ppbw Ca 40 ppbw Mg	5 ppbv As 200 ppbv Se 30 ppbv Cd	Not specified
Total Halides (Cl, F, Br)	5 ppmv	100 ppbv	50 ppbv
Total Alkalis (Na, K, Li)	100 ppbv	Not specified	Not specified

The goal of this research is to design and develop solid sorbents for use in polishing systems operating under warm-gas cleaning conditions (250°C–350°C) in steam and coal-derived water-gas shift mixtures. This polishing sorbent system must reduce concentrations of sulfur and its compounds from 1–20 ppmv, now achieved by existing sorbents, to very low ppbv (1–2 ppbv) levels. Dense palladium-catalyzed hydrogen transport membranes are envisioned for ultimate gas clean-up, eliminating essentially all impurities and purifying hydrogen to the highest possible levels while retaining CO_2 at very high concentration and pressure, up to 1000 psi (69 bar) for economic sequestration. However, for palladium-based membranes to function, impurities (especially sulfur, arsenic and phosphorus) that poison palladium, must be reduced, preferably to ppbv concentrations. Especially damaging to membranes is the formation of bulk Pd_4S.

Benefits of Warm Gas Cleanup with Hydrogen Transport Membrane

The economic benefits of using warm gas cleanup combined with Eltron's hydrogen transport membrane system for CO_2 capture in IGCC plants were outlined in a techno-economic analysis [7]. Three separate cases are shown in Table 2. The base case (Case 1) uses conventional cold gas (Selexol) technology for sulfur removal and CO_2 separation downstream of a high pressure General Electric (GE) gasifier, quench cooling and water-gas shift. Case 2 uses the GE gasifier, followed by conventional cold gas amine (MDEA) treating for sulfur removal, water-gas shift, and Eltron's membrane system for H_2/CO_2 separation. Radiant-convective cooling of syngas from the gasifier is used in Case 2 to avoid high moisture content in the feed to the amine absorber before the shift reactors. Thermal efficiency is improved in this case (partly due to the gasifier cooling method), while cost of electricity decreases slightly. A similar configuration is used in Case 3 (see Figure 1) as in Case 2, substituting warm gas cleaning for the desulfurization step. The cost of electricity in Case 3 decreases by $9.50 per MWh over the base case, with a 6% increase in thermal efficiency (see Table 2). Another benefit of the combination of warm gas cleaning and the hydrogen membrane is a significant increase in CO_2 capture from 90% to 99%.

Table 2. Summary of Economic Results for the Techno-Economic Analysis of CO_2 Capture in IGCC Plants

Pre-combustion Gas Cleaning & CO_2 Capture Method	Case 1: 2-Stage Selexol	Case 2: Cold Gas Cleaning & Eltron Membrane	Case 3: Warm Gas Cleaning & Eltron Membrane	Improvement of Case 3 over Base Case
Thermal Efficiency	27.4%	32.0%	33.4%	6.0%
% CO_2 Captured	90%	90%	99%	9%
Cost of Electricity ($/MWh)	115.5	114.5	106.0	9.5

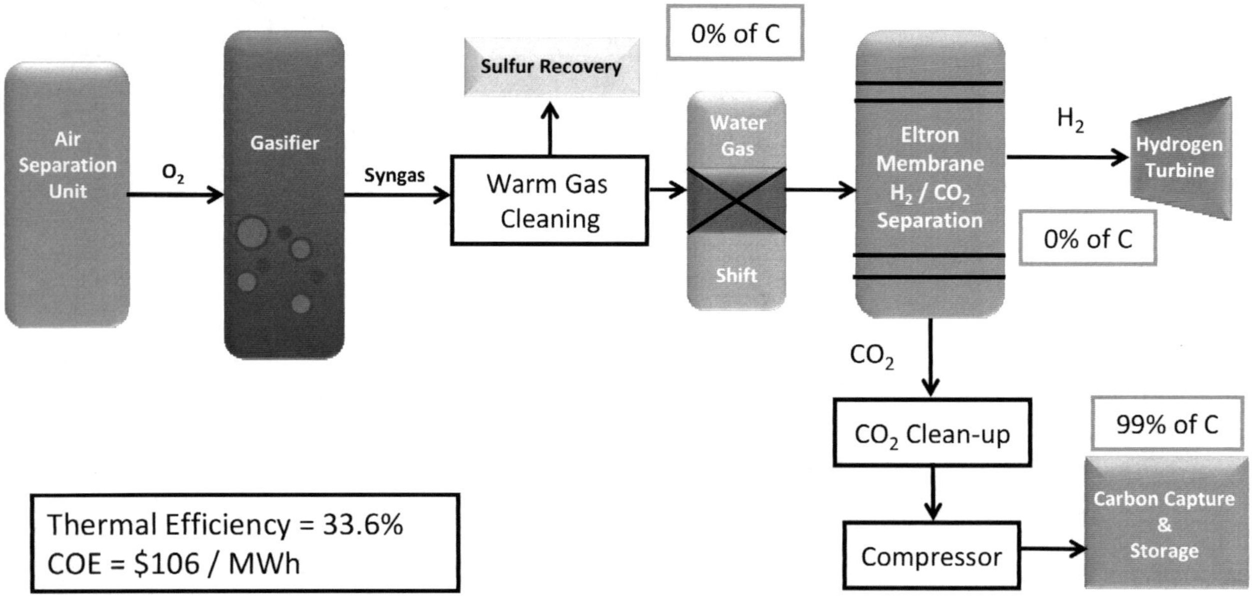

Figure 1. Case 3: CO_2 Capture with Warm Gas Cleaning and Eltron's H_2 Separation Membrane.

Drawbacks of the Zn-Based Sulfur Sorbent Systems

According to DOE guidelines [8], a baseline 640 MW_e IGCC electric power plant will consume 225 metric tons of Illinois #6 coal per hour and will produce 509 metric tons of CO_2 per hour. Using the recommended sulfur content (2.5 wt%) of Illinois #6 bituminous coal [8], a desulfurization system must remove 5.6 metric tons of sulfur per hour.

The sulfur polishing filters under development in the present research are intended to be placed downstream of a first-stage IGCC sulfur-removal system such as that now under development by Research Triangle Institute (RTI) and Eastman [9,10]. The RTI/Eastman system uses Zn-based sorbents to reduce sulfur concentration by 99.90% [10]. Therefore, 0.10% of the input sulfur ($\approx$ 6 kg/hr) will exit the RTI system—resulting in sulfur gases of sufficient concentration for the formation of sulfides that will rapidly destroy either fuel cells or hydrogen transport membranes. RTI reports H_2S levels exiting their system at 3–20 ppmv [10].

Zn-based sorbents such as Cu/ZnO are theoretically capable of bringing concentrations of sulfur to pptv concentrations under dry conditions. However, under high-pressure steam, H_2S is released by the equilibrium reaction: $ZnS + H_2O = ZnO + H_2S$ [5]. Breakthrough experiments conducted at Eltron Research & Development using commercial Cu/ZnO sorbents under water-gas shift conditions (40% steam by volume) show total sulfur concentrations in excess of 117 ppbv.

Sulfur Sorbent Screening

In order to rapidly screen new sorbent compositions, materials are exposed to sulfur as H_2S during TGA analysis. Using TGA, sorbent candidates are pre-reduced at 300°C in 10% H_2 (balance Ar) for 3 hours in order to remove a small percent of the lattice oxygen. Pre-reduction is followed by sulfur exposure at 300°C in 155 ppmv H_2S (balance Ar) for 10 hours. Since these tests are time-limited, none of these numbers reflect the full sulfur capacity of these sorbents. In these tests, a sorbent sample is placed into an 85 μL Al_2O_3 TGA crucible, open on top. The crucible is then placed into a Netzsch STA 449 F3 Simultaneous Thermal Analyzer for TGA tests. To date, 66 different screening tests have been performed on nearly 40 distinct materials. Long-term TGA testing has also been performed to determine the saturated sulfur capacity of sorbents. Eltron sorbents have been shown to capture over 7 wt% sulfur at 300℃.

Table 3 summarizes mass gain in terms of milligrams of sulfur sorbed per gram of sorbent ($mgS{\cdot}g^{-1}$) for select materials after 10 hours at 300°C in 155 ppmv H_2S (balance Ar). Values of sulfur uptake for some of the newly developed sorbents meet or exceed that of commercial Cu/ZnO. It should also be noted that values of mass gain listed in Table 3 exceed that which can be explained by surface adsorption—suggesting the formation of bulk sulfides which has also been verified by x-ray diffraction.

Table 3. Comparison of Unsaturated Sulfur Capacity at 300°C Before and After Attrition Milling.

Sorbent	Before Attrition Milling		After Attrition Milling	
	Surface Area ($m^2{\cdot}g^{-1}$)	Unsaturated Capacity (mg $S{\cdot}g^{-1}$ sorbent)	Surface Area ($m^2{\cdot}g^{-1}$)	Unsaturated Capacity (mg $S{\cdot}g^{-1}$ sorbent)
K53	0.3	15–18	10.1	28–29
K56	0.1	16	5.1	27
K57	0.1	23–38	6.2	45
K59	0.1	14–22	6.3	33
K60	0.2	16	7.9	30
K61	0.1	22–23	18.1	62
Commercial Sorbents			**As Received**	
Cu/ZnO	NA	NA	52.6	29
ZnO	NA	NA	40.3	30

Many of the Eltron sorbents have been prepared using solid state synthesis which results in low surface area prior to attrition milling (0.1–0.5 $m^2{\cdot}g^{-1}$). Despite the low surface areas, sulfur sorption was considerable for many of the sorbents. Eltron sorbents were attrition milled to reduce particle size and increase surface area (5–20 $m^2{\cdot}g^{-1}$). A corresponding increase in sulfur capacity was also observed (see Table 3). Sorbents of the same composition were also prepared using nitrate precipitation to yield high surface areas (> 50 $m^2 \cdot g^{-1}$). Further increases in sulfur sorption capacity were not observed, and these high surface area sorbents were shown to lose surface area upon exposure to high regeneration temperatures.

H_2S/COS Breakthrough Testing

Sulfur breakthrough tests were performed by exposing a packed bed of sulfur sorbent material to a simulated water-gas shift stream at 300°C containing by volume 40% H_2O (steam), 18% CO_2, 40% H_2, 3% CO, and 155 ppmv H_2S. Breakthrough was defined to be the point when total sulfur slippage first exceeded 1 ppmv. Gas delivery was from custom mixed tanks using mass flow controllers (MFCs). Water was admitted to the system via a liquid MFC and raised to make steam. The steam and the simulated water-gas shift streams were then combined and delivered to the sorbent test bed. In order to maximize the sulfur uptake prior to breakthrough, the sorbents were pre-reduced at 300°C under flowing H_2 at 300 sccm (STP at 0°C and 10^5 Pa) for 3 hours prior to sulfur exposure. By utilizing a small (5 mL) packed sorbent bed, a space velocity of 6000 hr^{-1} was achieved during sulfur exposure from a total gas delivery rate of 500 sccm. A high aspect ratio (length to cross-sectional area) bed geometry was used to minimize deleterious effects related to reaction kinetics.

The H_2S, COS and SO_x concentration were monitored throughout the breakthrough tests by continuous, automated GC sampling (500 μL Sulfinert® sample loop; Sulfinert® treated VICI injection valve). A Shimadzu GC-14A equipped with an FPD-14B detector and a Supelco (Bellefonte, PA) Model 12255-U packed GC column were employed to analyze the post-sorbent gas stream. The GC column was operated isothermally at 50°C (He carrier gas; 60 kPa)—resulting in a 2 minute chromatogram with adequate separation of H_2S, COS, and SO_x (not observed) within 2 minutes. A multi-point GC calibration was obtained via dilution of calibration standards for H_2S (1.1 ppmv and 0.42 ppmv) and COS (10.2 ppmv). All standards were mixed and diluted with UHP-grade Ar as the balance gas and supplied from Matheson Tri-Gas. Like the post-sorbent gas streams, the entire calibration assembly is Sulfinert® treated stainless steel to minimize sulfur losses throughout the system. Multiple injections were obtained at each calibration level in order to verify that a stable signal was obtained. This chromatography approach is not capable of handling super-saturated steam; therefore, a Nafion® gas dryer (Perma Pure, LLC) with counter-flowing nitrogen was utilized to reduce the moisture content of the effluent gas stream to a non-condensing level. Because H_2S and COS are water soluble gases, great care must be taken to avoid condensing H_2O at any point in the analysis stream. Based on this technique, quantitative GC analysis was possible for H_2S concentrations ≥ 74 ppbv and COS concentrations ≥ 10 ppbv.

Preliminary breakthrough testing indicated that in addition to the release of H_2S by reaction of steam with ZnS, commercial sorbents of Cu/ZnO release carbonyl sulfide (COS) under water-gas shift conditions. Additional equilibria exist for reactions of CO and CO_2 with ZnS that set the lower limits of COS concentration. Figure 2 shows breakthrough data collected downstream from a commercial Cu/ZnO sorbent exposed to H_2S contaminated water-gas shift mixtures at 300℃ and ambient press ure. At no time were the total sulfur or H_2S concentrations below 117 ppbv.

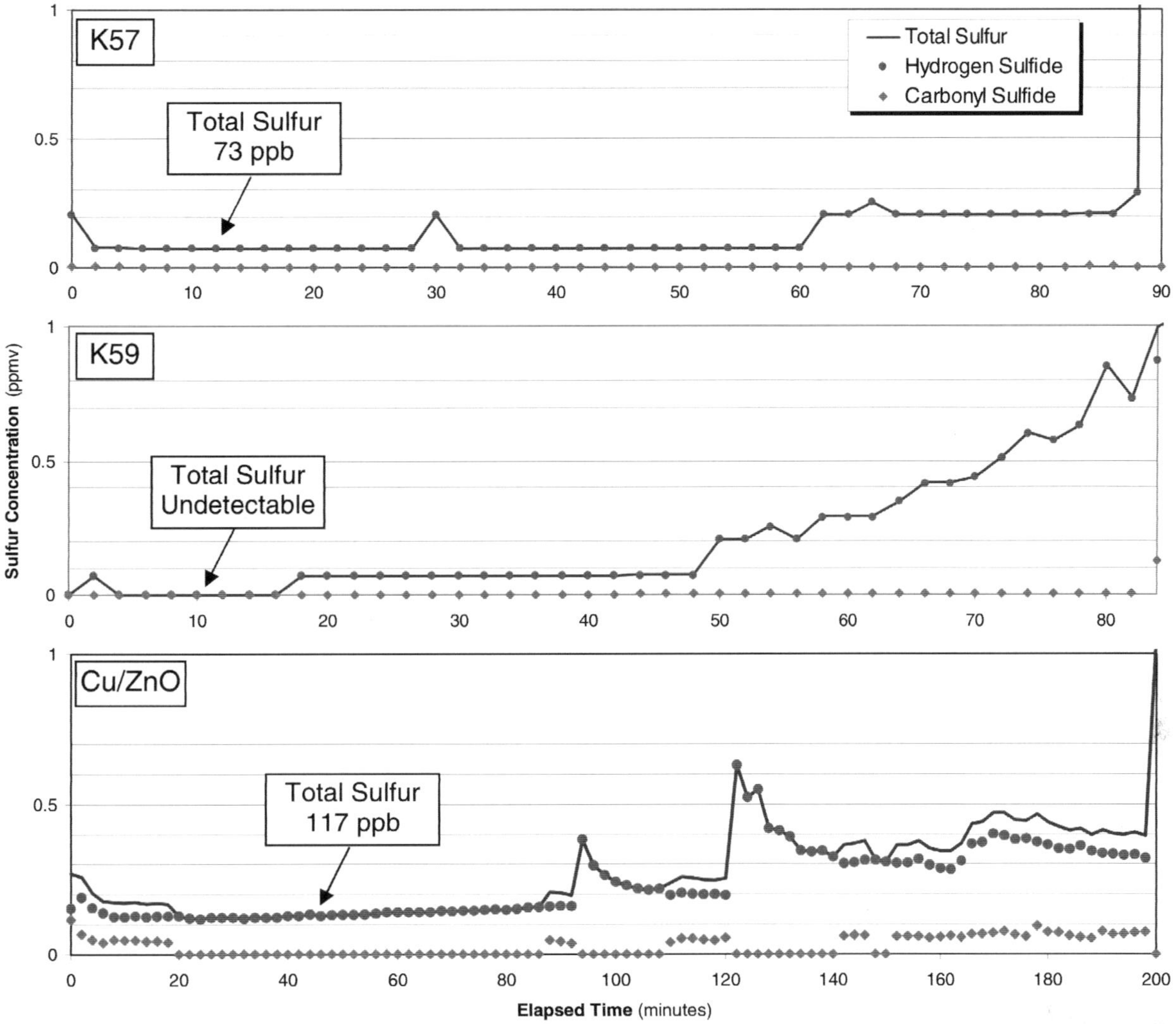

Figure 2. Breakthrough results for Eltron's K57 (top) and K59 (middle) sulfur sorbents and a commercial Cu/ZnO (bottom) sulfur sorbent. Eltron's sorbents show total sulfur at or below detection limits for nearly an hour. When testing the Cu/ZnO, at no time was the H_2S concentration below 100 ppb. Feed: 40% H_2O (steam), 18% CO_2, 40% H_2, 3% CO, and 155 ppmv H_2S at 300°C.

Several of Eltron's sulfur sorbents have been tested in the breakthrough reactor under the procedures described above. Two different Eltron sorbents (K57 and K59) have demonstrated sulfur removal capacities that exceed commercial Cu/ZnO (see Figure 2). Both sorbents show total sulfur at or below detection limits for nearly an hour. Unlike the Cu/ZnO, COS is not observed until the sorbents approach saturation.

Sulfur Sorbent Regeneration

Cyclic, high-temperature regeneration tests were performed on the Eltron sorbents in order to determine their potential for regeneration in hydrogen. Initial tests assessed the impact of regenerating the sorbents at various temperatures using a mixture of 2% H_2 in Ar in the TGA. The test began by pre-reducing the sorbents in 2% H_2 (balance Ar) for three hours at 300°C. Next the sorbents were cycled between H_2S exposure (155 ppmv, balance argon) at

300°C for 2 hours and various regeneration segments. As shown in Figure 3, the optimal regeneration step was determined by varying the dwell time (1–3 hr) and temperature (500–700°C) of the regeneration segment. At regeneration temperatures ≥ 600°C, the sorbent mass gain and loss was nearly the same for each segment. Regeneration is fairly rapid with the majority of the sulfur being liberated from the sorbent during the initial temperature ramp.

20-cycle tests incorporating 2-hour, 500°C regeneration segments were also performed with both the K57 and K59 sorbents. K57 showed 15.3 wt% net sulfur sorption at 300°C after 20 cycles with a 500°C regeneration seg ment, and K59 showed a net sulfur sorption of 11.1 wt%. In both cases, the 20-cycle tests indicated incomplete regeneration and a subsequent drop in sulfur sorption through the tests. By switching to a 1-hour, 600°C regeneration step, it is anticipated that the sorbents will show significantly increased net sulfur sorption and an extended lifetime.

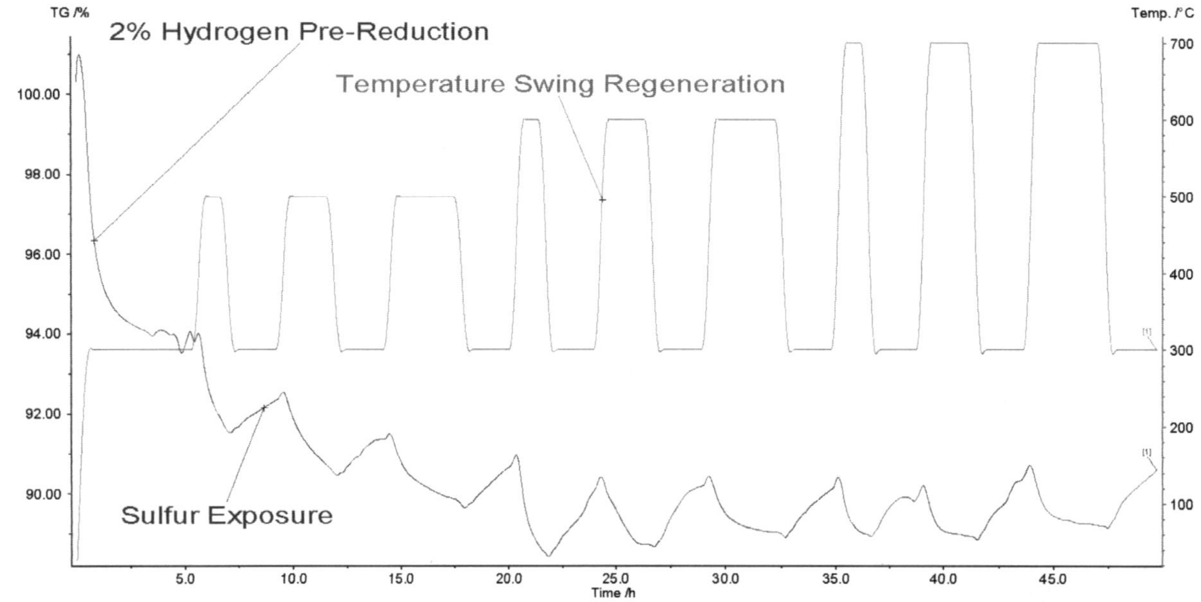

Figure 3. TGA results for Regeneration Test of K57. At regeneration temperatures ≥ 600°C, the sorbents gain and release nearly the same amount of sulfur in each segment.

As/P Sorbent Data

Eltron's proprietary sorbents have also been evaluated as arsenic and phosphorus sorbents. In these tests, small sorbent samples (approximately 3.5 mL) were pre-reduced at 300°C for 1 hour in H_2 and then exposed to arsenic (As_4) and phosphorus (P_4) vapors for 3 hours. The As_4 and P_4 vapors were generated by heating elemental arsenic and phosphorus in separate quartz crucibles. Exposure coupons were used both before and after the sorbent test bed in order to gauge the effectiveness of the sorbents in protecting easily poisoned nickel films. Coupons were prepared at Eltron from un-plated stainless steel VCR filter gaskets (Swagelok®) with a thin nickel coating (≈ 100 nm) applied to the front surface of the filter element. Initial and final masses were recorded for both the upper and lower coupons and the sorbent material. TGA data is used to estimate the expected mass loss due to pre-reduction. Appropriate evaporation temperatures were determined experimentally for As (525°C) and

P (450°C). These temperatures result in time-weighted average concentrations of 420 ppmv (As_4) and 250 ppmv P_4 throughout the 3-hour exposure period.

Eltron sorbent K57 strongly bound both arsenic and phosphorus with a total mass gain of nearly 5% in a 3-hour exposure test. Sorbent K59 was less successful in removing As, but showed increased P removal as compared to sorbent K57. Further analysis via Energy Dispersive Spectrometry (EDX) indicates that both arsenic and phosphorus were trapped by both sorbents (see Table 4).

Table 4. As and P Sorption Results

Sorbent	Total Weight Gain (%)	Elemental Analysis by EDX	
		As (wt%)	P (wt%)
K57	4.52	4.5	1.6
K59	0.82	2.2	2.5

References

[1] U.S. Department of Energy, Office of Fossil Energy and National Energy Technology Laboratory, Hydrogen from Coal Program for the Period 2008 through 2016, September 2008, p. 36.

[2] Energy Information Administration, Office of Integrated Analysis and Forecasting, U.S. Department of Energy, International Energy Outlook, www.eia.doe.gov/oiaf/ieo/index.html.

[3] W. Surdoval, Welcome and SECA Overview, Presentation at the 9th Annual SECA Workshop, Pittsburgh, Pennsylvania, Sheraton Station Square, August 5–7, 2008.

[4] M.V. Mundschau, X. Xie, C.R. Evenson IV, and A.F. Sammells, Advances in Hydrogen Separation Membrane Technology for the Separation of CO_2 and the Purification of Hydrogen Produced from Coal, in: Proceedings of the 30th International Technical Conference on Coal Utilization and Fuel Systems, April 17–21, 2005, Clearwater, FL.

[5] D. A. Gribble, Jr., S.L. Rolfe, and M.V. Mundschau, Perovskite Sorbents for Warm-Gas Removal of Sulfur, Proceedings of the Twenty-Sixth Annual International Pittsburgh Coal Conference, Pittsburgh, Pennsylvania, September 20–23, 2009, Session: Gasification: Synthesis Gas Cleaning, September 22, 2009.

[6] J.P. Trembly, R.S. Gemmen, D.J. Bayless, The Effect of IGFC Warm Gas Cleanup System Conditions on the Gas-Solid Partitioning and Form of Trace Species in Coal Syngas and Their Interaction with SOFC Anodes, Journal of Power Sources 163 (2007) 986–996, doi: 10.1016/j.jpowsour:2007.10.020.

[7] D.S. Jack, "Effective Utilization of Hydrogen Membranes for CO_2 Capture," The Clearwater Clean Coal Conference, May 31-June 4, 2009.

[8] U.S. Department of Energy, Cost and Performance Baselines for Fossil Energy Plants, DOE/NETL-2007/1281.

[9] A.A. Vierheilig, R.P. Gupta and B.S. Turk, Attrition Resistant, Zinc Titanate-Containing, Reduced Sulfur Sorbent and Methods of Use Thereof, U.S. Patent, 7,067,093 B2, Jun. 27, 2006.

[10] R. Gupta, B. Turk, M. Lesemann, J. Schlather and D. Denton, Status of RTI/Eastman Warm Gas Clean-up Technology and Commercialization Plans, Presentation at the Gasification Technologies Conference, October 5–8, 2008, Washington, D.C.

(117a) Post-Combustion Carbon Capture Research Progress through the DOE/NETL Existing Plants Program

Timothy E. Fout, Existing Plants Division, National Energy Technology Laboratory, Morgantown, WV

Jared Ciferno, U.S. Department of Energy, National Energy Technology Laboratory, Pittsburgh, PA

Lynn Brickett, Existing Plants Division, National Energy Technology Laboratory, Pittsburgh, PA

Coal-fired power plants have made significant progress in reducing emissions of sulfur dioxide, nitrogen oxide, particulate matter, and now mercury (Hg), since the passage of the Clean Air Act. However, significant Congressional activity has been underway on legislation that would require a reduction in carbon dioxide (CO_2) emissions. A potential solution for the reduction of emissions from coal fired power plants is the capture and subsequent storage of CO_2 in a geological formation (CCS). DOE/NETL's Existing Plants Program is focused on developing advanced CO_2 capture and compression technologies to aid in the development of the CCS process. The focus of the Existing Plants Program is driven by the fact that existing pulverized coal (PC) power plants produce a sizeable portion of current CO_2 emissions from all fossil-fuel-based sources, and that only about six gigawatts of the existing coal-fired electricity generating fleet is projected to retire by 2030. Current analysis indicates that state-of-the-art CO_2 capture technologies are not cost-effective technologies for capturing CO_2 from PC-based power plants. Also, these technologies have not been utilized on the scale required for CCS from a coal-fired power plant. The technologies being researched through the Existing Plants program aim at achieving a 90% capture rate while achieving no more than a 35% increase in the cost of electricity. This paper will give an update on the status of the program and recent progress of its post-combustion research projects.

See more of this Session: Post-Combustion Techniques for Capture/Separation of Carbon Dioxide I
See more of this Group/Topical: Topical 2: Advanced Fossil Energy Utilization

(117b) Vapor-Liquid Equilibrium (VLE) Study of CO2 with MEA and Ammonia Solvents

Donald Johnson, Power Generation and Utility Fuels, The Center for Applied Energy Research / University of Kentucky, Lexington, KY

Zhe Lu, Power Generation and Utility Fuels, The Center for Applied Energy Research / University of Kentucky, Lexington, KY

Joseph E. Remias, Power Generation and Utility Fuels, The Center for Applied Energy Research / University of Kentucky, Lexington, KY

Kunlei Liu, Power Generation and Utility Fuels, The Center for Applied Energy Research / University of Kentucky, Lexington, KY

James K. Neathery, Power Generation and Utility Fuels, The Center for Applied Energy Research / University of Kentucky, Lexington, KY

In the carbon capture and sequestration (CCS)scenario, vapor liquid equilibrium (VLE) curves represent the theoretical relationship between local gaseous CO2 vapor pressure and the respective carbon loading in solution for a given operating temperature. It determines the maximum carbon content in the lean solution to achieve 90% removal efficiency in the absorption process, and the minimum carrier gas flowrate required in the stripper for solvent regeneration at operating temperature. Unfortunately, the equilibrium between liquid and vapor phases can be very difficult to accurately quantify with simulation calculations, which use a combination of physical and thermodynamic equations along with post-experimental data for reference. Simulations can often yield results which are not a true reflection of what is observed in reality. In order to measure the VLE accurately, a CSTR equilibrium cell in conjunction with an on-line gas-sampling gas chromatograph (GC) for analyzing the vapor phase composition at equilibrium have been assembled. This apparatus will allow the engineer to test various solvents and determine the optimum operating conditions suitable for CCS. In this research effort, we report on initial gas equilibrium results that will be useful for designing CO2 scrubbers and associated stripping units using various aqueous ammonia and amine-based solvents. The results are reported at different CO2 loadings, temperatures, and pressures to obtain "true" vapor-liquid equilibrium curves for use in analysis and process design.

See more of this Session: Post-Combustion Techniques for Capture/Separation of Carbon Dioxide I
See more of this Group/Topical: Topical 2: Advanced Fossil Energy Utilization

Stripper Configurations for CO_2 Capture By Aqueous Amines

David H. Van Wagener, The University of Texas at Austin, Austin, TX; Dr. Gary T. Rochelle, The University of Texas at Austin, Austin, TX

Introduction

Amine-based absorption is considered an excellent option for CO_2 capture from coal-fired power plants. Compared to other options like oxycombustion and separation with membranes, absorption with amine solvents is advantageous because it has a relatively lower energy requirement for carbon capture and can be installed on the tail end of pre-existing coal plants. The majority of energy requirement for this process stems from solvent regeneration in the stripping section. A simple stripper is commonly used, but more complex configurations can help reduce the required energy input. It has been hypothesized that there is a relation between configuration complexity and performance that can be correlated. Complex configurations are desirable to improve stripper performance because of their ability to reduce driving forces. Fundamental work by Leites (2003) showed the benefit of reducing driving forces for generic chemical processes. Driving forces surface in several forms: temperature driving forces, mass transfer driving forces, and chemical reaction driving forces. Typical strippers for this application have very large driving forces in various locations. For example, the section of packing toward the bottom near the reboiler usually has a very large driving force, and the rest of the packing is nearly pinched, accomplishing the last of the reaction and mass transfer to approach equilibrium. Introducing more complexity to the by means of splits, recycles, and multiple pressure stages, the existing driving forces can be reduced to cut down on total exergy loss. In addition to reducing driving forces, more complex configurations with multiple pressure stages can improve the overall performance of the stripper and CO_2 compressor by collecting high-pressure CO_2, thereby reducing the total compressor work. High pressure stripping is beneficial to improve the selectivity of CO_2 while collecting at high pressure. Ambient pressure stripping is helpful to achieve a desired lean loading at lower temperatures, which can avoid excessive thermal degradation of the solvent.

Methods and Results

Complexity Modeling

Since complexity is a broad undertaking, the analysis started at the most simple separation method, a single heated flash, and progresses systematically to more complex configurations (Rochelle 2009). The mathematical definition of complexity is still loose, but it is expected that the number of process units in the configuration is roughly related to a complexity "value". The complexity is calculated as the total number of major pieces of equipment, excluding typical pumping and compression equipment. For example, a single stage heated flash has a value of 2 (a heater and separation vessel), and a simple stripper has a value of 3 (a heater, vessel, and packing). The complexity value of a configuration is decreased if an intercooler is eliminated by integrating a compression stage in the flowsheet. This is the case in the 2-Stage Multi-Pressure configuration. It is expected that more complex configurations will improve performance but have a diminishing return at higher complexity. It is also expected that different configurations with similar complexity values will have different performance according to how efficiently the process units are arranged. A systematic justification for the arrangement of process units as it relates to performance may be possible.

The analysis so far has focused on configurations using one, two, and three pressure stages. Each configuration was specified to run isothermally, except for special cases which included adiabatic sections. 9 m MEA was chosen as the solvent for this study due to the proven robustness of MEA. Additionally, 9 m MEA has improved solvent capacity over the industry standard, 7 m MEA. A thermodynamic model was developed for MEA by Hilliard (2008) which represents 9 m MEA fairly well; this Aspen Plus model using the electrolyte-NRTL method for MEA was used for this work. The different configurations can be separated into subcategories: multi-stage flash, and double matrix, and columns. The analysis began with the most simple separation method, a 1-stage flash, and built upward in complexity and total number of pressure stages. Each stage had a number of options: flash vs. packing, heated vs. adiabatic, and recompression of the vapor to the prior stage. The configurations that have been modeled are described in Figure 1 – 5. An additional configuration not shown is a stripper with an adiabatic lean flash. However, this flowsheet stems directly from the 2-stage multi-pressure configuration; the adiabatic lean flash configuration is a multi-pressure flowsheet with packing and the reboiler in the top section and an unheated flash in the bottom section.

Several of these configurations were simulated using multiple run specifications. The 2-stage flash was simulated for three cases: both stages heated isothermally, only the first stage heated, and only the second stage heated. These options were explored to determine the effectiveness of isothermal stripping. Next, the double matrix configuration was tested with a variety of options. First, it was simulated using only 3 flash sections as shown in Figure 5. This first case was also run with a uniform 20% split of the rich feed to the low-pressure flash. This value of split was used because it was found to be an optimum value in previous work. In the next case, this double matrix flash configuration was optimized for rich split at each lean loading. Finally, the double matrix configuration was also simulated with a section of packing in place of the adiabatic flash on the top of the low-pressure section. Another configuration not shown is the 3-stage flash, which follows the sequential progression of the multi-stage flash, similar to the 1- and 2-stage flash configurations. The vapor streams fed to the multistage compressor were pre-cooled to 40°C to knock out water vapor. Additionally, there was intercooling to 40°C between each stage of the compression trains.

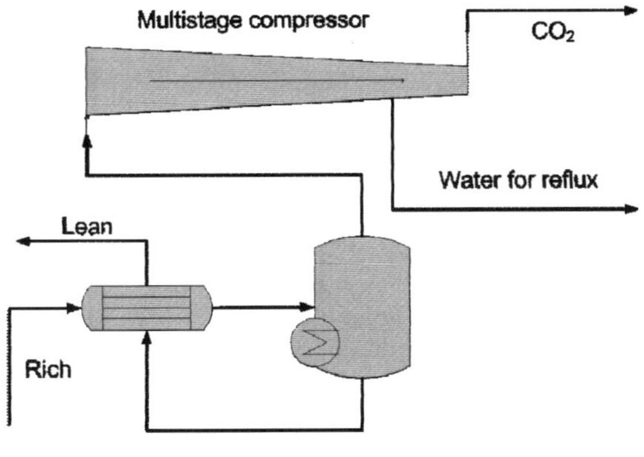

Figure 1. 1-Stage Flash

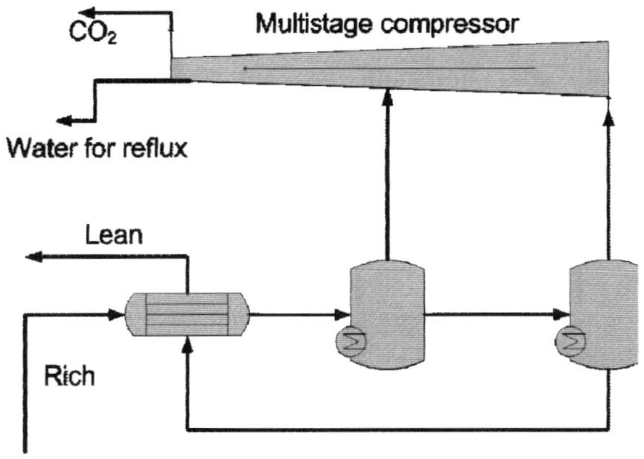

Figure 2. 2-Stage Flash

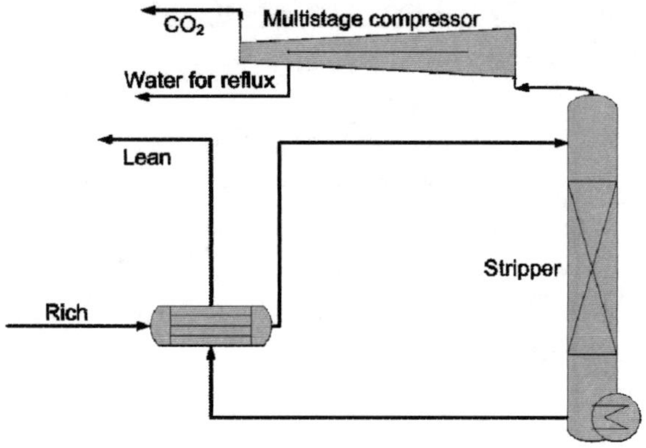

Figure 3. Simple Stripper

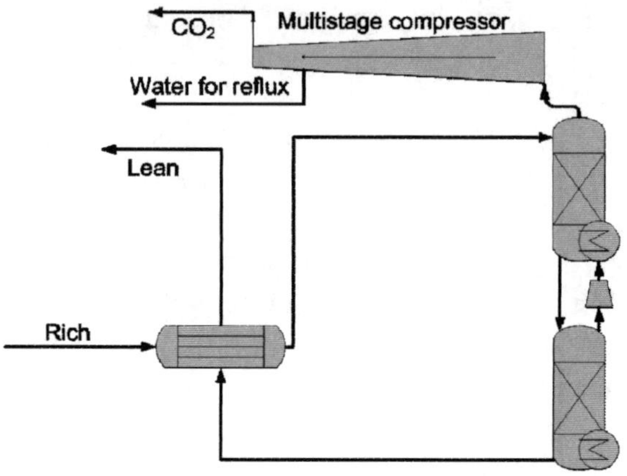

Figure 4. 2-Stage Multi-Pressure

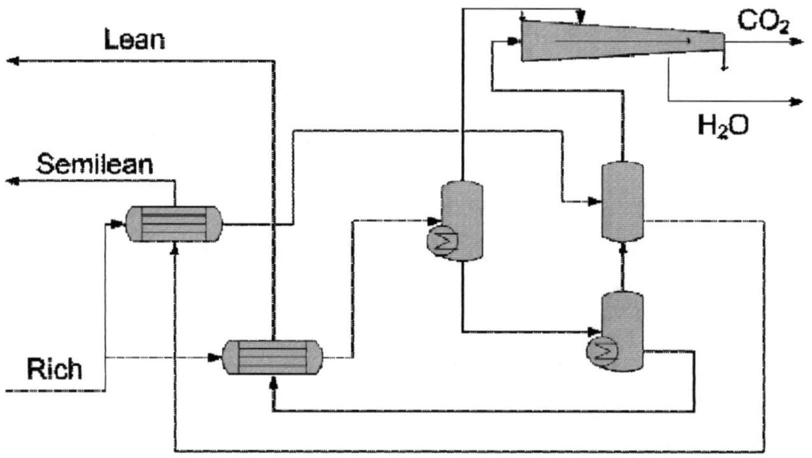

Figure 5. Double Matrix with Equilibrium Flashes

The simulations were performed with as many variables held constant as possible. With this goal in mind, the stripped CO_2 was compressed to 15 MPa using 7 stages in all simulations. This compression also included precooling and intercooling to 40°C with water knockout in each cooler. The rich pumps achieved the vessel pressure for a given configuration, and it also accounted for an additional 50 kPa for possible frictional losses as well as 250 kPa for gravitational losses if a packed column was used in exchange for a flash vessel. A constant rich loading of 0.5 was used, which corresponds approximately to a CO_2 partial pressure of 5 kPa in the solvent at 40°C, a typical rich end temperature for the absorber. A 5°C cold side approach was specified on all heat exchangers. A 10°C approach was specified on the steam reboilers. Equivalent work was calculated as follows:

$$W_{eq} = \sum_{i=1}^{n_{reboilers}} 0.75 * Q_i \left(\frac{T_i + 10K - T_{sink}}{T_i + 10K}\right) + W_{pumps} + W_{comps}$$

where Q_i and T_i are the duty and temperature for reboiler i. The equivalent work represents the total electrical penalty on a power plant assuming that electricity could be extracted from steam at a 75% Carnot efficiency rate. The lean loading was optimized to minimize total equivalent work for each configuration. The equivalent work was evaluated for lean loadings ranging from 0.2 to 0.45. The optimum lean loadings were all above 0.35, but the performance was evaluated down to lean loadings of 0.2 because many industrial users prefer these low lean loadings to reduce absorber capital cost.

Since isothermal operation should produce the most efficient thermodynamic results, all reboilers and heaters in each run were specified to have equal temperatures. The analyses were run for 100°C, 110°C, 120°C, and 130°C. Unless stated otherwise, the results presented in this report will be only for the 120°C runs. This temperature is expected to be the highest feasible operating temperature for MEA due to elevated thermal degradation rates above 120°C. The results from the different categories of configurations are charted below in Figure 6 - Figure 10. Additionally, the minimum equivalent work for each configuration is documented in Table 1. This table also lists the lean loading at the minimum equivalent work as well as the complexity rating, calculated by the basic method presented earlier.

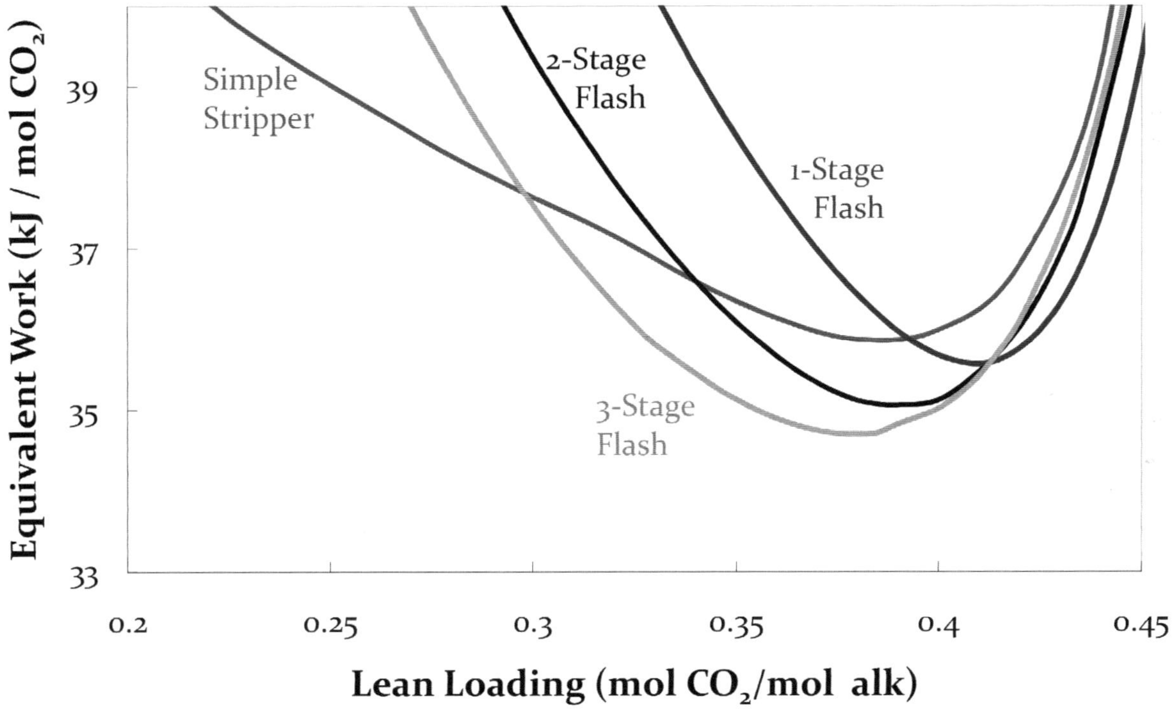

Figure 6. Multi-Stage Flash Results. 9 m MEA, 0.5 rich loading, 120°C reboilers, CO_2 compression to 150 atm, evenly distributed vapor generation per stage on mole basis.

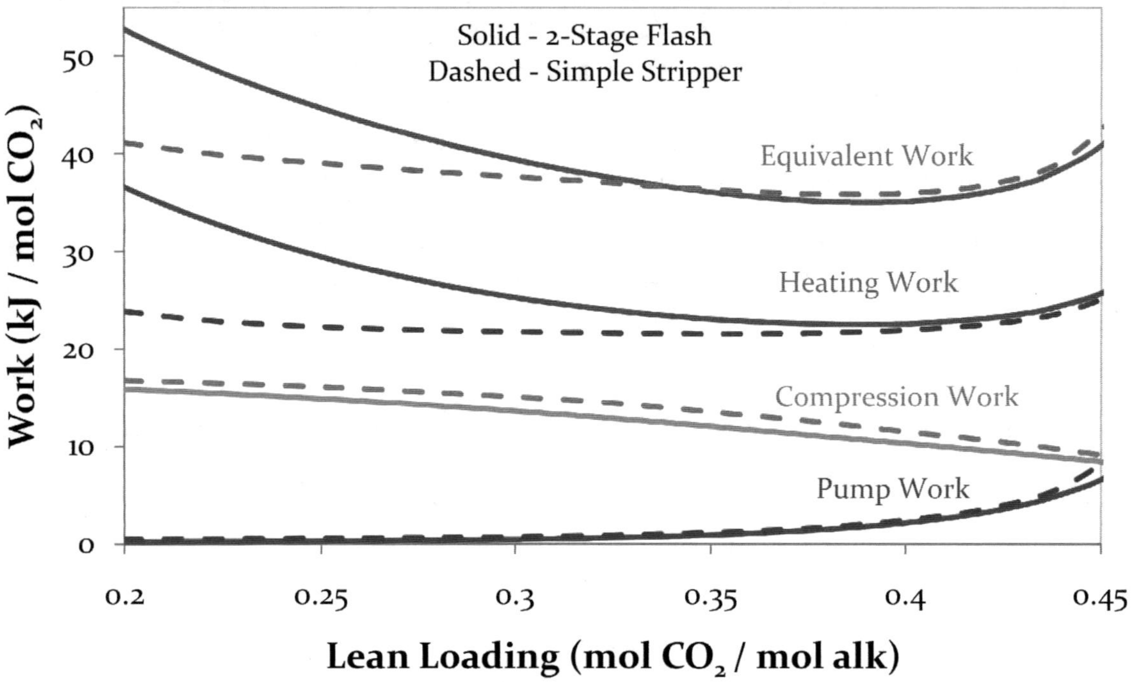

Figure 7. Contributions to Equivalent Work for 2-Stage Flash and Simple Stripper

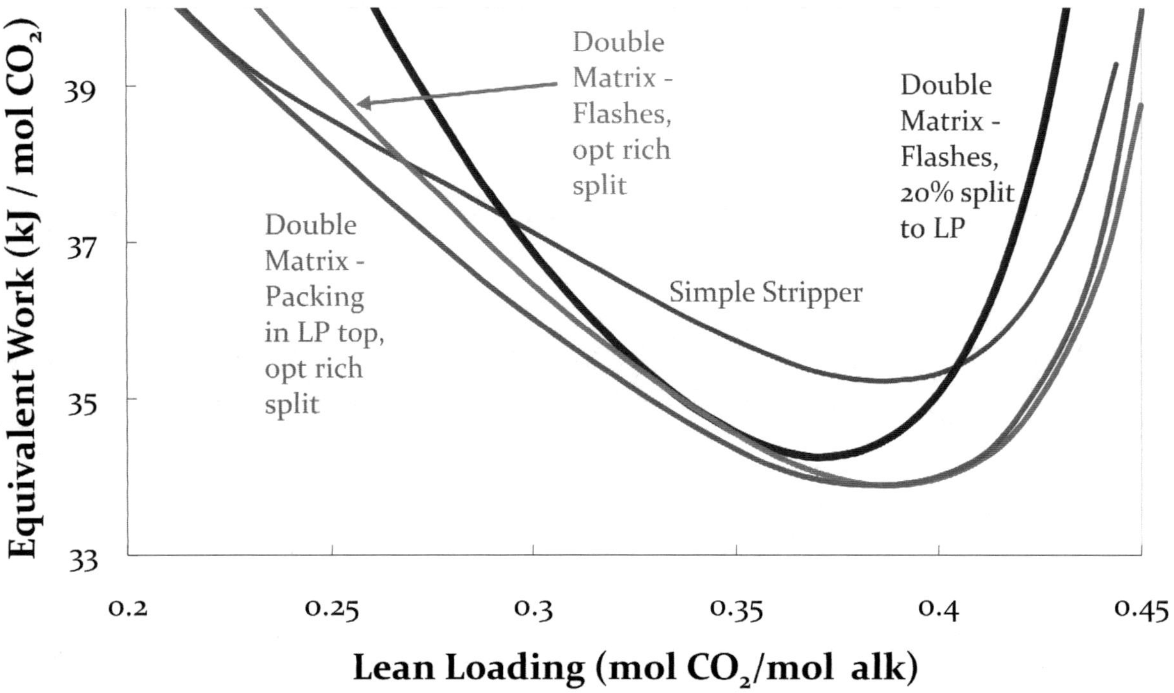

Figure 8. Double Matrix Results. 9 m MEA, 0.5 rich loading, 120°C reboilers, CO_2 compression to 150 atm, evenly distributed vapor generation per stage on mole basis.

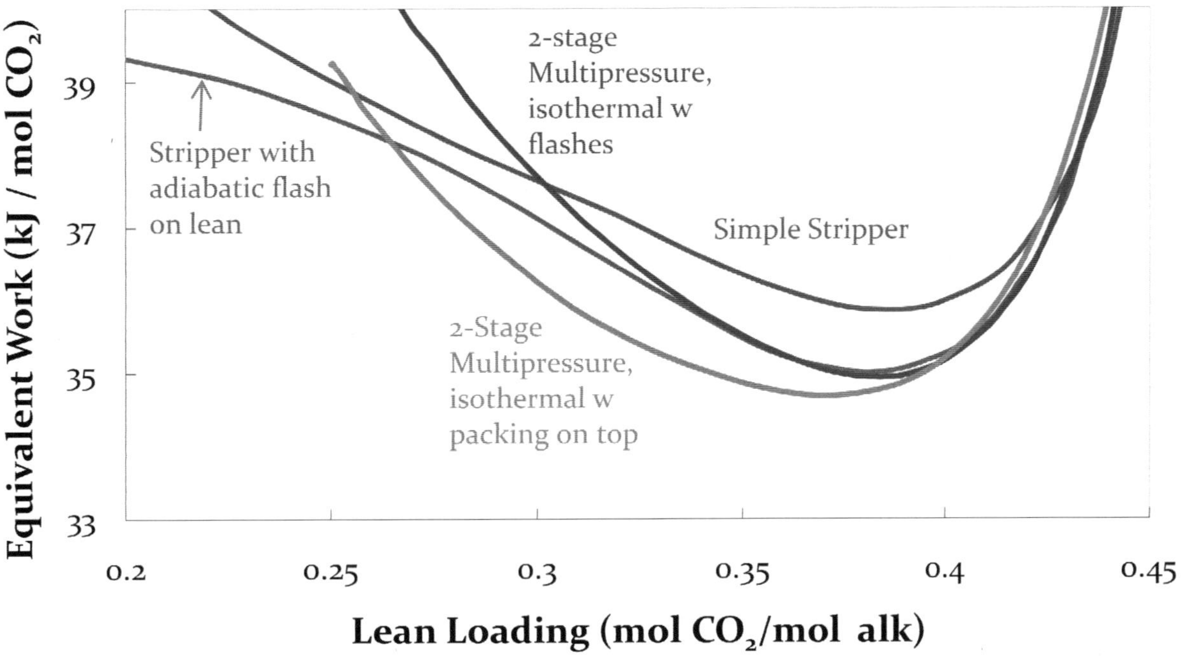

Figure 9. Column Results. 9 m MEA, 0.5 rich loading, 120°C reboilers, CO_2 compression to 150 atm, evenly distributed vapor generation per stage, mole basis (except Multipressure - 1.8 compression ratio).

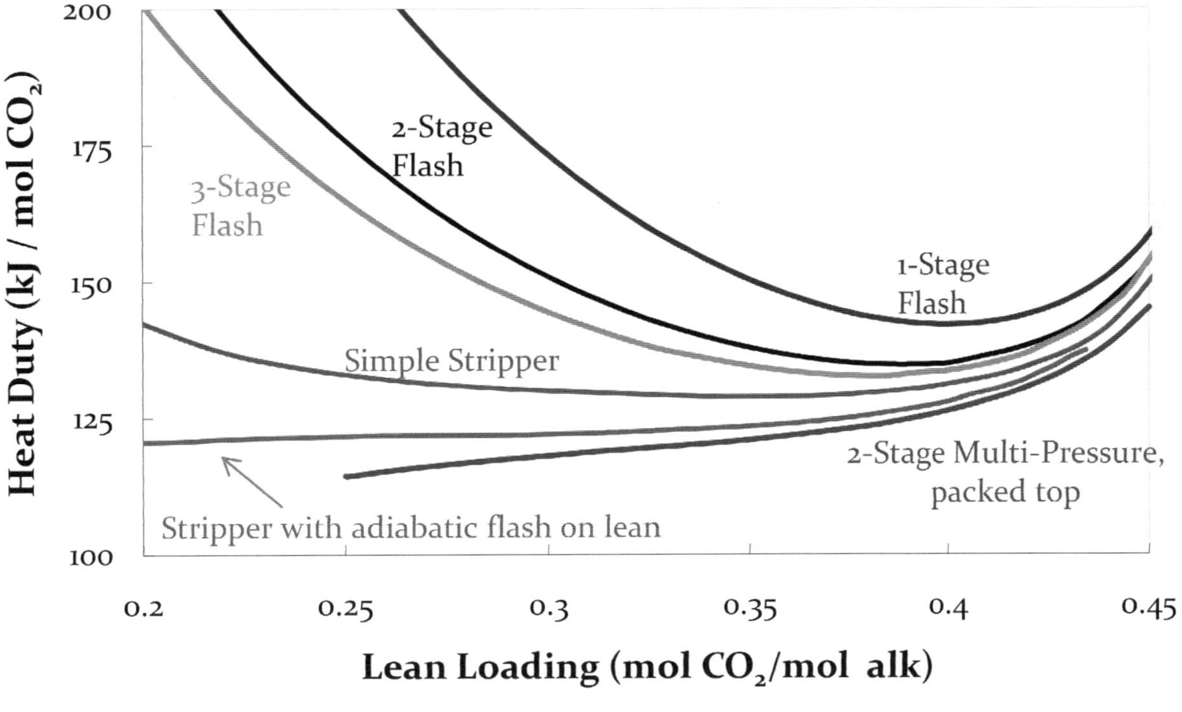

Figure 10. Heat Duties for Selected Configurations. 9 m MEA, 0.5 rich loading, 120°C reboilers, CO_2 compression to 150 atm, evenly distributed vapor generation per stage, mole basis, except multi-pressure.

Table 1. Results Summary for All Configurations

Configuration	Process units	Equivalent Work kJ/mol CO_2	Optimum Lean Loading
1-Stage flash	2	35.6	0.410
Simple Stripper	3	35.2	0.383
Stripper with adiabatic lean flash	3	35.0	0.387
2-Stage flash, 1st adiabatic	3	35.4	0.412
2-Stage flash, 2nd adiabatic	3	35.2	0.400
2-stage Multi-pressure, flashes	3	34.9	0.385
2-Stage flash	4	35.1	0.390
2-stage Multi-pressure, packed top	4	34.7	0.370
Double matrix, flashes	5	33.9	0.390
3-stage flash	6	34.7	0.380
Double matrix, packed LP top	6	34.0	0.370

The plots above demonstrate a clear improvement in stripper performance as complexity increased. This point is especially apparent in Figure 6, showing the difference between the various multi-stage flash configurations. Not only did the energy requirement decrease with more flash stages, but the optimum lean loading also decreases. This lower lean loading would be beneficial for the absorber to improve its performance and decrease the absorber packing height. The more complex configurations shown in the other figures decrease the energy requirement even more. In Table 1 the general trend of decreasing equivalent work with increasing complexity is displayed, though the trend did not fit a direct correlation with the number of process units. The improvement in performance also depended on two factors: use of vapor recompression and replacement of flashes with packing. Recompression of the low-pressure vapor improved performance by condensing some of its stripping steam to vaporize CO_2 at the higher pressure. Unlike vapor recompression, inserting packing sections in place of equilibrium flashes did not improve performance as much as expected. The effect of this addition was seen directly in the double matrix and 2-stage multi-pressure configurations. In the case of the double matrix, the baseline configuration with three equilibrium flashes was altered by replacing the top flash of the low-pressure section with a packed column. Adding packing in this location was expected to have the greatest effect on performance because a large amount of stripping steam was produced in the bottom flash of the low-pressure section, and the stripping steam required an adequate number of transfer units to fully condense and vaporize CO_2. The other flashes included reboilers which mostly eliminate the need for transfer units. Similarly, in the 2-stage multi-pressure configuration, packing was inserted in the high-pressure stage to better reach equilibrium with the inlet rich to reduce the amount of stripping steam in the vapor before it was cooled and compressed. As seen in Figure 8 and Figure 9, including packing in these two configurations had no effect at high lean loadings where the CO_2 capacity was low. However, packing improved performance in both configurations at low lean loadings when an increased number of transfer units was beneficial to achieve the increased amount of CO_2 liberation per unit solvent.

This analysis also demonstrated the significance of evaluating performance by equivalent work as opposed to heat duty. An analysis using heat duty neglected the effect of temperature on the value of steam. This heat duty analysis also overlooked the pumping and compression contributions to overall energy requirement. An example of this difference could be seen when comparing the 2-stage flash with the simple stripper in Figure 7. The 2-stage flash has a lower equivalent work but higher heat duty than the simple stripper. The improvement of the 2-stage flash over the simple stripper stemmed from reduction in compression work due to collection of CO_2 at high pressure in the first flash, so the heat duty analysis does not recognize this benefit. Additionally, Figure 10 shows that the optimum lean loading when evaluating heat duty was seemingly flatter than when evaluating equivalent work, which can be explained by the heat duty analysis neglecting the increased value of steam at high temperatures encountered when stripping to low loading.

Conclusions

Various stripper configurations with different levels of complexity have been analyzed in Aspen Plus® using the MEA model developed by Hilliard. 9 m MEA was the solvent. The multistage flash configurations demonstrated the improved performance with more pressure stages. More pressure stages typically yielded better performance due to more reversible operation. Replacing flash stages with packing sections has repeatedly demonstrated better performance at low loadings, though the improvement in performance at optimum loadings was generally insignificant.

The baseline simple stripper required 35.2 kJ/mol CO_2. The double matrix configuration performed best, and its optimum performance was nearly identical regardless whether packing was used at the top of the low-pressure section. The 3-stage flash and 2-stage multi-pressure configurations also performed well, but the multi-pressure had a significantly lower complexity level. The double matrix required 34.0 kJ/mol CO_2 with a complexity rating of 6, the 3-stage flash required 34.7 kJ/mol CO_2 with a complexity of 6, and the 2-stage multi-pressure required 34.7 kJ/mol CO_2 with a complexity of 4.

References

Hilliard, Marcus D. *A Predictive Thermodynamic Model for an Aqueous Blend of Potassium Carbonate, Piperazine, and Monoethanolamine for Carbon Dioxide Capture from Flue Gas.* Ph.D. Dissertation, The University Of Texas at Austin, 2008.

Leites, I L, D A Sama, and N Lior. "The theory and practice of energy saving in the chemical industry: some methods for reducing thermodynamic irreversibility in chemical technology processes." *Elsevier*, 2003: 55-97.

Rochelle, GT at al. "CO2 Capture by Aqueous Absorption, Third Quarterly Progress Report 2009." *Luminant Carbon Management Program.* The University Of Texas at Austin, 2009.

(117d) Enhancing Post-Combustion CO2 Capture in Power Plant – A Catalytic Approach

Reynolds A. Frimpong, Center for Applied Energy Research, University of Kentucky, Lexington, KY

Quanzhen Huang, Center for Applied Energy Research, University of Kentucky, Lexington, KY

Joseph E. Remias, Center for Applied Energy Research, University of Kentucky, Lexington, KY

James K. Neathery, Center for Applied Energy Research, University of Kentucky, Lexington, KY

Kunlei Liu, Center for Applied Energy Research, University of Kentucky, Lexington, KY

Improved solvent reaction kinetics and carbon capacity continue to be a major research interest for the scrubbing of CO2 in post-combustion processes. The DOE has set a new target for post-combustion CO2 capture of less than a 35% increase in the cost of electricity, while capturing at least 90% of the CO2 released during the combustion of fossil fuels in existing coal-fired power plants. In order to meet the DOE target, it will be necessary to promote the CO2 absorption rate in aqueous capture solutions. This work will discuss the use of catalysts to promote the reaction rate. Some current research efforts are targeted at developing new solvents or enhancing the absorption capabilities of some of the solvents already in use such as amine-based, ammonium-based, and alkali-carbonates. The use of catalytic additives in these solvents is one approach to promoting CO2 capture.

In this work, we describe the synthesis, characterization, and use of low cost non-toxic additives such as zeolites for CO2 capture rate enhancement. Ion-exchange reactions and metal impregnation were carried out on these additives to improve their catalytic properties. Results have shown that some zeolites provide catalytic enhancement in CO2 scrubbing when added to aqueous ammonium-based solutions.

See more of this Session: Post-Combustion Techniques for Capture/Separation of Carbon Dioxide I
See more of this Group/Topical: Topical 2: Advanced Fossil Energy Utilization

(117e) Design of a Continuous Foam-Bed Reactor System for CO2 Removal From Power-Plant Exhausts

Amit A. Gaikwad, Department of Chemical Engineering, Indian Institute of Technology, Delhi, New Delhi, India

Ashok N. Bhaskarwar, Department of Chemical Engineering, Indian Institute of Technology, Delhi, New Delhi, India

With renewed emphasis on reducing the carbon footprint in view of the United Nations Climate Change Conference in Copenhagen, reducing the greenhouse gas emissions is of prime importance. The average concentration of CO_2, the prominent of greenhouse gases, in the atmosphere today is about 385 ppm, which was about 280 ppm before the industrialization started. Intergovermental Panel on Climate Change (IPCC) says that the Earth's average temperature has risen by 0.74 degrees in the period from 1906 to 2005, and that the average temperature will continue to rise.

Various technologies are being developed to tackle this problem of CO_2 emissions. One such promising method is use of foam-bed reactors. The reactor offers very large gas hold-ups and low pressure drops, which is ideal for treatment of gaseous pollutants. This work focuses on treating CO_2 by reacting it with aqueous barium-sulfide solution in a continuous foam-bed reactor. The CO_2 required for this carbonation reaction could be obtained from smoke stack furnaces or power-plant exhausts, thereby reducing air pollution. H_2S gas is produced in the reaction, but it reacts faster with amines as compared to CO_2 and thus it can be removed with relative ease.

Some data for CO_2 removal using foam-bed reactors are available (Gaikwad and Bhaskarwar, 2006, 2007; Gaikwad et al., 2007) but they pertain to semi-batch mode of operation of the reactor. When continuous operation of a reactor is necessary, the conversion per pass is likely to be small, especially at high throughput rates of the liquid being processed. The conversion in the exit stream may, however, have to be quite high for the process to be acceptable from the point of view of meeting the required production level or environmental regulations. Multiple reactors will then appear, among others, as a feasible option.

Subramanyam et al. (1999) have developed a design procedure for treatment of the effluent from Kraft paper mills using foam-bed reactor. The primary type in multi-reactor configurations is the reactors-in-series. For the present case of CO_2 removal using aqueous barium-sulfide solution (gas absorption with chemical reaction and desorption), the graphical approach used in the design of a foam-bed reactor system was similar to the one proposed by Subramanyam et al. (1999). The modified single-stage model of a foam-bed reactor (Gaikwad and Bhaskarwar, 2007) was applied to the system of N foam-bed reactors, all of equal size in series, operated continuously.

The design of the foam-bed reactor system was done for the following specifications:

i. Effluent flow rate = 100 tons/ day

ii. Concentration of aqueous barium-sulfide solution in the inlet stream = 0.56 k mol/m^3

iii. Desired conversion = at least 95%

iv. Gas-flow rate = 8 x 10^{-3} m^3/s

v. Concentration of carbon-dioxide gas in the inlet stream = 1 x 10^{-2} k mol/m^3

A design chart has been prepared in the form of figure 1. In this chart, the height of a foam-bed reactor (H) has been plotted against the number of reactors required to be connected in series in order to achieve 95% conversion of the inlet liquid stream under the specified conditions (i-v). It is seen from the chart that the number of reactors and the height of each reactor are both smaller for a higher radius of the reactor (r_c). Such design plots have been made for different reaction velocities starting from 4300 m^3/k mol.s to 11000 m^3/k mol.s.

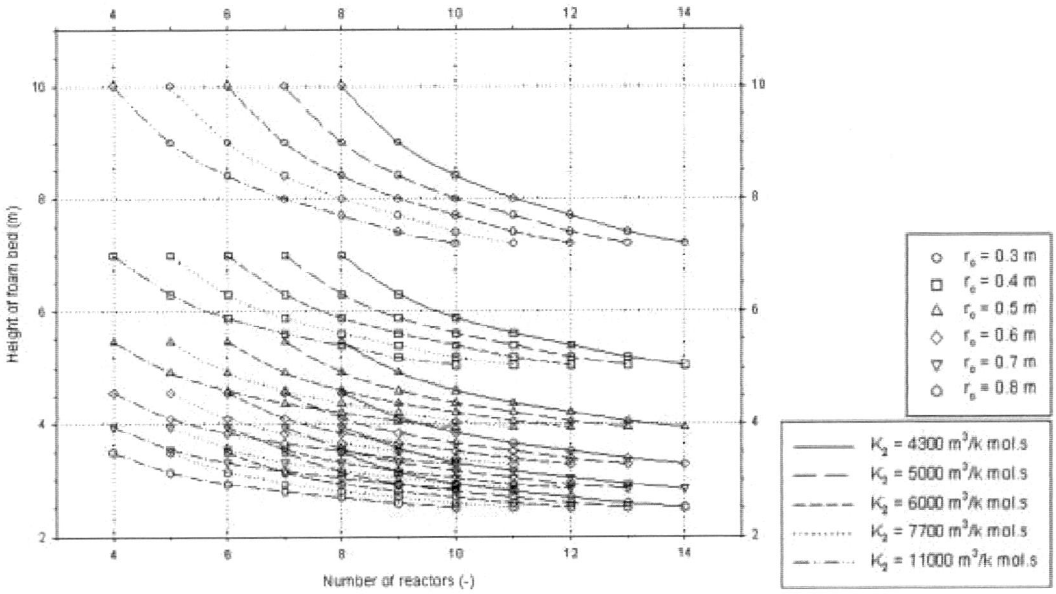

Figure 1. Variation of the height of foam bed with the number of reactors in (design chart I).

One can choose an aspect ratio (ratio of reactor height to reactor diameter) of say about 10, and find out the combination of H and r_c which gives a reasonably small number of reactors (say not greater than 10) for a given reaction velocity. Then, we have to turn to the second design chart shown in figure 2 wherein the average liquid hold-up in the reactor has been plotted against the number of reactors in series, again required for 95% conversion under the specified conditions (i-v). Having found out H and r_c in the previous step, one can now locate the point on the chart corresponding to the average liquid hold-up existing under the actual operating conditions by suitably interpolating for the corresponding value of the parameter r_c^2H, if not directly readable from curves shown. The number of foam-bed reactors actually required in series can then be directly read off from the corresponding abscissa. This number, if it is a fraction, will have to be rounded off to the nearest higher integer to ensure a conversion greater than 95%.

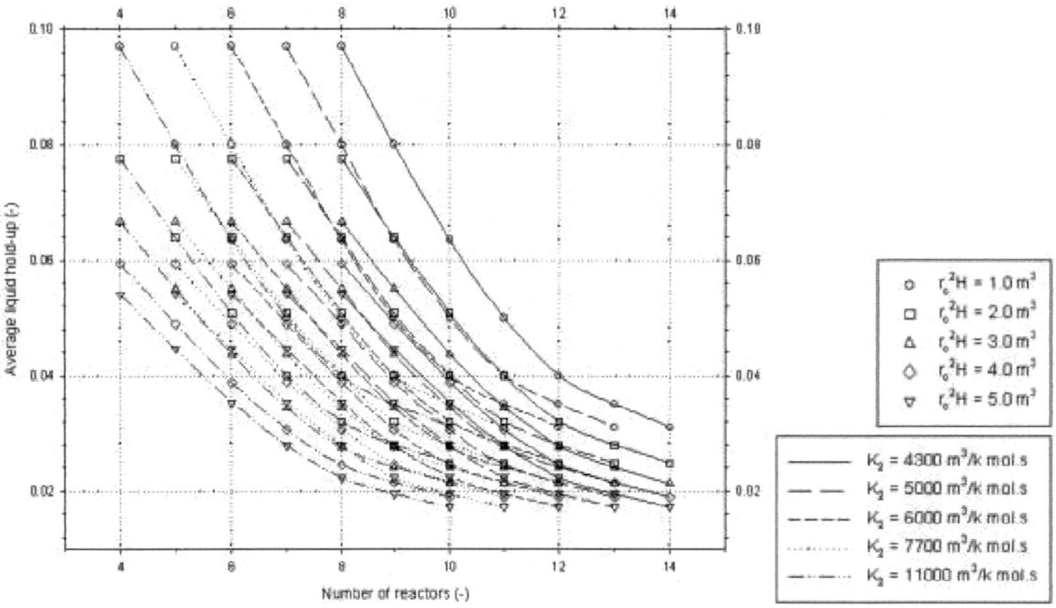

Figure 2. Variation of the average liquid hold-up with the number of reactors in series (design chart II).

Thus using these design charts one can obtain the dimensions and number of foam-bed reactors required for a continuous operation to remove CO_2. The design process covers broad range of reaction velocities giving a flexibility of choosing a reactant liquid other than aqueous barium-sulfide solution. Following the procedure outlined in this work, similar plots can be generated for different gas velocities, conversions, effluent flow-rates, and liquid and CO_2 concentration to obtain the required design parameters.

See more of this Session: Post-Combustion Techniques for Capture/Separation of Carbon Dioxide I
See more of this Group/Topical: Topical 2: Advanced Fossil Energy Utilization

(117f) An Incremental, Multi-Technology Approach to Retrofitting Existing Pulverized Coal Power Plants for Carbon Capture

David C. Miller, Collaboratory for Process & Dynamic Systems Research, National Energy Technology Laboratory, U.S. Department of Energy, Morgantown, WV

Global climate models indicate that acting rapidly to reduce carbon dioxide emissions will provide greater benefits to the environment than delaying action; however, new technologies, currently under development, may be significantly less costly than those currently ready to be deployed. This tradeoff between time to deployment and cost makes it difficult to determine the best course of action. This paper will explore scenarios under which partial post-combustion capture is achieved with aqueous amine absorption in the near term and additional capture capability is achieved in the future through newer technologies, such as advanced sorbents or membrane processes. Results from process simulations will be presented to demonstrate the potential for a multi-technology approach to achieve synergistic reductions in the cost of carbon capture.

See more of this Session: Post-Combustion Techniques for Capture/Separation of Carbon Dioxide I
See more of this Group/Topical: Topical 2: Advanced Fossil Energy Utilization

Hydrogen production by ethanol reforming driven by solar energy

Marcello De Falco[1], Fausto Gallucci[2]*

[1.] *Università di Roma La Sapienza, Dip. Ing. Chimica e Materiali, Via Eudossiana 18, 00184 Roma, Italy*
[2.] *Fundamentals of Chemical Reaction Engineering Group, Faculty of Science and Technology, IMPACT, University of Twente, Enschede, The Netherlands*

marcello.defalco@uniroma1.it , f.gallucci@tnw.utwente.nl

ABSTRACT

In this paper hydrogen production via reforming of ethanol has been studied in a novel hybrid plant consisting in an ethanol reformer and a concentrating solar power (CSP) plant by means of a molten salt stream. The heat required by the reforming of ethanol has been supplied to the system by molten salts produced by solar energy. The temperature drop in the molten salt is few K (less than 20K) making the molten salt stream still suitable for steam and electricity production in the co-generative plant.

A 2D mathematical model has been formulated and used in order to simulate the performance of the reformer coupled with heat integration through molten salt. The results show that the hybrid plant proposed here is a good solution for sustainable hydrogen production. An optimised set of operating conditions have been found. 99% ethanol conversion is achieved with 4.5 m long reactor, while the simulations indicate that both reactor diameter and molten salt flow rate need to be further optimized in order to find the best heat management. A small temperature drop (17 K) in the molten salt stream has however been observed which makes the integration of hydrogen production inside a solar power system really attractive.

On the other hand there is a compromise between temperature and CO content in the exhaust stream. High temperature is required for achieving high ethanol conversion and to convert the methane produced in the first part of the reactor. The high temperature makes the CO shift reaction not complete so that a downstream low temperature shift reactor is needed. A solution to be studied in the future can be the use of membrane solar ethanol reforming.

INTRODUCTION

Production of hydrogen from ethanol is very attractive since bio-ethanol is a renewable material mainly produced from biomass fermentation [1,2]. Most literature on ethanol reforming is devoted to the study of suitable catalysts. Usually, catalysts based on Rh, Ru, Pd, Pt, Ni, Co and Cu are used on supports of Al_2O_3, SiO_2, MgO and La_2O_3. The reaction conversion and selectivity of the products (mainly H_2, CO, CO_2 and in minor part, CH_4, CH_3CHO and C_2H_4) depend on the catalyst used as well as on the operating temperature.

An alternative and more efficient way to produce hydrogen from ethanol is the oxidative steam reforming reaction, where the ethanol reforming and partial oxidation are coupled in order to reduce the amount of energy required by the system and, eventually, to achieve the autothermal reforming [3-5]. However, the addition of oxygen could result in high temperature hot spots which are detrimental for the catalyst stability.

Other possibilities to obtain autothermal reforming of ethanol have been presented in recent literature. For example, recently Gallucci *et al.* [6] propose a fluidized bed membrane reactor in which part of the hydrogen produced is recovered through Pd membranes and converted with air to supply the energy required by the ethanol reforming reaction. Aicher *et al.* [7] proposed a complete portable system for producing hydrogen for

fuel cells from ethanol consisting in an autothermal reformer, high and low temperature shift reactors, a selective methanation reactor, and a tail gas combustor. Casanovas *et al.* [8] obtained the autothermal reforming of ethanol by thermally coupling the ethanol reforming and the ethanol oxidation reactions in a two side micro-reactor. However, the energy required by the ethanol reforming can be supplied in another way. In fact, as recently proposed for methane reforming, the energy needed for the endothermic reaction can be supplied via coupling of reformer and molten salts plant [9,10]. The molten salts act as energy carrier between the reformer and a solar plant.
The objective of this work is to study for the first time the hydrogen production via reforming of ethanol in a hybrid system where the heat supply is achieved via molten salts. The aim is to produce hydrogen from ethanol keeping the molten salt still at high temperature thus suitable for the steam/electricity production.

CSP MOLTEN SALT PLANT

The concentrating solar power (CSP) plant basically consists of a solar collector field, a receiver, a heat transfer fluid loop and a heat storage system (Fig. 1). The mirrors of the solar field concentrate the direct solar radiation on the solar receiver set at the focal line. The heat transfer fluid (e.g. molten salts) removes the high temperature solar heat from the receiver and is afterwards collected into an insulated heat storage tank to be pumped, on demand, to the heat users (usually steam generators for electricity production by means of steam-turbine) where it releases its sensible heat. Finally, the heat carrier fluid is stored into a low temperature tank (290°C) ready to restart the solar heat collection loop.
A proper dimensioning of the heat storage system allows driving the process 24 h/24 h in continuous at the designed working conditions, thus making easier and more convenient the coupling of CSP with a chemical plant as ethanol steam reformer.

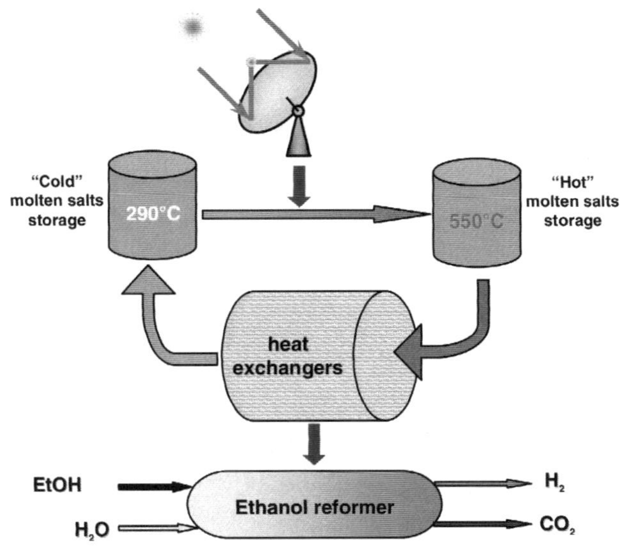

Figure 1 - Simplified scheme of parabolic trough power plant with two-tank molten salt storage

The idea to match the CSP plant with ethanol steam reforming reactor derives from the consideration that thermal level reached by molten salt stream (550°C) is adherent with thermal requirements of reactor (preferred operating temperature equal to 450-550°C). The molten salts present a very high energy capacity thus coupling the low temperature reforming with molten salts heating system could result in a few K temperature drop in the molten salts stream which remains suitable for steam/electricity production. By this way, the plant proposed is able to produce both hydrogen and electricity using a renewable energy source.

The core of the process presented and evaluated in this study is the steam-reforming reactor heated up by the molten salt. Figure 2 shows the section of reactor system simulated in the following. The reactor is composed by a bundle of 4 reformers assembled in a heat exchanger and molten salt is fed in the shell side.

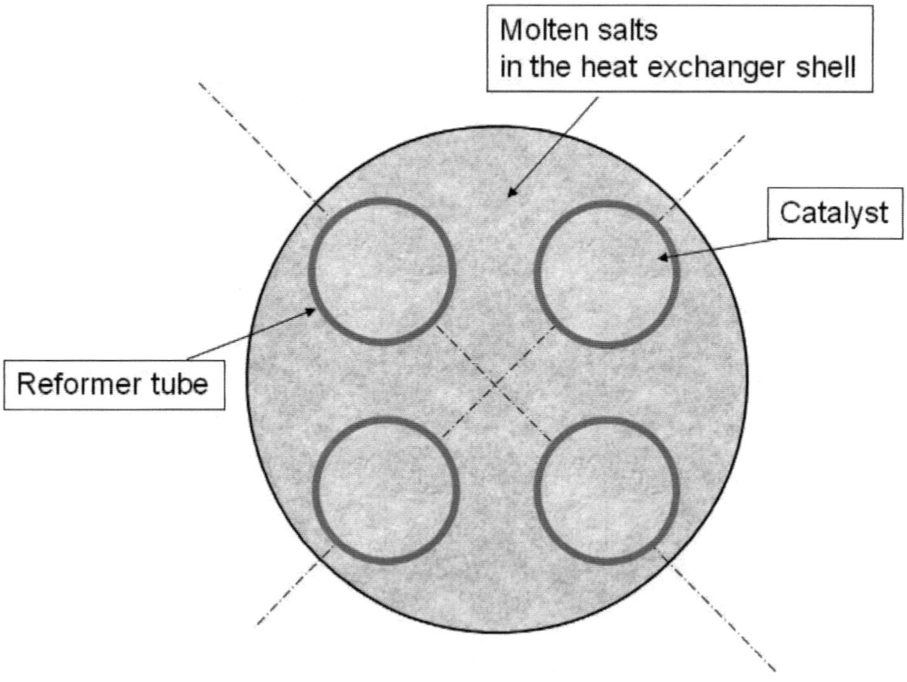

Figure 2 – Tubes-and-shell reactor configuration

REACTOR MODEL

The ethanol steam reformer heated up by the molten salt stream is modeled in detail by a two-dimensional mathematical model based on mass, energy and momentum balances with the intrinsic kinetic equations reported by Mas *et al.* [11]. The catalyst packed in reactor is Ni-Al. The catalyst used, working in steady state, does not produce by-products as acetaldehyde or ethylene, thus the amount of hydrogen produced per mole of ethanol converted is maximized. Moreover, since acetaldehyde or ethylene formation is avoided, the carbon deposition is also avoided being both these two by-products carbon precursors. The ethanol reaction system consists on the following 4 reactions:

$C_2H_6O \rightarrow CH_4 + CO + H_2$ (1)
$C_2H_6O + H_2O \rightarrow CH_4 + CO_2 + 2H_2$ (2)
$CH_4 + H_2O \Leftrightarrow CO + 3H_2$ (3)
$CH_4 + 2H_2O \Leftrightarrow CO_2 + 4H_2$ (4)

The following assumptions have been made:

- steady-state conditions;
- negligible axial dispersion and radial convection;
- ideal gas behavior;
- a single tube representative of any other tube;
- carbon deposition reactions neglected.

The comparison between experimental literature values and model results shows that the model predicts fairly well the experimental results at high temperatures where the error is almost always lower than 5%.

RESULTS AND DISCUSSION

By applying the set of suitable operating conditions reported in the following Table 1, the reactor was simulated with the target of 99% ethanol conversion. A reactor length of 4.5 m is required for achieving the target conversion which also corresponds to a total pressure drop of about 2 bar.

Table 1 Optimized parameters considered for reactor design

Parameter	Value imposed
Pressure	10 bar
Steam to carbon ratio	5
Temperature	823 K
GHSV	1115 h^{-1}

As reported in Fig. 3(A) the ethanol conversion increases sharply in the first part (first centimetres) of the reactor while is then increasing linearly till the last part of the reactor where it reaches slowly 100% conversion.
As a consequence of this behaviour the heat flux required is higher in the first part of the reactor while is then constant and decreases in the last part of the reactor. This heat required results in temperature profiles as indicated in Fig. 3(B). The lowest temperature is achieved in the first part of the reactor while in the last part of the reactor the temperature slightly increases. It is also quite evident that the molten salts feed rate and their heat capacity are high enough so that the temperature of the molten salt decreases of few degrees along the reactor. In particular, for the 4.5 m reactor the molten salts temperature drop, for reaction heat duty and process steam generation, is as low as 17.2 K.
The figure also reports the temperature at different radial positions in the reactor. It can be noted that the difference between the reactor wall and the reactor axis is about 45 K. Moreover most of the heat is lost in the first half of the reactor radius. This figure also indicates that both the molten salt feed rate and the reactor radius can be optimised in order to optimise the energy management.

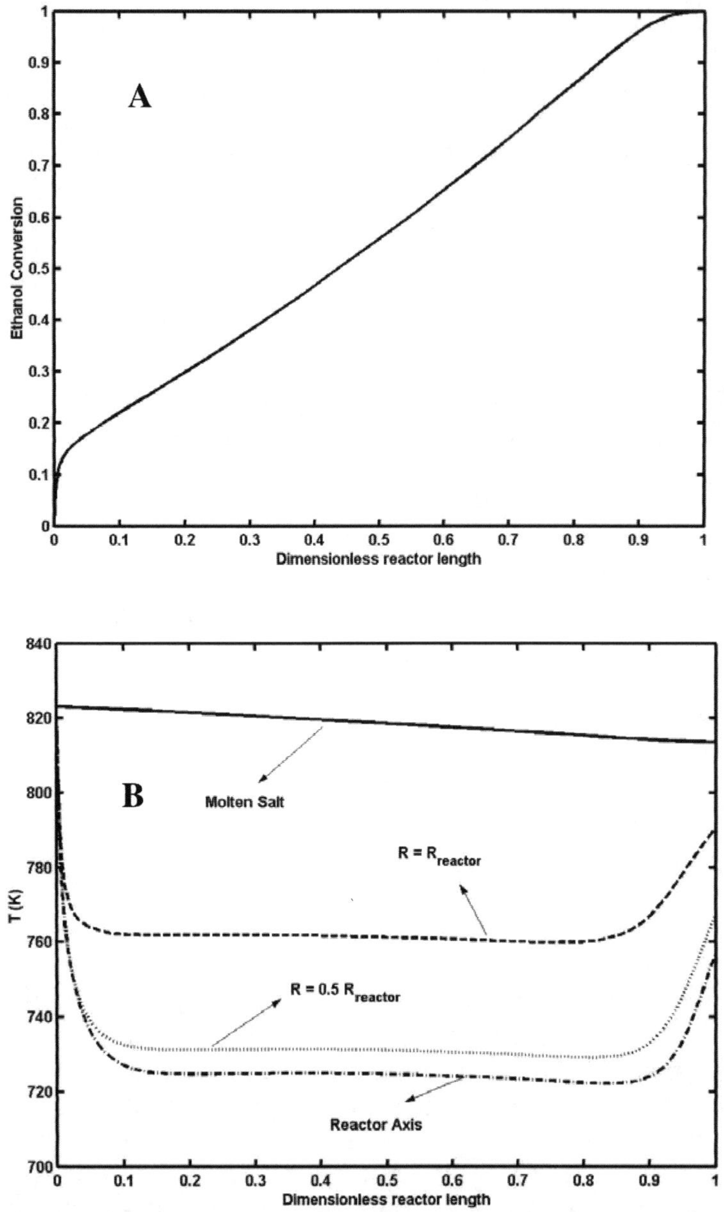

Figure 3. Ethanol conversion (A) and temperature profiles at different reactor radial position (B) as a function of reactor length

Figure 4 reports the gas molar fractions inside the ethanol reformer as a function of the axial position. It is important to notice that in the first part of the reactor a quantity of methane is produced; however, this methane is then converted to additional hydrogen via methane reforming (reactions 3 and 4). On the other hand, at the end of the reactor the molar fraction of CO and CO_2 are almost equal while the water content is still quite high. This suggests that the high temperatures needed for the ethanol reforming make the water shift reaction (an exothermic reaction) not effective, so that an additional low temperature shift reactor is needed in order to decrease the CO content and increase the hydrogen content in the stream. A solution can be the use of membrane reactor for the solar reforming. This will be studied in a future work.

This first reactor assessment for ethanol reforming via solar energy carried by molten salts demonstrate the feasibility of the reforming process. It has been demonstrated that the maximum temperature drop in the molten salts can be restricted to about 17 K, which allows to carry out the reforming of ethanol with a conversion of 99% which corresponds to a hydrogen rich stream containing about 107 Nm^3/h of hydrogen. The molten salts are still at high temperature and can be effectively used for steam/electricity production in the co-generative plant.

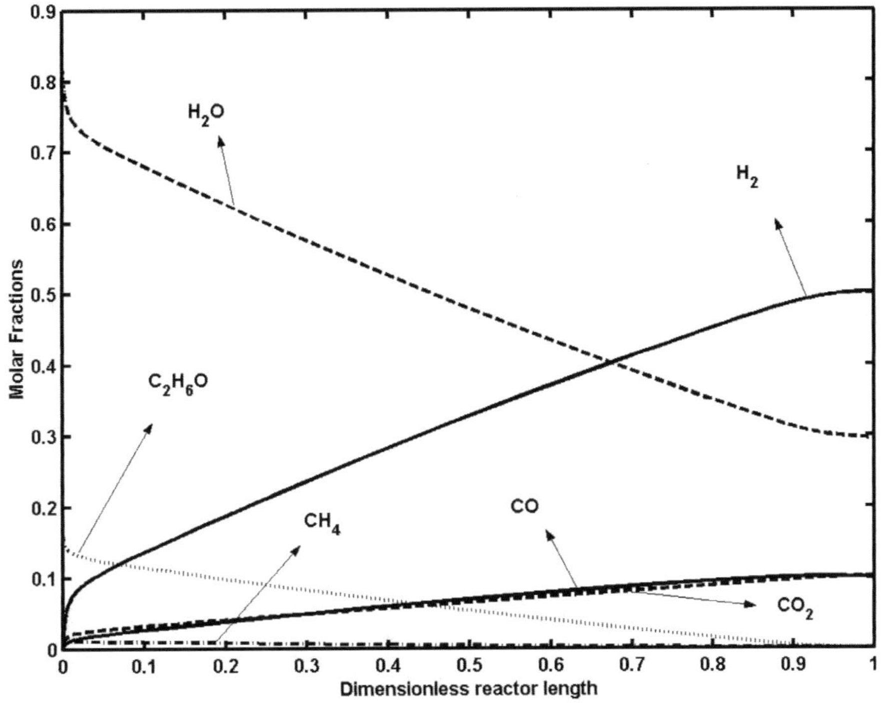

Figure 4. Gas composition profiles as a function of reactor length

CONCLUSIONS

The ethanol steam reforming driven by molten salt heated by solar energy has been studied and an optimised set of operating conditions have been found. A value of ethanol conversion equal to 99% is achieved with 4.5 m long reactor, while the simulations indicate that both reactor diameter and molten salt flow rate need to be further optimized in order to find the best heat management. A small temperature drop (17 K) in the molten salt stream has however been observed which makes the integration of hydrogen production plant inside a solar electrical power production system really attractive.

On the other hand there is a compromise between temperature and CO content in the exhaust stream. High temperature is required for achieving high ethanol conversion and to convert the methane produced in the first part of the reactor. The high temperature makes the CO shift reaction not complete so that a downstream low temperature shift reactor is needed. A solution to be studied in the future can be the use of membrane solar ethanol reforming.

References

[1] Badmaev SD, Snytnikov PV, Hydrogen production from dimethyl ether and bioethanol for fuel cell applications, *Int J Hydrogen Energy*, 2008;33:3026–30.

[2] Ni M, Leung DYC, Leung MKH, A review on reforming bioethanol for hydrogen production, *Int J Hydrogen Energy* , 2007;32:3238–4

[3] Lin W.-H., Hsiao C.-S., Chang H.-F., Effect of oxygen addition on the hydrogen production from ethanol steam reforming in a Pd-Ag membrane reactor, *J. Membrane Science*, 2008;322(2): 360-67

[4] Srisiriwat N., Therdthianwong S., Therdthianwong A., Oxidative steam reforming of ethanol over Ni/Al_2O_3 catalysts promoted by CeO_2, ZrO_2 and CeO_2-ZrO_2, *Int J Hydrogen Energy* 2009 34 (5), pp. 2224-2234

[5] S. Cavallaro, V. Chiodo, A. Vita and S. Freni, Hydrogen production by auto thermal reforming of ethanol on Rh-Al_2O_3 catalyst, J. Power Source, 123 (2003) 10-16

[6] Gallucci, F., Van Sint Annaland, M., Kuipers, J.A.M., Pure hydrogen production via autothermal reforming of ethanol in a fluidized bed membrane reactor: A simulation study, *Int. J. Hydrogen En.*, (2010) DOI: 10.1016/j.ijhydene.2009.12.014

[7] Aicher, T., Full, J., Schaadt, A., A portable fuel processor for hydrogen production from ethanol in a 250 Wel fuel cell system, *Int. J. Hydrogen En.*, 34 (19), (2009) 8006-8015

[8] Casanovas, A., Saint-Gerons, M., Griffon, F., Llorca, J., Autothermal generation of hydrogen from ethanol in a microreactor, *Int. J. Hydrogen En.*, 33 (7), (2008) 1827-1833

[9] De Falco, M., Giaconia, A., Marrelli, L., Tarquini, P., Grena, R., Caputo, G., Enriched methane production using solar energy: an assessment of plant performance, *Int. J. Hydrogen En.*, 34(1), (2009) 98-109

[10] M. De Falco, A. Basile, F. Gallucci, Solar membrane natural gas steam reforming process: evaluation of reactor performance, *Asia Pacific J Chem. Eng.*, in press

[11] V. Mas, M.L. Bergamini, G. Baronetti, N. Amadeo, M. Laborde, A kinetic study of ethanol steam reforming using a nickel based catalyst, *Topics in Catal.*, 2008;51(1-4):39-48

Autothermal reforming of ethanol in a fluidized bed membrane reactor for ultra-pure hydrogen production

Fausto Gallucci, Martin Van Sint Annaland, J.A.M. Kuipers*

Fundamentals of Chemical Reaction Engineering Group, IMPACT, Faculty of Science and Technology, University of Twente, P.O. Box 217, NL-7500 AE Enschede, The Netherlands

f.gallucci@tnw.utwente.nl , M.vanSintAnnaland@tnw.utwente.nl
J.A.M.Kuipers@tnw.utwente.nl

ABSTRACT

In recent years, many efforts have been made to develop processes and reactor technologies for the production of cheap, ultra-pure hydrogen that can be used in efficient PEM fuel cells. Nowadays, hydrogen is mainly produced by steam reforming of natural gas in multi-tubular reactors. The main drawback of natural gas reforming is that this reaction leads to a H_2 rich gas mixture also containing carbon oxides and other by-products. Consequently, in order to produce pure H_2, chemical processes are carried out in a number of reaction units (typically high temperature reformer, high and low temperature shift reactors) followed by separation units (mostly pressure swing adsorption units are used). An attractive way to produce hydrogen is the reforming of renewable fuels (*e.g.* bio-ethanol) inside membrane reactors, where hydrogen production and hydrogen separation through selective membranes are integrated in one apparatus.

In this paper the production of ultra-pure hydrogen via autothermal reforming of ethanol in a fluidized bed membrane reactor has been studied. The steam reforming of ethanol is an endothermic reaction requiring a great amount of heat. The needed energy for the steam reforming is obtained by burning part of the hydrogen recovered via the hydrogen perm-selective membrane. In this configuration, the air used for the combustion is never in contact with the reacting mixture, which makes thus autothermal reforming with integrated CO_2 capture feasible. Simulation results based on a phenomenological model show that it is possible to obtain overall autothermal reforming of ethanol while 100% of hydrogen can in principle be recovered at relatively high temperatures and at high reaction pressures. At the same operating conditions, ethanol is completely converted, while the methane produced by the reaction is completely reformed to CO, CO_2 and H_2.

Keywords: Membrane Fluidized Bed, Ethanol Reforming, Autothermal Reforming, Hydrogen, Membrane Reactor.

* Corresponding author

FLUIDIZED BED MEMBRANE REACTOR MODEL

In our fluidized bed model ethanol and steam enter the reactor where the catalyst is mildly fluidized. The hydrogen produced is directly recovered as an ultra-pure stream via perm-selective dead-end Pd-based membranes. Part of the hydrogen is recovered in different (U-shaped) Pd-based membranes where it reacts with oxygen (fed as air) producing heat. The heat is transferred to the bed, supporting the reforming reaction. A scheme of the fluidized bed used is reported in Fig. 1.

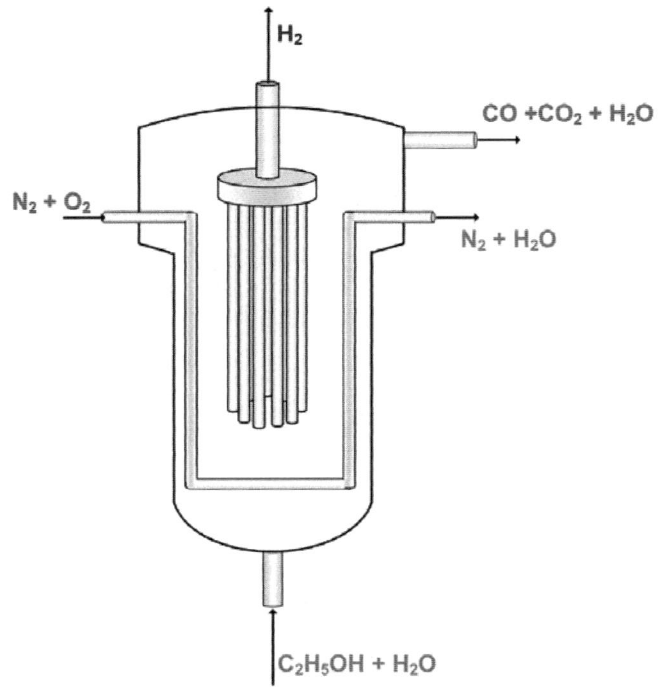

Figure 1. Schematic of the fluidized bed membrane reactor for autothermal reforming of ethanol

A typical phenomenological two-phase model for a membrane assisted fluidised bed reactor is considered here, where the number of CSTRs in the cascade and the sizes of the CSTRs are directly related to the extent of gas back-mixing in each phase. The overall (bubble and emulsion phase) component mass conservation equations have been formulated, accounting for chemical transformations in the emulsion phase and a net gas production due to the chemical reactions and gas withdrawal (hydrogen) via membranes. The overall energy balances are also accounting for possible energy exchange via the sweep gas. The exact number of CSTR in both emulsion phase and bubble phase need to be determined by comparing the model results with experimental results. However, no experimental data on fluidized bed reforming of ethanol are available at the moment, so the number of CSTR in both phases were kept equal to the number of CSTRs determined for the methane reforming (N_b =5 and N_e =35) [1].

Empirical correlations for the mass transfer coefficients and fluidization properties of the fluidized beds have considered. The kinetic rate expressions for ethanol steam reforming have been taken from Mas *et al.* [2]. Selective removal of H_2 using Pd-based membranes has been modelled with a lumped type flux expression, using experimental data from [3]. Pure component physical data have been taken from Daubert and Danner [4], while mixture properties have been computed following Reid et al. [5].

The described reactor model is subsequently used to compare the reactor performance as a function of the operating conditions in terms of the following parameters:

Ethanol conversion $= \dfrac{\phi_{C2H5OH,IN} - \phi_{C2H5OH,OUT}}{\phi_{C2H6O,IN}}$

Hydrogen recovery $= \dfrac{\phi_{H2,extracted}}{\phi_{H2,produced}}$

X Selectivity $= \dfrac{\phi_X}{\phi_{CO} + \phi_{CO2} + \phi_{CH4}}$ (where X is CO, CO_2 or CH_4)

A direct validation of the fluidized bed membrane reactor for ethanol reforming is not possible at the moment since no experimental results are available for this reactor type. However, the model can simulate a typical plug flow reactor, if the number of CSTRs considered is sufficiently high and permeation through membranes is ignored. By simulating a reactor with 200 CSTRs and no membranes, the results in terms of ethanol reforming agree fairly well the experimental results presented by Mas *et al.* [2]. The results of the validation are reported elsewhere [6].

RESULTS AND CONCLUSIONS

The theoretical feasibility of ultra-pure hydrogen production via autothermal reforming of ethanol in a fluidized bed membrane reactor has been demonstrated through a theoretical investigation. The model used here does not include deactivation by carbon deposition as for the kinetic model presented by Mas *et al.* [19]. The high steam to ethanol ratio used (6) is considered high enough to prevent carbon formation. Moreover, since acetaldehyde or ethylene formation is avoided by using an adequate catalyst, the carbon deposition is also avoided being both these two by-products carbon precursors. Al the simulations related to the FBMR have been performed using a reactor volume of 0.1 m^3.

When the reaction is carried out in a membrane reactor (Fig.2), also complete conversion of ethanol and methane can be achieved at high temperatures, but in addition at temperatures above 860 K more CO is shifted into CO_2 due to the extraction of hydrogen from the reaction mixture through the membranes. As a result, due to the hydrogen permeation also the hydrogen mole fraction is much lower in the membrane reactor in comparison to the outlet composition of the reactor without membranes.

The hydrogen extraction can be increased (as indicated in Fig 2) by increasing the membrane area installed inside the reactor. By increasing the membrane area 5 times the (almost) complete hydrogen recovery can be achieved at high temperature.

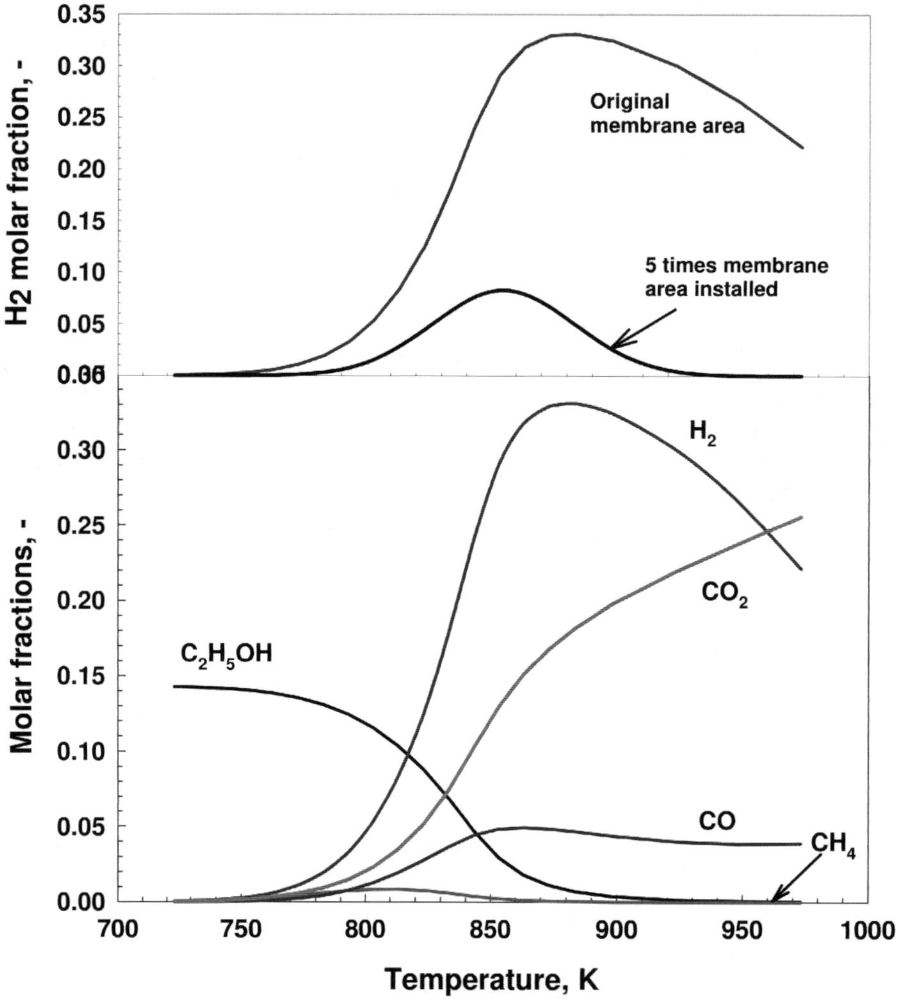

Figure 2. Product distribution in a membrane reactor as a function of temperature (reaction pressure 5 bar, permeation pressure 0 bar). Hydrogen molar fraction as a function of membrane installed (top view)

Effect of pressure and temperature on FBMR performance

The permeation of hydrogen through the membrane is an activated process which increases with the temperature and increases with the difference in hydrogen partial pressure between reaction side and permeation side. It is thus worth investigating the effects of both temperature and pressure on the reactor performance. As shown in Fig. 3, the ethanol conversion increases by decreasing the reaction pressure at each temperature investigated. This is because the ethanol reforming system, proceeds with an increase in the number of mole. On the other hand, the endothermic reaction system requires high temperatures, so that the conversion increases with temperature for each pressure investigated. Complete conversion is achieved at temperatures exceeding 950 K for each reaction pressure investigated. However, at low pressures (< 3 bar) complete conversion is already attained at 875 K. Higher ethanol conversion can also be achieved at high operating pressures and low temperatures through installing additional membrane area.

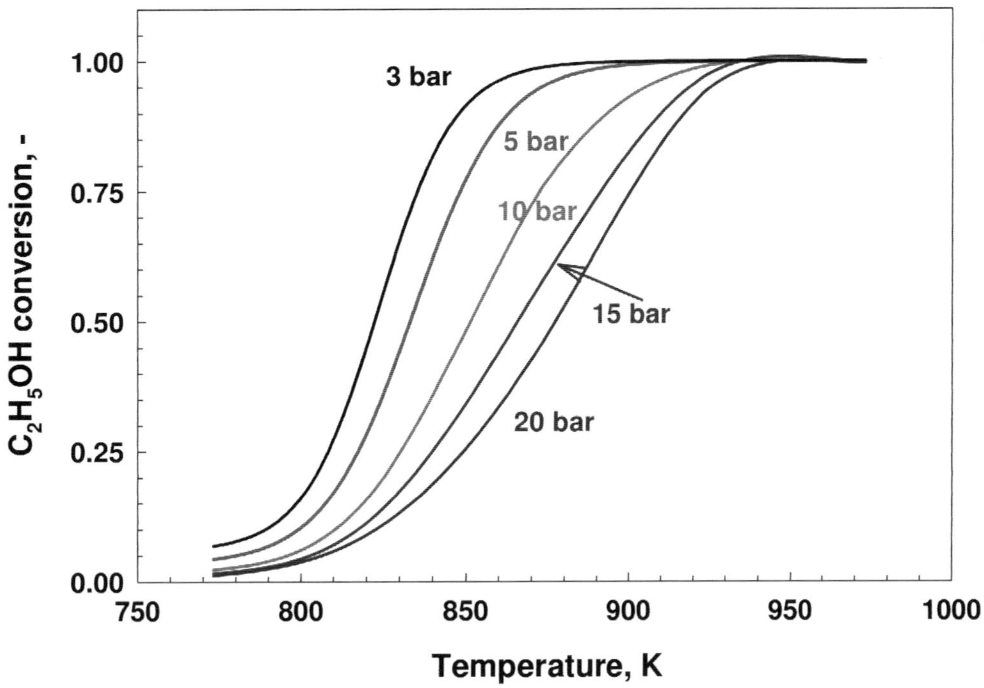

Figure 3. Ethanol conversion in a fluidized bed membrane reactor as a function of temperature at different reaction pressures (permeation pressure 0 bar).

The ethanol reforming is carried out in membrane reactors in order to recover ultra-pure hydrogen through the Pd-based membranes. As indicated in Fig.4, at low reaction pressures (< 3 bar), the pure hydrogen recovery hardly reaches 30% even at higher temperatures. On the other side, at high pressure (20 bar) and high temperature, more than 75% of the hydrogen is recovered as pure hydrogen. These results indicate that the membrane reactor performs best at high pressures and high temperatures. The hydrogen recovery depends on the difference of hydrogen partial pressure between the reaction side and the permeation side. In our simulation it has been considered vacuum in the permeate side. If the aim is to recover the same amount of hydrogen at higher pressure in the permeate side, a higher pressure in the reaction side need to be used in order to obtain the same hydrogen partial pressure difference as in the case presented here.

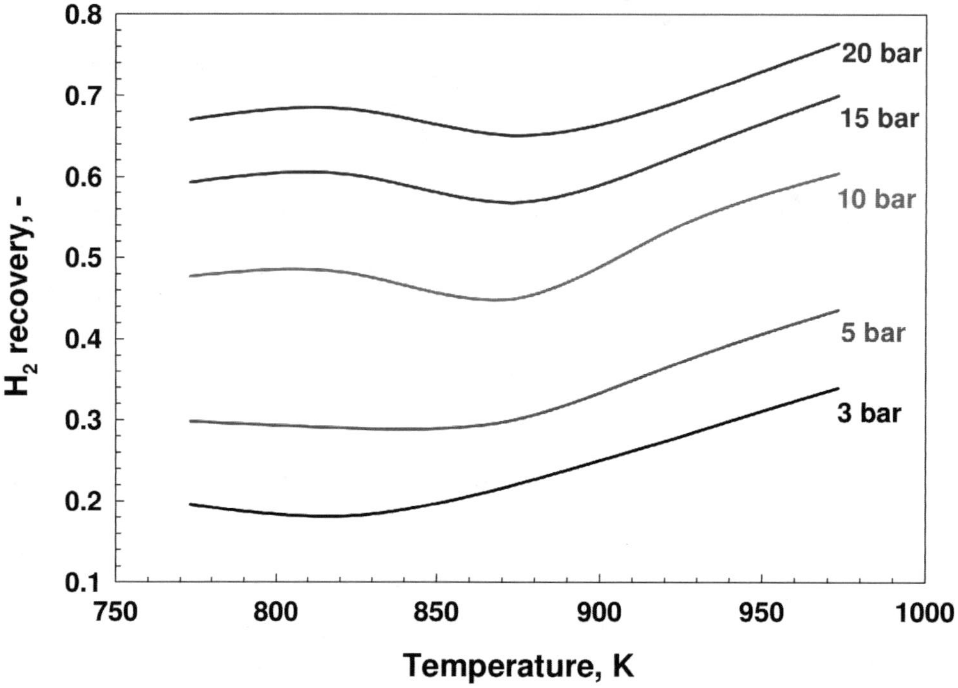

Figure 4. Hydrogen recovery in a fluidized bed membrane reactor as a function of temperature at different reaction pressures (permeation pressure 0 bar).

The model is now applied to a case of 1 Nm^3/h of pure hydrogen production. The temperature considered is 973 K, while a high pressure (20 bar) is considered in order to increase the amount of hydrogen recovered via the membranes.

CONCLUSIONS

The simulation results with a phenomenological two-phase model showed that pure hydrogen can be recovered through Pd-based membranes, realizing complete ethanol conversion while the heat needed for the reforming reaction can be easily supplied by burning approximately 15% of the hydrogen recovered, obtaining overall autothermal reforming. Mixing between air and the reaction mixture is circumvented so that the CO_2 separation after the reactor is easier, since the exhaust gas of the reactor consists of a mixture of CO, CO_2 and water. The simulations also show that a critical point to be experimentally studied in the fluidized bed membrane reactor is the extent of bubble-to-emulsion phase mass transfer limitations. In the worst case (limitations calculated as in a fluidized bed without membranes) the mass transfer limitations inside the reactor result in an increase of the membrane area of 88% with respect the case in which no mass transfer limitations prevail.

REFERENCES

[1] Gallucci F., van Sint Annaland M., Kuipers J.A.M., Autothermal Steam Reforming of Methane in a novel Fluidized Membrane Reactor. Part 2 Comparison of reactor configuration, *Topics in Catalysis*, 2008; 51(1-4), 133-45

[2] Mas V., Bergamini M.L., Baronetti G., Amadeo N., Laborde M., A kinetic study of ethanol steam reforming using a nickel based catalyst, *Topics in Catalysis*, 2008;51(1-4):39-48

[3] Gallucci F., van Sint Annaland M., Kuipers J.A.M., Autothermal Steam Reforming of Methane in a novel Fluidized Membrane Reactor. Part 1 Experimental Demonstration, *Topics in Catalysis*, 2008; 51(1-4), 133-45

[4] Daubert T.E., Danner R.P., Physical and Thermodynamic Properties of Pure Chemicals; Core Edition, Taylor & Francis: London (1998)

[5] Reid R.C., Prausnitz J.M., Poling B.E., The Properties of Gases and Liquids. McGraw-Hill Book Company: New York (1988)

[6] Gallucci F., Van Sint Annaland M., Kuipers J.A.M., Pure hydrogen production via autothermal reforming of ethanol in a fluidized bed membrane reactor: A simulation study, *Int. J. Hydrogen En.*, (2010) doi:10.1016/j.ijhydene.2009.12.014

(137c) Stability Study of Dense Cermet Membranes for Hydrogen Separation

Yunxiang Lu, Energy System, Argonne National Laboratory, Argonne, IL

S.E. Dorris, Energy System, Argonne National Laboratory, Argonne, IL

T.H. Lee, Energy System, Argonne National Laboratory, Argonne, IL

C.Y. Park, Energy System, Argonne National Laboratory, Argonne, IL

U. Balachandran, Energy System, Argonne National Laboratory, Argonne, IL

ABSTRACT

Argonne National Laboratory is developing dense cermet (i.e. ceramic-metal composite) hydrogen transport membranes (HTMs) for separating hydrogen from mixed gases, particularly product streams generated during coal gasification and/or methane reforming. Hydrogen separation with these membranes is nongalvanic (i.e., it does not use electrodes or an external power supply to drive the separation), and hydrogen separated from the feed stream is of high purity, so post-separation steps are unnecessary.

These cermet membranes (60-70vol% Pd with Y_2O_3-stabilized ZrO_2) with thickness of 10-20 mm were prepared by conventional ink/paste painting and co-sintering techniques on a porous substrate (alumina or zirconia). Nongalvanic hydrogen permeation flux in the temperature range of 400°C-600°C were measured. The chemical stability of membranes was tested in various gas mixtures that contained H_2, CO, CO_2, and/or H_2S. The dense HTM demonstrated high selectivity for H_2 over CO and CO_2. When tested in a gas mixture that contained 30% CO_2 and 1% CO at 400-600 °C. HTM cermet membrane showed continuous slow performance degradation; however, the observed flux degradation can be fully reversed. Because H_2S reacts with Pd-containing membranes to form palladium sulfide (Pd_4S), which impedes hydrogen permeation through the membranes, the Pd/Pd_4S phase boundary was determined in the temperature range of 400-600°C using various feed gases with 10-50% H_2 and ≈5-100 ppm H_2S. Tests in H_2S-containing atmospheres showed that the membrane is stable at 500°C in gases that contain 10% H_2 and 15-20 ppm H_2S for 200 hrs and at 550°C in gases that contain 10% H_2 and 25-30 ppm H_2S for 100 hrs. These results suggest that Pd-containing cermet membranes are promising candidate for hydrogen separation/purification under coal gasification conditions. The present status of HTM development at Argonne will be presented in this talk.

Work supported by the U.S. Department of Energy, Office of Fossil Energy, National Energy Technology Laboratory's Clean Fuels Program, under Contract DE-AC02-06CH11357.

See more of this Session: Fuel Processing for Hydrogen Production II
See more of this Group/Topical: Topical 2: Advanced Fossil Energy Utilization

Hydrogen Enriched Natural Gas (HENG): Generation and Applications in Combined Heat and Power Generation and as a Fuel for High Efficiency Boilers

B. Allen, T. Boutot, J. Bullerwell, J. Liu, S. Norton and T. Whidden, Atlantic Hydrogen Inc., Fredericton, New Brunswick, Canada E3B 5A3

We describe a pre-commercial process for the in line production of hydrogen and solid, nanoscale carbon from a natural gas feedstock that avoids concomitant generation of CO_2. The process is pre-commercial scale (25-50 Nm^3/hr) and exhibits acceptable energy efficiencies (2-3 kW-hr/Nm^3 H_2) and reliable long-term performance. In this report we review the long term operational characteristics of the Hydrogen Enriched Natural Gas (HENG) pilot plant (the "CarbonSaver™") and describe the combustion characteristics of HENG as a fuel in real-world applications. HENG is used to fuel a 75 kW Tecogen combined heat and power (CHP) generator employed in a net metering environment and providing electricity and heat to our facility. Performance characteristics and emissions profiles of the internal combustion engine in the CHP unit are described when using both HENG and natural gas fuels. Similarly, the performance and emissions characteristics of a New York Thermal residential heating boiler and of a high efficiency, low NOx regenerative burner are described and contrasted for natural gas vs. HENG fuels.

INTRODUCTION

Natural gas is recognized as an important part of the solution to climate change, as it has the smallest carbon footprint among fossil fuels and can be used with high efficiency. In principle, natural gas fuel has numerous environmental advantages. For instance, when used under lean burn conditions in internal combustion engines, natural gas offers e.g. better control of fuel feed and mixture ratios, reduced knock, lowered equipment costs (i.e. no requirement for catalysts in the exhaust line) and significant reductions in GHG and NO_x emissions (1). However, there are a number of difficulties associated with the use of natural gas in low emission operations for ICE's and other combustion systems. For example, slower flame speeds and incomplete combustion under lean conditions contribute to problems in ICE's such as misfires, lower engine efficiencies and increased emissions when NG fuel is used. In other types of combustion systems, extremely lean operation can lead to increased particulate and unburned hydrocarbon emissions. A solution to this problem lies in the substitution of natural gas by mixtures of hydrogen in natural gas (4-20% v/v = Hydrogen Enriched Natural Gas, HENG). At these concentrations, HENG is physically compatible with existing natural gas transmission and distribution infrastructure, as well as end-use equipment. Moreover, codes and standards in many jurisdictions treat HENG with less than 20 per cent hydrogen as natural gas, which facilitates its initial deployment into gas networks. The adoption of HENG fuel has a number of potential benefits:

- HENG decreases the carbon intensity of natural gas. For every tonne of carbon removed before combustion, approximately 3.7 tonnes of CO_2 are avoided when HENG is burned.
- HENG increases the efficiency of natural gas conversion into useful energy. Adding even small amounts of hydrogen leads to more complete combustion of the fuel, including CO, methane and other hydrocarbons in the gas stream. This can improve engine efficiency and lower emissions of harmful pollutants.

- HENG helps avoid the formation of thermal NO_x, because it allows stable combustion at leaner gas mixtures to achieve lower flame temperatures than is possible with conventional natural gas.

Conceivably, the widespread adoption of HENG fuel could reduce CO_2 emissions in the United States by over 111 million tonnes per year. This is equivalent to the emissions from 20 million passenger vehicles or 10 million homes per year. In addition, HENG can be highly cost-competitive with many energy-related CO_2-reduction strategies under consideration. On the basis of cost per tonne of CO_2 abated, HENG costs less than most projections for carbon capture and storage technologies.

Unfortunately, the straightforward substitution of HENG in NG applications is complicated by the necessity for transporting and blending hydrogen with natural gas at dispensing facilities for the fuel. No infrastructure currently exists for the widespread distribution of hydrogen. Consequently, much of the potential advantage in using HENG fuel, at least in terms of reduced emissions, may be lost due to the necessity of road or rail transport of the gas. An in-line process for the production of HENG from natural gas that does not incur GHG penalties is therefore needed. Conventional steam reforming for hydrogen production processes is not desirable in this application, since these processes produce copious amounts of co-product CO_2. Rather, a process that can directly treat natural gas to produce HENG without generating any greenhouse gases would be ideal, since such a process leverages existing natural gas pipelines and local delivery systems, and takes advantage of the growing recoverable gas reserves in North America and LNG supplies globally.

In this report we describe equipment and a process for such an in-line production process for HENG. The process uses high frequency, high pressure plasma treatment of natural gas feedstock. We describe the product and process characteristics from lab and pilot scale processing, including the nanoscale structure of the carbon product. Emissions reductions observed when HENG is substituted for natural gas as the feed to a 75 kW combined heat and power (CHP) unit, residential hot water boiler and a conventional high efficiency industrial boiler are described.

EXPERIMENTATION

Our system employs an array of proprietary pulsed plasma torches which, together with ancillary equipment, form the "CarbonSaver™" system. Our pilot scale system is capable of processing up to 50 Nm^3/hr of natural gas to produce HENG targeted at ca. 10% hydrogen v/v and solid carbon. Figure 1 shows a high pressure (10 bar) lab-scale torch system (Figure 1a) and an atmospheric pressure pilot scale CarbonSaver™ (Figure 1b). In the lab-scale system, gas flows to the plasma torch were controlled by Unit Instruments UFC-1000 mass flow controllers (MFCs) calibrated to NIST traceable standards and having a URS-100 power supply. The pilot system used more conventional industrial flow control approaches. Gas analyses were performed using an Agilent 6890N gas chromatograph paired with an Agilent 5973 mass spectrometer. The system was configured with two thermal conductivity detectors (TCDs), a mass spectroscope detector (MSD), a 30 m silica gel column, a 0.09 m packed molecular sieve column and a 30 m Plot Q column. Gas flows, gas switching, and oven temperature and ramps were set up to permit fast hydrogen/methane analyses by TCD,

(a) **(b)**

Figure 1. CarbonSaver™ natural gas reformer; (a) high pressure laboratory scale reactor; (b) atmospheric pressure pilot plant

followed by optional full GC/MS analyses of the heavier molecular species. The GC/MS was calibrated for all major species of interest using certified gas mixtures from Praxair. In-line mass spectral analyses were performed using a Stanford Research Systems RGA200 residual gas analyzer.

In a typical HENG production test, the system was cycle purged using nitrogen until no oxygen was detected in the in-line GC/MS scans of the gas throughput. Natural gas (Praxair, >85% methane, balance propane/nitrogen with low levels of heavier organics and trace mercaptans) or methane (Air Liquide, 99+%) was supplied to a torch at rates of up to 10 Nm3/hr and the reactor purged without plasma present. A baseline sample of the natural gas was analyzed by GC/MS. The plasma was then initiated using the proprietary HV- pulsed power supply while maintaining the natural gas flow. The product was analyzed using in-line GC/MS at regular periods through the process run. Replicate samples for fast analyses of methane and H_2 were collected every 4 minutes while samples for slower GC/MS scans of heavier molecular weight species were collected at less frequent intervals. The solid carbon product was collected from the interior surfaces of the reactor system and from the solids filtration system following each process run. The carbon product was characterized using TEM (JEOL 2011 STEM), SEM (JEOL 2400), and Raman Spectroscopy (Renishaw inVia Confocal Research Raman Microscope).

Engine emissions were analyzed using a Horiba Multi-gas Analyzer Unit, model VA-3000. The analyzer contained a chemiluminescence detector for the quantitative determination of nitric oxide and a non-dispersive infrared detector that was used to measure carbon dioxide concentrations. Each detector was zeroed with HP+ Grade nitrogen and calibrated with a Certified Standard Grade gas mixture prior to the testing. A Baldwin Classic Peltier Cooler, model M115, from Perma-Pure LLC was used to cool and dry the flue gases to 5°C dew point prior to entering the analyzer. A water slip sensor downstream of the cooler monitored the dew point of the sample to ensure that the setpoint was not exceeded. A model P085 Iwaki air pump was used to draw the sample from the flue and supply it to the analyzer. The sample was regulated to a constant positive pressure of 0.7 psig at the analyzer inlet.

PRODUCTION OF HENG

The reformed HENG product from both the atmospheric pressure CarbonSaver™ and our lab scale high pressure system had similar chemical speciation. Table 1 shows a typical analysis of the HENG produced in these systems along with representative process parameters. In this particular case, the process was carried out at approximately 10 bars. Beyond methane and hydrogen, the reforming process produces a low concentration of acetylene as the primary contaminant. Acetylene is not an issue in the combustion processes tested. The presence of a low level of inerts in the feed and product gases (i.e N_2) is not shown. Energy requirements for the process are typically of the order of 350 to 400 watts/Nm^3 HENG. This value may be compared with those for steam methane reforming (ca. 200 watts/Nm^3 HENG equivalent once ancillary process energies are considered) and electrolysis (450-500 watts/Nm^3 HENG equivalent). The CarbonSaver™ system has demonstrated stable operation with steady production of ca. 10% hydrogen over run times of up to 36 hours.

	Test #1	Test #2	Units
Hydrogen	10.73	11.15	% Of Volume
Acetylene	1.54	1.92	% Of Volume
Ethylene	0.34	0.46	% Of Volume
Benzene	0.06	0.08	% Of Volume
Propylene	0.02	0.03	% Of Volume
1,3-Butadiene	0.04	0.06	% Of Volume
Methylacetylene	0.05	0.06	% Of Volume
Methane	84.60	83.49	% Of Volume
Input Power	2641	2841	Watt
Exhaust Gas Temperature	231.5	232.7	˚C
Pressure	146.3	148.6	PSI
Plasma Efficiency	312	392.6	W/m^3 @ 10% HENG
Carbon Yield	12.3	13.2	Grams/m3@10%

Table 1. Product analysis and process conditions, CarbonSaver™ high pressure process.

COMBUSTION TESTING OF HENG

The HENG produced in the CarbonSaver™ was evaluated as a fuel for a 75 kW Tecogen CHP unit run under extreme lean burn conditions. Figure 2 shows a graph in which the stabilizing influence of HENG fuel under lean burn conditions is apparent. The plot at the top of the chart in Figure 2 shows the instantaneous power output of the Tecogen, the bottom plot is hydrogen content in the gas. The power output is plotted while the air to fuel ratio (AFR) in the feed is varied from relatively rich to ultra-lean (denoted in the labels and divisions at the top of the graph) with and without hydrogen addition to the HENG feed. At the beginning of the scan, the system is operated at lean conditions with no hydrogen in the natural gas feed. Our Tecogen equipment configuration is not calibrated for a direct read-out of the AFR; however, the value of this parameter is directly correlated with a measurable millivolt signal. In this study, the AFR signal varied from 4600 mV for lean burn conditions to 4920 mV for extreme lean burn conditions. It can be seen from Figure 2 that the power output is relatively stable, albeit with some noise. As the graph progresses to the right, the AFR is gradually increased under fuel conditions of approximately 10% HENG. The presence of the hydrogen in the fuel permits steady operation to an AFR setting of 4920 mV. In order to confirm that the hydrogen was stabilizing the operation of the engine at this extreme lean condition, the hydrogen was reduced to 0% at an AFR of 4920 mV. It can be clearly seen from the graph in Figure 2 that the engine operations became unstable as soon as the hydrogen was removed from the fuel.

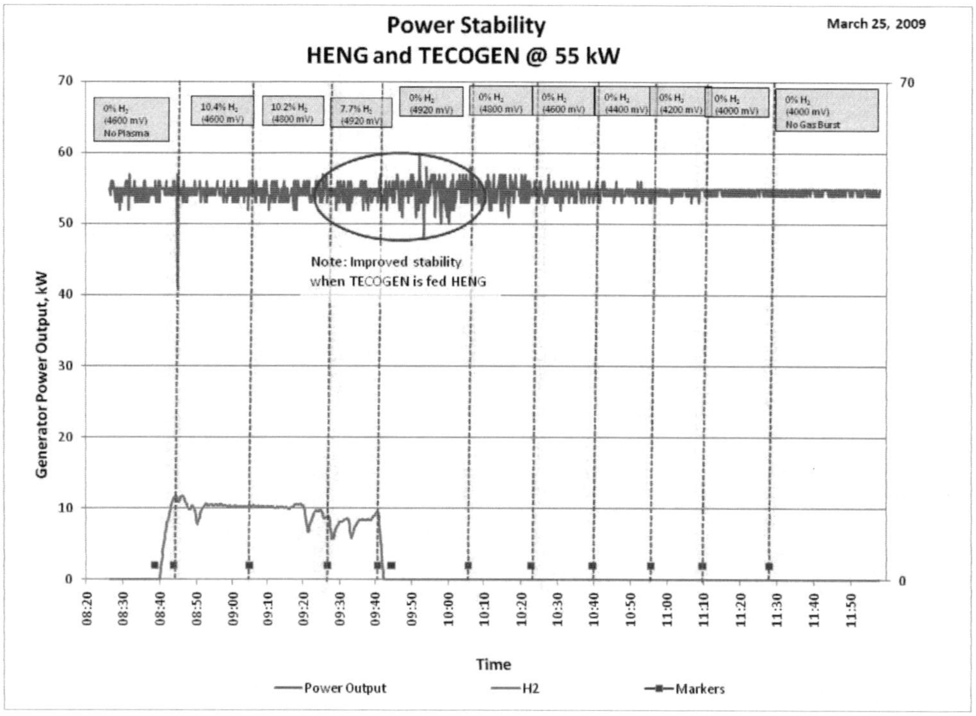

Figure 2. Power stability variations with AFR and Hydrogen concentration in HENG

The effect of hydrogen content in the HENG on the engine efficiencies was also evaluated. Figure 3 shows the impact of variations in hydrogen content with engine efficiency as measured by the amount of fuel required for constant power output operation. It can be

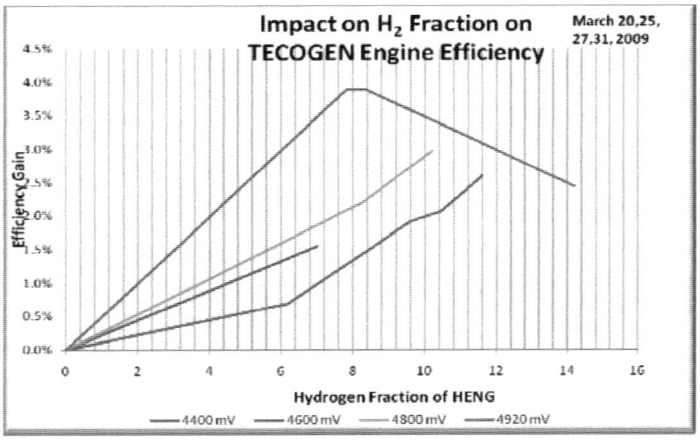

Figure 3. Engine efficiency variation with hydrogen content in HENG fuel.

seen that, at all AFRs, engine efficiency was improved by the presence of hydrogen in the fuel. The greatest improvement in engine efficiency was observed under the leanest burn conditions, with an efficiency improvement of ca. 4% seen at an AFR signal of 4920 mV. This greater efficiency is presumably due to improvements in combustion within the ICE which can be traced to the combustion characteristics of the added hydrogen.

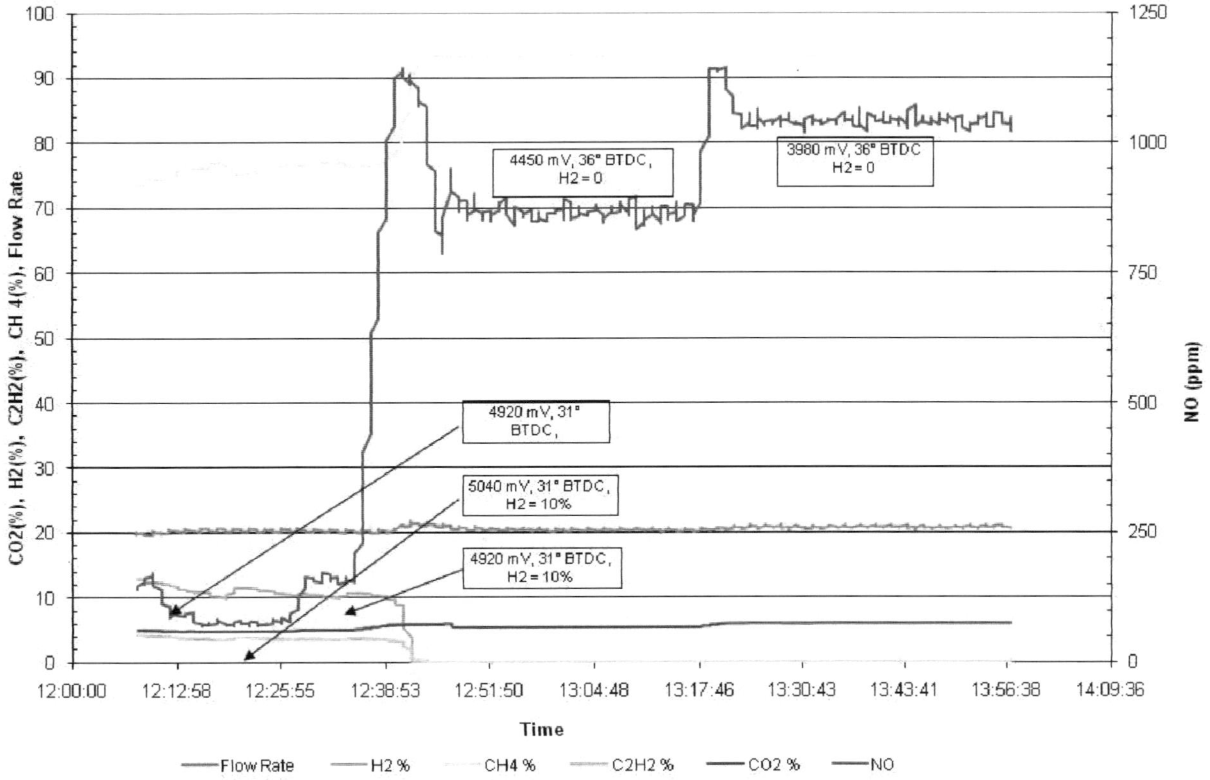

Figure 4. The effect of hydrogen content on NOx emissions of the Tecogen CHP.

The addition of hydrogen to natural gas produces a dramatic reduction in the NO_x emissions from the Tecogen engine. Figure 4 shows a graph of the measured NO_x emissions during a test run in which the percentage of hydrogen in HENG and the AFR were varied. It can be seen that the chart of the engine emissions shows relatively low (150 ppm) NO_x levels at the beginning of the test period when the hydrogen concentration in the fuel is approximately 12% and the AFR indicates extreme lean conditions (4920 mV). As the test progressed, the AFR was first increased to a reading of 5040 mV, resulting in a reduction in the NO_x emissions from approximately 150 down to 75 ppm. Resumption of the AFR indicator to 4920 mV produces a repeatable return to 150 ppm NO_x emissions. When the hydrogen is removed from the fuel, the NO_x fuel under lean burn (4920 mV) conditions, the NO_x is seen to rise rapidly to approximately 1125 ppm (the slight mismatch in the timing between the hydrogen trace and the NO_x trace is an artefact of the gas sampling arrangement). Without hydrogen the system behaved as might be expected, with higher NO_x levels observed under richer burn conditions.

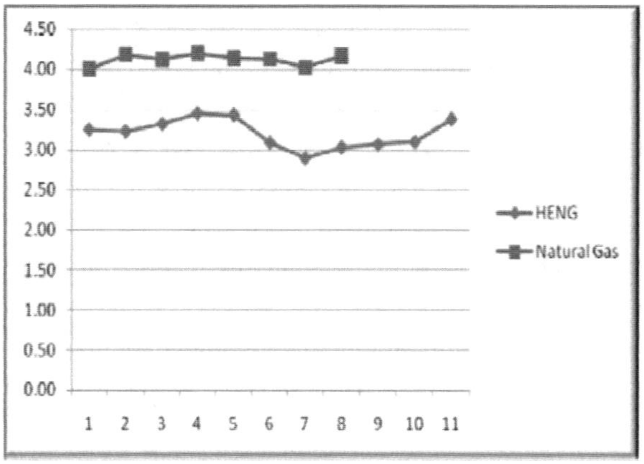

Figure 5. Emission reductions from HENG use in a residential boiler

HENG was also evaluated as a fuel for open combustion systems. Figure 5 shows the results for the use of HENG in a New York Thermal Trinity Ti150 boiler with a Sentry 2100 controller set to operate with natural gas under automatic control. Samples for gas analysis were drawn from the exhaust flue immediately downstream of the boiler. The boiler exhibited relatively low levels of NO_x in when operated with natural gas fuel (ca. 4-4.25 ppm). When HENG was substituted as a fuel for the system, a small but consistent reduction in the NO_x emissions was observed, to a level of ca. 3.25 ppm. More extensive testing was performed using a research tunnel furnace at the Natural Resources Canada Canmet laboratory. Figure 6 shows a representative chart of NOx reductions when HENG was employed as a fuel for this system. More extensive discussion of the results for the RTF system is available in a companion presentation at this conference by P. Gogolek (2). In both tests of HENG as a fuel for open flame combustor systems, the substitution of natural gas by HENG produced a small but significant reduction in NO_x emissions.

CARBON PRODUCED IN THE CARBONSAVER™ PROCESS

The carbon produced by the CarbonSaver™ process is a mixture of amorphous and nanostructured materials. The carbon consists of highly aggregated nanoscale particles with diameters between 20 and 50 nm. The carbon exhibited a relatively high surface area, as ca 200 m^2/g. TEM analysis (Figure 7a) showed the presence of significant concentrations of highly ordered, nanoscale graphitic structures, typically ca. 100 nm in diameter. High resolution micrographs suggest that these structures consist of approximately 7 – 10 stacked graphene

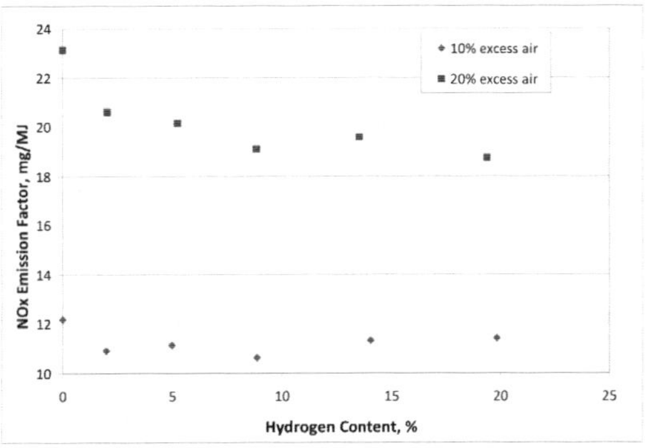

Figure 6. RTF furnace test of HENG: NO_x emissions on energy basis.

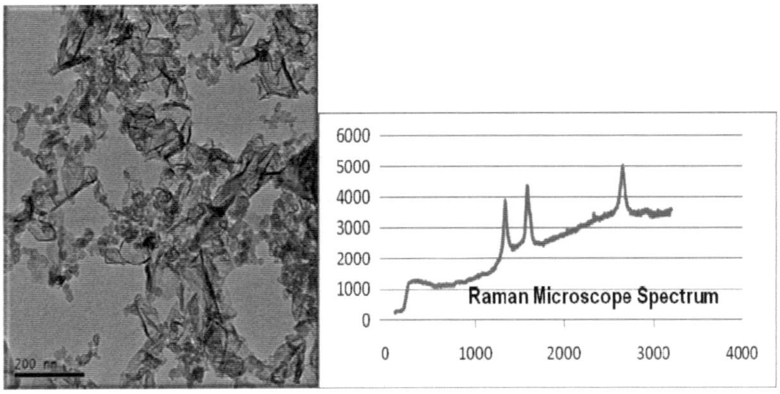

Figure 7. (a) TEM of the carbon produced in the CarbonSaver™ system and (b) the Raman spectrum of this carbon.

layers. Raman spectra of the carbon (Figure 7b) are consistent with significant concentrations of ordered carbon nanostructures.

Conclusion

Our results show that the CarbonSaver™ system is a viable, GHG reformer for various hydrocarbon feedstocks, especially natural gas. The specific energy requirements of the process, although not as efficient as those of existing chemical hydrogen production processes such as SMR, are competitive with other non-GHG producing hydrogen processes. We have demonstrated the utility of HENG as an ultra-low emission fuel suitable for use in internal combustion engines and open flame combustors. The carbon produced has significant concentrations of interesting nanostructure carbon species which may have practical use in composite engineering materials.

References:

1. "Methane-Hydrogen Mixtures as Fuels", G. A. Karim, I. Wierzba and Y. Al-Alousi, *Int. J. Hydrogen Energy*, 21, 625 (1996).
2. "Combustion Testing of Hydrogen-Enhanced Natural Gas", P. Gogolek, A. Caverly, P. Hughes and T. Whidden, AIChE Spring Meeting March 2010, San Antonio, TX.

Ag/ZnO one dimensional catalyst for the Steam Reforming of Methanol by Hydrogen production

R. Pérez-Hernández[1, δ], A. Gutiérrez-Martínez[1], A. Mayoral[2], F. Leonard Deepak[2,3], Ma. E. Fernández-García[1], G. Mondragón-Galicia[1], M. Jose-Yacaman[2].

[1]Instituto Nacional de Investigaciones Nucleares, Carr. México-Toluca S/N La Marquesa, Ocoyoacac, Edo. de México C. P. 52750, México.
[2]One UTSA Circle. Department of Physics and Astronomy. The University of Texas at San Antonio, Texas, 78249, USA
[3]International Iberian Nanotechnology Laboratory, Avda Mestre Jose Veiga, Braga 4715, Portugal.

Dr. M. Jose-Yacaman
Department of Physics and Astronomy, The University of Texas at San Antonio, Texas, 78249, USA. E-mail : miguel.yacaman@utsa.edu
[δ] Should be addressed: Dr. R. Pérez-Hernández: E-mail: raul.perez@inin.gob.mx

1. Introduction

The technologies for the production of high-quality transportation fuels from non-petroleum hydrocarbon resources have attracted considerable attention by scientists and chemical companies in accordance with the recent environmental demands and the decrease of fossil energy reserves. The fuel cell technology offers the potential to reduce the consumption of energy resources and the emissions of environmental pollutants from automobiles and electric power plants. In particular, polymer electrolyte fuel cell (PEFC) systems using hydrogen as an energy source, have attracted much attention due to their low operating temperature [1]. Conventional hydrogen production methods use gasoline or natural gas reforming. However, from the different feed stock fuels available, alcohols are very promising candidates because they are easily decomposed in the presence of water and generate a hydrogen-rich mixture at relatively low temperatures.

Methanol has been identified as a highly suitable liquid fuel for H_2 production, because of its high hydrogen/carbon ratio which favors energetically the steam reforming of methanol (SRM), as well as, the absence of carbon–carbon bonds which reduces soot formation, which otherwise may lead to deactivation of the catalyst [2, 3] and ease of handling. Oxidative Steam Reforming of Methanol (OSRM) and Steam Reforming of Methanol (SRM) for hydrogen production on Cu, Ni, Au and Pd supported on CeO_2, ZrO_2 and Al_2O_3 has been extensively studied in recent years [4-11]. Iwasa et al. [12] reported that ZnO base-catalysts have exceptionally high activity and selectivity to CO_2 and H_2 on the steam reforming of methanol and methanol synthesis [13] with low selectivity toward methane [14, 15]. Copper zinc oxide catalysts are important commercially in the conversion or preparation of synthetic gas [16]. Although several new systems were proposed for different reactions with ZnO-base catalysts, no decisive novelty appeared in the last years. In this context, new highly active and selective catalysts are necessary in order to advance the industry of the high-quality transportation fuels. For this reason we prepared Zinc oxide

(ZnO) 1D nanorods and, used them as a support for silver nanoparticles. The growth morphology of these 1-D catalyst was studied in detail using electron microscopy techniques, and their catalytic activity was studied in the steam reforming of methanol for H_2 production.

2. Experimental Method

2.1 Synthesis of the ZnO nanostructures and Catalysts preparation (Ag/1D-ZnO)

All reagents including zinc nitrate ($Zn(NO_3)_2{\bullet}6H_2O$) and sodium hydroxide (NaOH), were analytical grade purchased from Aldrich and used as received. The ZnO 1D nanorods were synthesized using the method reported previously [17-19]. The synthesized ZnO 1D nanorods were impregnated with an aqueous solution of $AgNO_3$ at an appropriate concentration to yield 1.5 and 5 wt % of Ag in the catalysts. The samples were then dried at 100°C for 1h and calcined at 400 °C for 2h in static air and finally reduced with a H_2 stream at 400 °C for 1h before further characterization. The labeling of different catalysts will be referred as xAg/ZnO, where x = 1.5 and 5 wt % of metal on the catalysts respectively.

2.2 Characterization

The morphology of the nanorods was analyzed by Scanning Electron Microscopy (SEM) which was performed in a FEG Hitachi S-5500 ultra high resolution electron microscope (0.4 nm at 30kV) with a BF/DF Duo-STEM detector. Transmission electron microscopy (TEM), selected area electron diffraction (SAED) and weak beam dark field (WBDF) analysis were performed using a Philips Tecnai G2, 200Kv, TEM equipped with Schottky-type field emission gun, ultra-high resolution pole piece (Cs=0.5 mm), and a scanning transmission electron microscope (STEM) unit with high angle annular dark field (HAADF) detector operating at 200 kV. X-ray diffraction (XRD) powder patterns were recorded in a Siemens D-5000 diffratometer, using Cu Kα (λ=0.15406 nm). The method to measure the total surface area and TPR was reported previously [4, 5, 20-23]. The steady-state activity in the SRM reaction was performed in a conventional fixed-bed flow reactor (8 mm i.d.) using 0.1 g of the catalyst in a temperature range from 225 to 400 °C with steps of 25 °C with 7 h of stabilization time at each temperature and atmospheric pressure on an automatic multitask unit RIG-100 from ISR INC. The catalyst was first activated in a stream of H_2 (60 ml/min) from room temperature to 400 °C with a heating rate of 10°C/min and held at this temperature for 1h. For the SRM reaction, He (50 ml/min, GHSV= 30,000 h^{-1} based on the total flow) was added by means of two mass flow controller (RIG-100) and bubbled through a tank containing water and the other methanol, the partial pressure of CH_3OH and H_2O was 75 and 12.75 Torr respectively. The molar ratio in the steam was CH_3OH (1.95μmol)/H_2O (1.97 μmol) = R ≈ 1.0. The effluent gas of the reactor was analyzed by gas-chromatography (Gow-Mac 580 instrument) equipped with a two columns system (molecular sieve 5 Å and Porapack Q columns), double injector controlled by Clarity software V.2.6.04.402 and TCD. The spent 1.5Ag/ZnO catalyst was tested again in a second cycle of catalytic activity. So at the end of the reaction (400 °C) the catalyst was cleaned with He for 30 min. Then the sample was brought up to the reaction temperature in He and the reaction mixture was introduced and the catalytic activity for the second cycle was determined.

3. Results and discussions

ZnO with 1D morphology was growth modifying the concentration of the precursors during the synthesis. Using the specific concentration of the [OH^-] and [Zn^{2+}] resulted in the formation of cactus-like ZnO nanostructures. Secondary growth produced agave-like crystals from new crystal growth on the edges. The SEM images (Figure 1a) reveals that the flowers consisted of nanoneedles with lengths of about 7 μm and diameters of about 900 nm and other near to 150 nm. Each flower has a central part as a common growth point for all of the constituent nanoneedles. Inset in Figure 1a, showed that the ZnO 1D crystal obtained by the hydrothermal method belongs to the hexagonal crystal system. However, when the sample was seeing in transversal section (Figure 1b) the top of the rod showed a sharp tip. The reason for the alignment of the ZnO structure is due to the matching lattice structure through the polar nature of the ZnO surface [17]. Initially, the ZnO crystalline grains mainly grow along the <0110> direction due to epitaxy. However, as a result of the polar characteristic of the ZnO, the growth along the <0001> direction has a gradual increasing velocity [24]. As in most cases, the surfaces of the metal oxide nanoparticles formed in hydrothermal conditions were terminated by hydroxyl groups [25]. We noted that the average diameter of ZnO nanorods proportionally increased as the concentration of the precursor solution was increased. Due to the higher amount of the growing units of ZnO, the diameters of ZnO nanorods swelled, effect attributed to the radial growth along the <01$\bar{1}$0> direction [18]. The formation of hexagonal prism and pyramid like ZnO crystals is attributed to the difference in the growth velocities of the various crystal facets. The growth velocities of ZnO under hydrothermal conditions along the different directions has been already reported to be (0001) > (1011) > (1010) [17, 26, 27].

Figure 1. a) SEM image of the 1D ZnO microrods. Inset Figure in (a), showed a top view of the nanorod. b) Transversal section of the microrods showed a sharp tip.

Detailed structural characterization of the zinc oxide nanorod before Ag impregnation was performed by TEM analysis. Figure 2 shows a low magnification TEM image of a typical ZnO nanorods obtained in this work, the diameter of this nanorod was about 150 nm. All of these nanorods terminate in a triangular truncated phase which has a diameter of 30 nm. Inset in Figure 3a shows the SAEDP electron diffraction pattern in [$\bar{1}$ 2 $\bar{1}$ 0] direction of the entire wire in which we can observe forbidden spots along the planes {0001}. In addition, the SAEDP revealed a single crystal structure along the entire rod which was also confirmed with HRTEM images indicating that it grows free of defects.

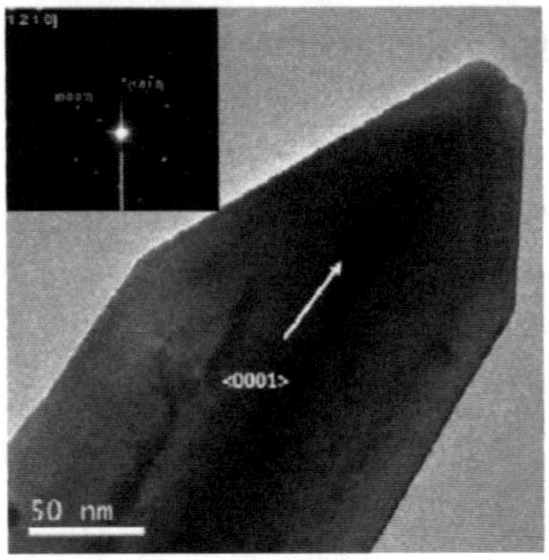

Figure 2. Low magnification TEM images of one of the nanorods that growth along the [0001] directions. Inset image showed the experimental electron diffraction pattern in the [$\bar{1}$ 2 1 0] direction.

We used the ZnO 1-D nanorods for silver impregnation and tested as catalysts for catalytic applications. We observed Ag nanoparticles with quasi-spherical morphology and average size about 15-30 nm (inset on Figure 3a) homogeneously distributed along the nanorods on the 5Ag/ZnO catalyst. The TEM images of the 5Ag/ZnO 1D catalyst showed good dispersion of the silver nanoparticles over the entire ZnO nanorod. Figure 3b shows the HRTEM image of silver nanoparticles deposited on the ZnO 1D nanorod, and was indexed assuming *Fm*-3*m* symmetry obtaining a unit cell value of 4.09 Å in agreement with the data reported for the JCPDS card number 065-2871. On the other hand, the nanoparticles showed lattice fringes with interplanar spacing of 0.235 nm corresponding to the {111} planes of fcc Ag. The Fast Fourier transform (FFT) pattern (inset in Figure) also confirms fcc metallic Ag along the [110] zone axis.

a) b)

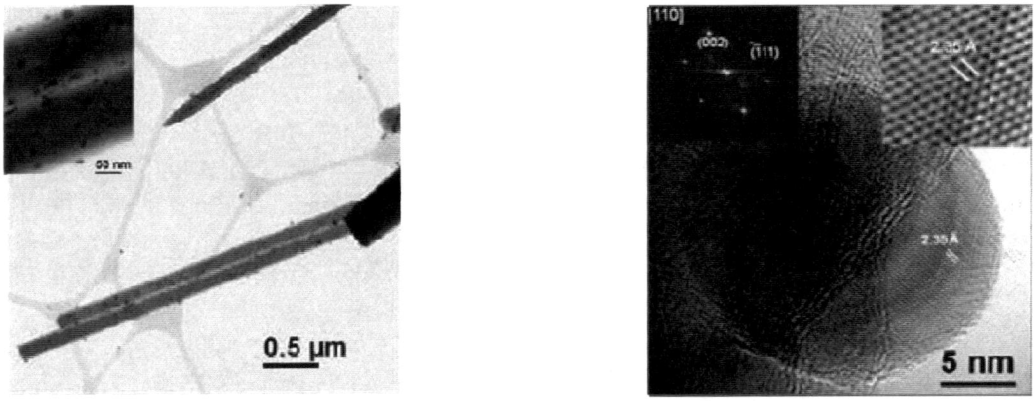

Figure 3. **a)** Low magnification TEM image of the 5Ag/ZnO 1D catalyst. Inset Figure showed the higher magnification of the one dimensional catalyst. **b)** Aberration corrected HRTEM image of a 5Ag/ZnO catalyst which was recorded along [110] orientation. The d-spacing measured between two adjacent planes of atoms was measured to be ≈ 2.35 Å corresponding to the {111} planes. The particle size was about 22.5 nm. The FFT is shown inset.

The crystalline structure of the ZnO, 1.5Ag/ZnO and 5Ag/ZnO catalysts powder was investigated by XRD. We observed that the peak identified as metallic Ag increase as a function of silver loading (Figure not showed). The average crystal size of the Ag measured by Scherer equation was 28.3 nm and 33.3 nm for 1.5Ag/ZnO and 5Ag/ZnO catalysts respectively. TPR profiles of the calcined samples and catalysts after catalytic reaction did not show a positive peak from r.t. to 500 ºC, associated with the consumption of hydrogen necessary for the reduction of the silver oxide or some reducible species present on the catalyst (Figures not show). This result was attributed to that all the silver oxide probably is reduced at below ambient temperature.

Table 1 showed the temperature dependence of the catalytic performance on methanol steam reforming over Ag/ZnO one dimensional catalysts from 225 to 400 °C. The catalytic performance of the steam reforming of methanol reaction of the bare ZnO showed tiny catalytic activity at low temperature but, when it was increased may display excellent SRM performance. Methanol conversion over the Ag-free ZnO support at 350 °C was 5 % and at 400 °C was 64 %. Koga et al. [28] studied the auto thermal reforming of methanol reaction on ZnO whiskers. They showed that the ZnO whiskers used in their study did not present any catalytic activity towards the methanol reforming reaction. However, when Cu was deposited on it, methanol conversion was observed at 250 °C. Mo et al. [29] studied the POM reaction over different CeO_2 and ZnO powders. They observed 85 % of the methanol conversion on CeO_2 sinthesized in their laboratory and 0 % for commercial CeO_2 and ZnO respectively. On the other hand, when 1.5 wt. % of silver was supported on the ZnO 1D, the sample showed better performance on the steam reforming of methanol reaction than bare ZnO 1D. The light-off temperature for the 1.5Ag/ZnO 1D sample was approximately at 275 °C while on the 5Ag/ZnO 1D sample it was near to 300 °C. The methanol conversion at 350 °C was ~ 39 and ~ 16 % for 1.5Ag/ZnO and 5Ag/ZnO one dimensional catalysts respectively. The catalytic activity of the samples at the maximum reaction temperature showed the following order: 1.5Ag/ZnO > 5Ag/ZnO > ZnO 1D samples. This finding could be correlated with the Ag particle size present on the samples. The catalyst with small Ag particle size showed high catalytic activity than sample with big Ag particle size. Croy et al. [30] studied the H_2 production through methanol decomposition on Pt/TiO_2 catalysts. They observed high catalytic activity on the catalysts with small particle size and diminish as the particle size increase. On methane combustion was observed better catalytic activity on the catalyst with low gold loading than the one with high gold loading, this result was associated with the dispersion of Au and the atomic ratio of Au^{3+}/Au^{0} [31, 32]. On the other hand, the catalyst that showed best performance in the catalytic activity was evaluated on a second cycle of the reaction, and it was found that the overall activity was slightly higher than the first cycle of the reaction. This finding could be associated at the particle size as was reported on XRD section for this sample [33].

Table 1. Temperature dependence of SRM activity of the Ag/ZnO one dimensional catalysts. Partial pressure of CH_3OH and H_2O was 75 and 12.75 Torr respectively. GHVS=30,000 h^{-1}. H_2O/CH_3OH = 1.0 mole ratio.

Temperature	ZnO	1.5Ag/ZnO	1.5Ag/ZnO 2° cicle	5Ag/ZnO
225	0	0	1.34	0
250	0	0	6.81	0
275	0	3.45	12.22	0
300	0	5.99	15.72	0
325	0	17.44	24.24	2.86
350	3.06	38.93	44.83	15.56
375	29.92	65.01	76.07	36.84
400	63.38	97.05	100	71.98

High loading of silver nanoparticles can reduce the surface active for the catalytic reaction dropping the reaction, as was observed on the 5Ag/ZnO one dimensional catalyst. Because, high content of Ag on the nanorods diminish the methanol adsorption over zinc oxide and hence avoid the dissociative adsorption of methanol. Those reports and our results of the catalytic activity, suggest that the base-ZnO samples could be suppressed the undesirable CO by-production [33]. This is known to act as a catalytic poison for Pt anode electrocatalysts in fuel cells. According to the SEM and TEM results we propose a similar behavior to occur on our Ag/ZnO one dimensional catalyst. In other words, we suggest that the low rim/edges sites ratio of the ZnO

nanorods is responsible for the catalytic activity and selectivity toward H_2 and low CO production, because, nanorods have larger edge area for the catalytic reaction than rim sites (Figure 4) [33]. The next steps (2-5) could be occur on the edges sites of our Ag/ZnO one dimensional catalyst via the Mars and van Krevelen (MVK) mechanism then the role of Ag is to accept the hydrogen to be released to the gas phase (Figure 4).
The main products of the SRM reaction of the Ag/ZnO 1D catalyst were H_2, CO and CO_2; among them; H_2 and CO_2 were the principal products. At the beginning of the reaction where tiny methanol is consumed only CO_2 and H_2O were produce. Table 2 showed the H_2 yield (μmol) as a function of temperature on the SRM reaction of the ZnO, 1.5Ag/ZnO and 5Ag/ZnO 1-D catalysts. The H_2 yield, defined as μmole H_2/μmole feeding CH_3OH, is negligible at low conversions and increases as reaction temperature was raised. Between them, the 1.5Ag/ZnO one dimensional sample had the best H_2 yield than ZnO and 5Ag/ZnO 1D samples. The H_2 yield observed on the 1.5Ag/ZnO 1D sample was 2.48 μmole at 400 °C but on the second cycle of the reaction the value increase until 2.73 (the theoretical value for SRM is 3). On the other hand, it is cleared that the H_2 production on the first and second cycle of the reaction of the 1.5Ag/ZnO 1D catalyst was practically the same in whole reaction temperature. Bare ZnO and 5Ag/ZnO one dimensional samples showed approximately the same H_2 production, between them the last sample had lowest selectivity toward hydrogen.

Table 2. H_2 yield as a function of the reaction temperature. Partial pressure of CH_3OH and H_2O was 75 and 12.75 Torr respectively. GHVS=30,000 h^{-1}. H_2O/CH_3OH = 1.0 mole ratio.

Temperature	ZnO	1.5Ag/ZnO	1.5Ag/ZnO 2° cicle	5Ag/ZnO
225	0	0	0	0
250	0	0	0	0
275	0	0.12	0.07	0
300	0.06	0.18	0.19	0
325	0.24	0.48	0.53	0.28
350	0.84	1.34	1.30	0.68
375	1.69	2.42	2.44	1.52
400	2.52	2.48	2.73	2.44

$CH_3OH_{(g)} + MOM \rightarrow CH_3OM + MOH$ (M = Al, Zn, Ce) [7] (2)
$H_2O_{(a)} \rightarrow H_{(a)} + OH_{(a)}$ (3)
$CH_3O_{(a)} + OH_{(a)} \rightarrow CO_2 + 2H_2$ (4)
$CH_3OH + 2OH_{(a)} \rightarrow CO_2 + H_2O + 2H_2$ (5)

4. Conclusions

The ZnO 1D nanorods were impregnated with silver by the classic impregnation technique. TEM results showed that is possible to prepare good silver nanoparticles dispersed over entire surface of the ZnO 1D nanorods. The average Ag crystallite size determined by Scherrer equation in DRX analysis was 28.3 nm for the 1.5Ag/ZnO sample and 33.3 nm for the 5Ag/ZnO catalysts respectively. These Ag/ZnO 1D nanorods catalysts showed good catalytic activity on the steam reforming methanol reaction for hydrogen production. Among these catalysts, the sample with small Ag particle size showed best performance in methanol conversion than on the catalyst with big Ag particle size. Good stability during the reaction was observed on the 1.5Ag/ZnO 1D sample (second cycle of reaction). According to the results of SEM and TEM techniques the catalytic activity observed on the ZnO one dimensional catalyst occurs in the edge sites rather than in the rim sites. H_2 yield reaches the value of 2.48 in the first cycle of the reaction and increase up to 2.78 in the second cycle of the reaction at 400 °C which is very close to the theoretical one. On the other hand, the controllable growth of ZnO nanostructures described here may open up the possibility of exploring novel applications in the areas of catalysis in addition to high volume applications.

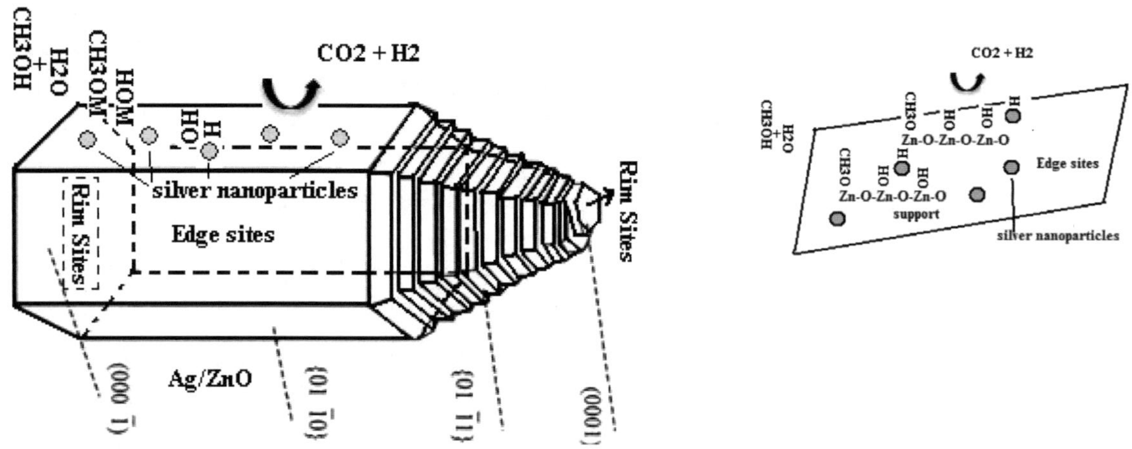

Figure 4. Rim and edge sites for Ag/ZnO one dimensional catalyst and reaction mechanism from eq. 2-5.

Acknowledgements

The authors are indebted to the Welch foundation through grant funding, Peñoles Company in Mexico and the projects ININ-CA-711, CA-009 and CONACyT (México) J-48924 for financial support. Thanks to D. Olmos MSc and Dr. D. Ferrer for their technical support.

References

[1] A.A. Mirzaei, H.R. Shaterian, S.H. Taylor, G.J. Hutchings, Catal. Letters 87 (2003) 103.

[2] B. Lindstrom, L.J. Pettersson, Int J Hydrogen Energy. 26 (2001) 923.

[3] L.J. Pettersson, R. Westerholm, Int J Hydrogen Energy. 26 (2001) 243.

[4] R. Pérez-Hernández, G. Mondragón Galicia, D. Mendoza Anaya, J. Palacios, C. Angeles-Chavez, J. Arenas-Alatorre, Int J Hydrogen Energy. 33 (2008) 4569.

[5] R. Pérez-Hernández, A. Gutiérrez-Martínez, C.E. Gutiérrez-Wing, Int J Hydrogen Energy. 32 (2007) 2888.

[6] M. Turco, C. Cammarano, G. Bagnasco, E. Moretti, L. Storaro, A. Talon, M. Lenarda, Appl. Catal. B: Env. 91 (2009) 101.

[7] M. Turco, G. Bagnasco, C. Cammarano, P. Senese, U. Costantino, M. Sisani, Appl. Catal. B: Env. 77 (2007) 46.
[8] S. Fukahori, T. Kitaoka, A. Tomoda, R. Suzuki, H. Wariishi, Appl. Catal. A: Gen. . 300 (2006) 155.
[9] T.L. Reitz, S. Ahmed, M. Krumpelt, R. Kumar, H.H. Kung, J. Mol. Catal. A: Chem. . 162 (2000) 275.
[10] T.L. Reitz, P.L. Lee, K.F. Czaplewski, J.C. Lang, K.E. Popp, H.H. Kung, J. Catal. . 199 (2001) 193.
[11] R. Pérez-Hernández, L.C. Longoria, J. Palacios, M.M. Aguila, V. Rodríguez, Energy Materials. 3 (2008) 152.
[12] N. Iwasa, T. Mayanagi, N. Ongawa, K. Sakata, N. Takezawa, Catal. Lett. 54 (1998) 119.
[13] J. Strunnk, K. Kahler, X. Xia, M. Comitti, F. Schuth, T. Reinecke, M. Muhle, Appl. Catal. A: Gen. . 359 (2009) 121.
[14] H. Sakurai, M.H. 361., Catal. Today 29 (1996) 361.
[15] H. Sakurai, M. Haruta, Appl. Catal. A: Gen. 127 (1995) 93.
[16] M.S. Spencer, Catal. Lett. . 66 (2000) 255.
[17] E. De La Rosa, S. Sepulveda-Guzman, B. Reeja-Jayan, A. Torres, P. Salas, N. Elizondo, M. Jose-Yacaman, J. Phys. Chem. C. 111 (2007) 8489.
[18] B. Reeja-Jayan, E. De la Rosa, S. Sepulveda-Guzman, R.A. Rodriguez, M.J. Yacaman, J. Phys. Chem. C 112 (2008) 240.
[19] S. Sepulveda-Guzman, B. Reeja-Jayan, E.d.l. Rosa, A. Torres-Castro, V. Gonzalez-Gonzalez, M. Jose-Yacaman, Mater. Chem. Phys. 115 (2009) 172.
[20] R. Pérez-Hernández, F. Aguilar, A.Gomés-Cortés, G. Díaz, Catal. Today. 10-108 (2005) 175.
[21] R. Pérez-Hernández, A. Gómez-Cortés, J. Arenas-Alatorre, S. Rojas, R. Mariscal, J.L.G. Fierro, G. Díaz, Catal. Today. 107-108 (2005) 149.
[22] R. Pérez-Hernández, D. Mendoza-Anaya, M.E. Fernández, A. Gómez-Cortés, J. Mol Catal A: Chemical 281 (2008) 200.
[23] C. Gutiérrez-Wing, R. Pérez-Hernández, G. Mondragón-Galicia, G. Villa-Sánchez, M.E. Fernández-García, J. Arenas-Alatorre, D. Mendoza-Anaya, Solid State Sci. . 11 (2009) 1722.
[24] Z.L. Wang, J. Phys.: Condens. Matter 16 (2004) R829.
[25] T. Trinidade, J.D.P. DeJesus, P. O'Brien, J. Mater. Chem. . 4 (1994) 1611.
[26] W.J. Li, E.W. Shi, W.Z. Zhong, Z.W. Yin, J. Cryst. Growth 203 (1999) 186.
[27] Z. Zhang, J. Mu, J. Colloid Interface Sci. . 307 (2007) 79.
[28] H. Koga, T. Kitaoka, H. Wariishi., Chem. Commun. 43 (2008) 5616.
[29] L. Mo, X. Zheng, C.-T. Yeh, ChemPhysChem. 6 (2005) 1470.
[30] J.R. Croy, S. Mostafa, J. Liu, Y.-H. Sohn, B.R. Cuenya, Catal. Lett. 118 (2007) 1.
[31] Y. Zhang, J. Deng, L. Zhang, W. Qiu, H. Dai, H. He, Catal. Today 139 (2008) 29.
[32] X. Zhang, H. Shi, B.Q. Xu, Catal. Today 122 (2007) 330.
[33] R. Pérez-Hernández, A. Gutiérrez-Martínez, A. Mayoral, F.L. Deepak, M.E. Fernández-García, G. Mondragón-Galicia, M. Jose-Yacaman, submitted to Catal. Today (Dec-2009).

(137f) Comparison of Sulphur–Iodine, Copper–Chloride and Hybrid-Sulphur Thermochemical Cycles for Hydrogen Production

Nesrin K. Demir, Mechanical Engineering, UOIT, Oshawa, ON, Canada

Ibrahim Dincer, Mechanical Engineering, UOIT, Oshawa, ON, Canada

Greg F. Naterer, Mechanical Engineering, UOIT, Oshawa, ON, Canada

The rapid increase in worldwide energy demand, caused by the rise in global population and living standards, has led to higher greenhouse gas emissions and diminishing fossil fuel reserves. To address these challenges, there has been growing R&D for new alternative energy solutions to address global energy requirements and environmental problems. Nuclear energy is one of the more promising low-carbon sources for large scale energy production. Renewable energy sources have been limited due to their cost, reliability and availability.

Hydrogen is one of the promising, clean and effective energy carriers. Hydrogen can be produced using either nuclear or renewable energy sources. Many past studies have focused on developing efficient and economic pathways for hydrogen production. Thermochemical cycles are one of the most promising methods for large-scale hydrogen production, and these cycles are usually classified according to the chemicals used in the processes.

The sulphur-iodine, copper-chloride and hybrid-sulphur water splitting cycles are promising methods of thermochemical hydrogen production. The sulphur–iodine cycle involves three thermochemical steps. The copper–chlorine cycle involves five steps, of which four are thermally driven chemical reactions, and one is an electrochemical reaction. The hybrid-sulphur cycle involves two steps, of which one is a thermochemical reaction and the other is an electrochemical reaction. This paper compares these three cycles from the perspectives of energy, exergy, efficiency and cost. The major engineering advantages and disadvantages of these cycle variations are analyzed and discussed.

Keywords: Hydrogen, energy, exergy, sulphur-iodine, copper-chloride, hybrid-sulphur, water splitting cycles, thermochemical cycles.

See more of this Session: Fuel Processing for Hydrogen Production II
See more of this Group/Topical: Topical 2: Advanced Fossil Energy Utilization

(137g) Enhancing the Water Gas Shift Reaction Using Synthetic Sorbents for the Capture of CO2

Saquib Sultan, Department of Chemical Engineering and Biotechnology, University of Cambridge, Cambridge, United Kingdom

Christoph R. Müller, Department of Mechanical and Process Engineering, ETH Zurich, Zurich, Switzerland

John S. Dennis, Department of Chemical Engineering and Biotechnology, University of Cambridge, Cambridge, United Kingdom

In order to capture the carbon content of coal, one option would be to gasify it with mixtures of, *e.g.* steam, CO_2 and O_2, to produce synthesis gas: the water-gas shift (WGS) reaction, $CO + H_2O \leftarrow\rightarrow CO_2 + H_2$, could then be used to increase the yield of hydrogen. In principle, if the WGS reaction is operating close to thermodynamic equilibrium, the yield of H_2 could be enhanced by the simultaneous removal of the CO_2 from the gas-phase using a suitable sorbent based on CaO, which is carbonated in: $CaO_{(s)} + CO_{2\,(g)} \rightarrow CaCO_{3(s)}$. This would move the WGS equilibrium to the right, thereby producing H_2 with only a small contamination by CO_2. The captured CO_2 would subsequently be recovered by removing the carbonate and calcining it to generate a pure stream of CO_2 which would be ready for sequestration, *e.g.* in deep saline aquifers. The regenerated calcareous sorbent would then be recycled to the water-gas shift reactor.

This paper describes the results of experiments in which the WGS reaction was conducted in the presence of a CaO-containing sorbent. A major limitation of such a scheme is the degradation of the conversion of the CaO over progressive cycles of carbonation and calcination, particularly with natural sorbents, such as limestone. In our experiments, the sorbent employed was a synthetic one, namely calcium magnesium acetate (CMA, Ca : Mg = 7 : 3, molar basis), the performance of which did not degrade substantially with cycling. The enhanced WGS reaction was conducted at temperatures of 500, 550, 600 and 650°C, with calcination at 800°C, in a packed bed. The gases fed to the reactor were obtained from cylinders (in the cases of CO and inert N_2) and steam was generated from deionised water.

From the experiments performed, it was shown that CaO-containing sorbents can affect the WGS in two ways: (i) by catalysis of the reaction itself and (ii) by altering the equilibrium position through abstraction of CO_2 from the gas phase. It was found that a trade-off has to be made between the production of H_2 and the contamination of the product gas with CO_2 (the CO_2 slip). For the WGS performed at 600°C, equilibrium conditions were reached while at the lowest experimental temperature of 500°C, there was very little CO_2 slip.

See more of this Session: Fuel Processing for Hydrogen Production II
See more of this Group/Topical: Topical 2: Advanced Fossil Energy Utilization

HYBRID MEMBRANE ABSORPTION PROCESS FOR POST COMBUSTION CO_2 CAPTURE

S. James Zhou and Howard Meyer, Gas Technology Institute, Des Plaines, IL

Ben Bikson and Yong Ding, PoroGen Corporation, Woburn, MA

Abstract

Gas Technology Institute (GTI) and PoroGen Corporation are jointly developing a novel hybrid membrane – absorption process for pre- and post-combustion CO_2 capture (Carbo-Lock™ process). The novel Carbo-Lock™ process combines advantageous features of both absorption and membrane processes to generate cost-effective separation and capture of CO_2 from various emission sources. Together the organizations are making extensive laboratory and bench-scale measurements to bring this technology development to commercialization.

The Carbo-Lock™ process is a hybrid of membrane and the conventional absorption processes. CO_2-containing gas passes through small membrane tubes (hollow fibers with porous walls) while a CO_2 selective solvent (typically an amine solution) flows on the shell side of the membrane tubes. CO_2 passes through the nano-porous membrane and is absorbed in the selective solvent. The CO_2 rich solvent can be regenerated in a second membrane module operated in a reverse process.

The Carbo-Lock™ process uses a novel hollow fiber membrane technology patented by PoroGen. This novel membrane is made from a chemically and thermally stable commercial engineered polymer poly(ether ether ketone) or PEEK. The PEEK membrane contactor can provide a platform for solvent-based systems beyond conventional amines. The reduced size requirements translate to lower solvent inventories, less metal exposure to corrosive liquids, and lower visual impact (smaller footprint) for siting at congested power plants. Since the gas and liquid phases are not directly in contact, the solvent purity levels are expected to remain high and solvent losses, from foaming and carryover, are expected to remain low. Utilizing the membrane contactor in the regeneration section will greatly reduce energy usage for desorbing CO_2 from the solvent. The system operates at very low gas pressure drops, comparable to those achieved with very large diameter columns, and much lower than those of conventional membrane systems.

The preliminary test and computer simulation results of hybrid Carbo-Lock™ membrane-absorption process are presented in this paper.

Introduction

The hybrid membrane/absorption process as shown in Figure 1 is enabled by poly(ether ether ketone), PEEK, based hollow fiber membrane contactor that can replace any column-based mass transfer device in numerous industrial applications such as the column-based devices currently employed in the amine absorption process for CO_2 separation and capture. The membrane is in the hollow fiber configuration and is produced by a patented environmentally benign melt extrusion process. PoroGen is the only manufacturer of the porous PEEK membranes worldwide. The hybrid membrane/absorption device is a contactor that utilizes solvent absorption mechanism with the membrane as the phase boundary between the gas and the liquid. Advanced concepts of our hybrid process employ selective membranes that enhance separation efficiency by functioning as a "selective valve", enabling fast transport of CO_2 while blocking gas components that cause degradation from reaching the absorbent (for example, blocking oxygen from reaching amine) and/or preventing excessive loss of absorbent by evaporation.

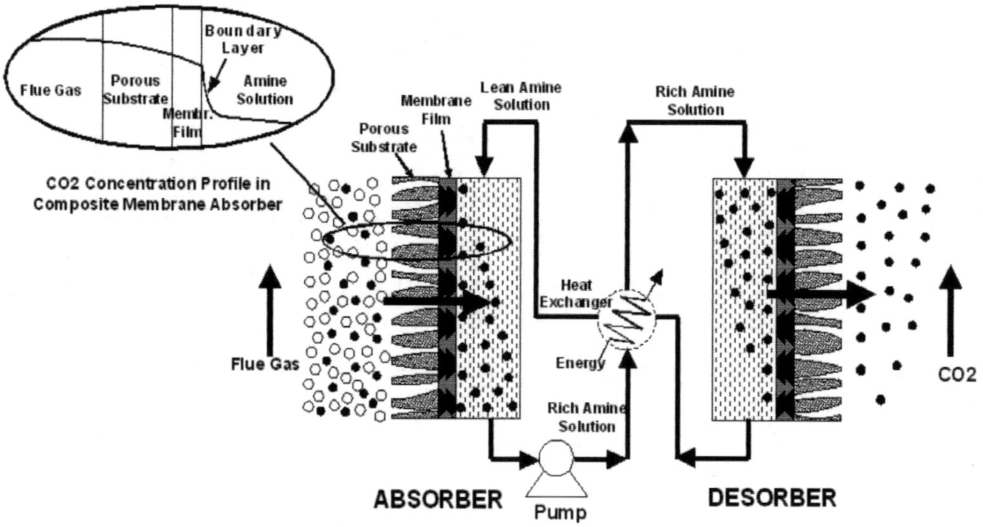

Figure 1 **Integral Membrane Absorption Process for CO2 Capture from Flue Gas**

Process Benefits

The interfacial area is a key variable that directly influences the overall mass transfer and thus the size and performance of the processing equipment. In the PEEK membrane contactor, the interfacial area will be increased by an order of magnitude or greater over conventional packed or tray columns by using super-hydrophobic porous hollow fiber membrane modules. The hollow fiber configuration will allow for high membrane surface area per unit volume and will result in a much more compact absorption system of dramatically reduced size and weight which will also reduce the footprint of the carbon capture island, especially at land-limited sites. For example, the membrane contactor CO_2 removal plant for a 350MW facility is only 11% the size of a conventional packed column based MEA plant.[1] The modular design of membrane contactors will enable the deployment of shop pre-fabricated equipment rather than field fabricated equipment, and will decrease the necessary foundation/footing requirements for structures. Additional benefits include, but are not limited to, the elimination of equipment corrosion due to the inert materials of construction used in hollow fiber contactors (polymeric materials only), decreased solvent losses because there is no carryover nor foaming, lower solvent degradation, and decreased parasitic fan loads because of the decreased pressure and pressure drop required for the hollow fiber membrane contactor vs. conventional membranes and columns.

The membrane is a nanoporous, super-hydrophobic hollow fiber membrane. The novel hollow fiber membrane technology is based on the chemically- and thermally-resistant, commercial engineering polymer, PEEK. The PEEK membrane material utilized is a high temperature engineered plastic that is virtually non-destructible under the operating conditions encountered in typical flue gas and other industrial applications. It can withstand contact with essentially all common treating solvents. This is a major distinguishing feature and major benefit of our technology. The broad benefits of the membrane contactors are well recognized and numerous attempts have been made to deploy them in combination with amine based solvents for CO_2 capture. However, all tested hollow fiber contactors were destroyed or their performance otherwise degraded in contact with the absorption solvent used for CO_2 capture due to the aggressive nature of such solvents. This led to development of specially formulated, less aggressive solvent systems that can be placed into contact with available membranes. However, to date, these specialized systems resulted in inferior CO_2 capture performance.

The Membrane Modules

PGC constructs hollow fiber membrane modules utilizing computer controlled helical winding process. This methodology enables construction of contactor modules with exceptionally uniform and controlled packing density. The structured packing insures optimal mass transfer without channeling or bypassing, low pressure drop and minimizes concentration polarization in liquid boundaries.

Figure 2. Structured packing generated by computer controlled helical winding

Membrane contactor module represents enabling component of this technology. The module consists of the membrane cartridge, o-rings and an external housing. The critical components of membrane module are the hollow fiber membrane and the epoxy tube-sheets. All materials of cartridge construction have been initially tested at high temperature in aggressive solvent environment. To enable efficient counter-current flow, the packing density in the cartridge must be uniform and flow bypassing and entrance and exit effects must be minimized.

Mass Transfer Coefficients

The porous membrane in the membrane contactor typically operates in a non-wetting mode with the liquid side pressure below the breakthrough pressure. Under such conditions given the hydrophobic nature of the membrane and the small pore size, the pores are considered to be completely filled with gas. The gas/liquid interface is thus located at the liquid side pore openings. Depending on the pore size, gas phase mass transfer occurs by molecular diffusion with or without contributions from Knudsen diffusion. The overall mass transfer in the contactor system is controlled by the diffusion rate in the liquid layer.

The mass transfer rate through a membrane contactor is known to be controlled by the liquid side mass transfer due to the 10,000 fold lower CO_2 diffusivity in the liquid phase. Previous data showed that the mass transfer coefficient in the contactor increased with increasing temperature.[2] The mass transfer coefficient is further a strong function of CO_2 partial pressure.[2] In particular, the CO_2 mass transfer coefficient is higher at lower CO_2 partial pressures. It is, therefore, beneficial to operate the membrane contactor at higher temperatures and lower CO_2 partial pressures, conditions that match those of the flue gas. The PEEK membrane can operate at temperatures up to 200°C continuously. This temperature is well in excess of the typical absorption process condition. The high temperature operating capability is of a critical importance in the high temperature regeneration process. To enable CO_2 collection at an elevated pressure in the de-sorption step that will lower the sequestration compression cost, a much higher temperature of regeneration is required.

Initial experiments were carried out to measure mass transfer coefficients as a function of temperature, solvent composition and feed gas flow rates using PEEK hollow fiber contactors. The test module was constructed by computer controlled helical winding and featured structured fiber packing. Representative size modules with 10 ft^2 of membrane area were employed. Synthetic flue gas of a typical flue gas CO_2 concentration and MEA solvent (28%) were utilized. Results are summarized in Table 1.

Table 1. Measured mass transfer coefficients of PEEK hollow fiber contactor

Feed Gas Flow Rate, SCFH	Solvent Rate, L/min	Solvent T, F	Inlet CO_2 Partial P (Kpa)	Kg, mol/(m^2.s.kPa)
25.08	0.74	85	4.11	2.96E-04
25.13	0.755	82.2	3.23	2.93E-04
25.15	0.4	84	3.30	2.74E-04
25.12	0.397	84.4	4.31	2.72E-04
25.16	0.422	124.3	4.71	2.83E-04
25.12	0.41	124.5	3.73	2.83E-04
25.16	0.73	121.7	3.65	2.77E-04
25.06	0.75	115.3	4.51	2.72E-04
34.97	0.585	115.5	5.87	2.62E-04
34.98	0.57	115.5	4.88	2.63E-04
25.16	1.07	122	5.61	4.15E-04
35.1	1.1	120.3	5.68	6.98E-04

These results show that high mass transfer coefficients, Kg, can be attained. The mass transfer coefficients attained are comparable to the best results reported previously.[2,3] Increase in the mass transfer coefficient is expected upon further optimization of hollow fiber morphology and module structured packing design.

Process Economics

The transformational nature of the technology is rooted in its ability to reduce energy and capital costs across a broad spectrum of separation applications. The hybrid membrane/absorption technology combines advantageous aspects of both technologies and cuts across entire spectrum of applications currently served by absorption separations. The technology is a generic technology that enables high mass transfer between gas and liquid streams and can be applied to numerous industrial applications. The technology is compatible with essentially entire spectrum of absorbents from commercial physical and chemical absorbents to experimental advanced absorbents like ionic liquids. In the later case our technology becomes an enabling technology since conventional columns do not function well in contact with viscous ionic liquids. The hybrid membrane/absorption technology carries the promise of dramatic capital and operating cost reduction for CCS.

The process stability and reliability has been one of the key barriers of solvent absorption processes for large-scale CO_2 separation and capture. With the combination of membrane and solvent absorption, the phase boundaries are established by the physical boundaries of the membrane and solvent, making the combined process inherently more stable and reliable than the conventional solvent process.

To estimate the impact that the membrane contactor could have on the carbon capture, GTI used the Integrated Environmental Control Model[4] and Illinois #6 coal for a net 550 MW pulverized coal power plant with constant 2007 dollars as shown in Table 2. This data show that with the conventional column-based MEA process for post-combustion CO_2 separation and capture, the levelized cost of electricity (LCOE) increased by 85.6%; whereas if the membrane contactor based technology is used, the increase is only 42.2% based on membrane module cost of $\$100/m^2$. The LCOE analysis used the same parameters and assumptions used in reference 5.

Table 2. Comparison of new pulverized coal plants

	Pulverized Coal Boiler		
	PC Subcritical		PC Subcritical with Membrane Contactor for CO_2 Removal
	Case 9[3]	Case 10[3]	
CO_2 Capture	No	Yes	Yes
Gross Power Output (kWe)	583,315	679,923	625,000
Auxiliary Power Requirement (kWe)	32,870	130,310	75,000
Net Power Output (kWe)	550,445	549,613	550,000
LCOE (mills/kWh)[1]	64.0	118.8	91.0[2]

[1] Based on an 85% capacity factor

[2] Value is based on membrane cost of $\$100/m^2$

[3] Case 9 and 10 are from reference 5

The sensitivity of LCOE to membrane module price is shown in Figure 3. As expected, the LCOE decrease with the decrease in membrane cost.

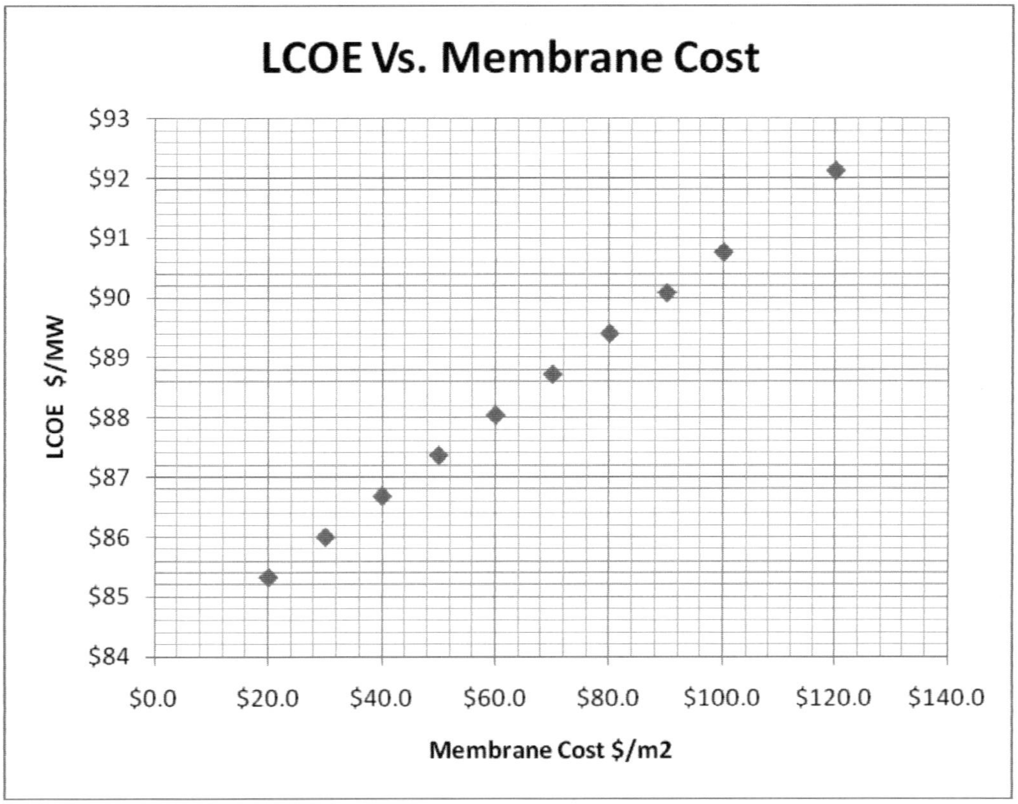

Figure 3. Sensitivity of LCOE to Membrane Price

To reach the goal of 90% CO_2 capture at no more than 35% increase in cost of electricity, the LCOE needs to be less than $87 ($/MW); this represents a membrane cost of less than $\$40/m^2$ as shown in Figure 4. Alternatively, the target can be met by increasing mass transfer coefficient by 150%.

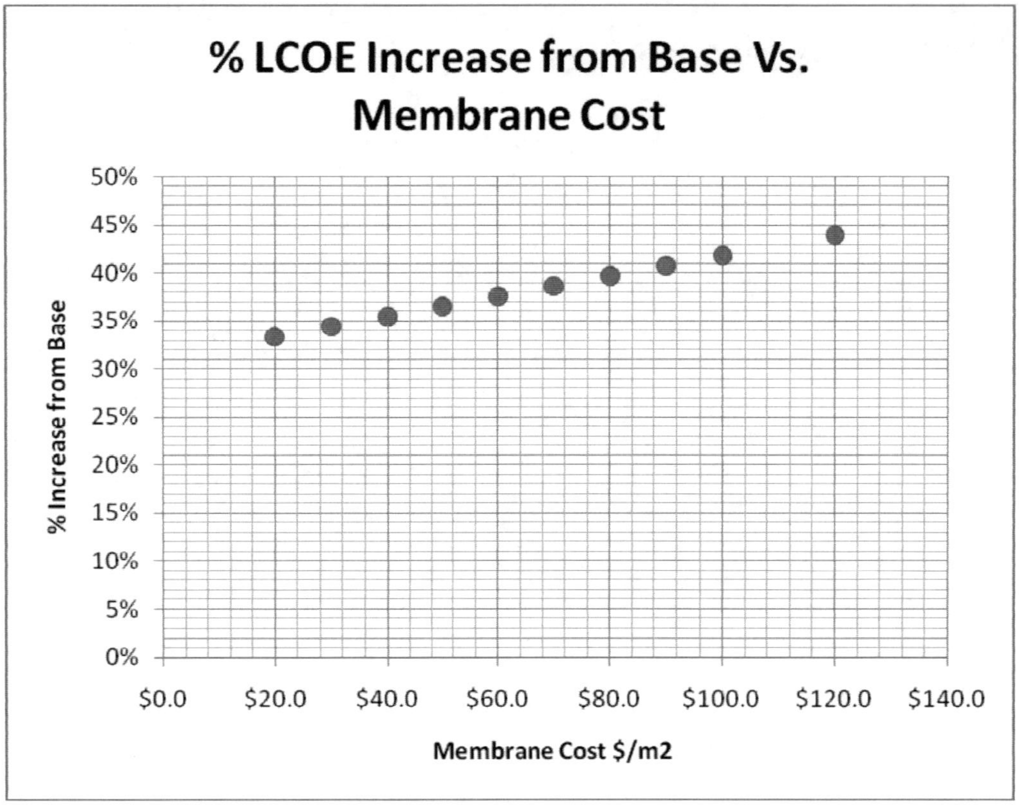

Figure 4. Percent increase in LCOE from PC base Plant(without CO_2 capture) as a function of membrane cost.

Conclusions

Based on the mass transfer measurements and the engineering estimate, the hybrid membrane/absorption process is a promising technology for carbon capture applications with numerous advantages over the current tray or packed column based technology. The new technology carries the promise of attaining the goal set by DOE of the 90% carbon capture at no more than 35% increase in.the cost of electricity generation.

Acknowledgement

GTI gratefully acknowledges the financial support provided by the Illinois Department of Commerce and Economic Opportunity through the Office of Coal Development and the Illinois Clean Coal Institute.

References

1. David C. Thomas, "Carbon Dioxide Capture for Storage in Deep Geologic Formation – Results from the CO_2 Capture Project," Elsevier, 2005, 133-155.
2. Karl A. Hoff, "Modeling and Experimental Study of Carbon Dioxide Absorption in a Membrane Contactor," PhD Thesis, Norwegian University of Science and Technology, March 2003.
3. Shuiping Yan, et. al. "Comparative Analysis of CO_2 separation from flue gas by membrane gas absorption technology and chemical absorption technology in China," Energy Conv. And Manag. **49** (2008) 3188-3197.
4. http://www.iecm-online.com/
5. "Cost and Performance Baseline for Fossil Energy Plants," DOE/NETL-2007/1281.

(142b) Post-Combustion CO2 Capture Using An Efficient Membrane Process

Xiaotong Wei, Membrane Technology and Research, Inc., Menlo Park, CA

Timothy C. Merkel, Membrane Technology and Research, Inc., Menlo Park, CA

Haiqing Lin, Membrane Technology and Research, Inc., Menlo Park, CA

Bilgen Firat, Membrane Technology and Research, Inc., Menlo Park, CA

Jenny He, Membrane Technology and Research, Inc., Menlo Park, CA

Karl D. Amo, Membrane Technology and Research, Inc., Menlo Park, CA

Richard Baker, Membrane Technology and Research, Inc., Menlo Park, CA

Production of power from coal inevitably generates carbon dioxide (CO2) as a by-product. To mitigate global climate change, this CO2 must be captured and sequestered. With current technologies, separating CO2 from flue gas accounts for about 70% of the total cost for CO2 capture and sequestration. Among a number of emerging technologies that are being evaluated for CO2 capture, membrane technology is attractive because of its inherent advantages such as high energy efficiency, a small footprint, environmentally friendly operation (no chemicals), mechanical simplicity, and proven reliability.

Working with NETL, MTR has developed new membranes and process designs to recover CO2 from power plant flue gas. The membranes have CO2 fluxes ten times higher than standard commercial membranes, combined with useful CO2/N2 selectivities. Based on our best current process design, these membranes offer the potential to capture 90% of the CO2 in the flue gas using about 12% of the power produced by the plant, at a cost of $20-$30/ton of CO2 captured. Successful commercialization will provide a technology that can be used in the near-term to reliably retrofit existing coal-fired power plants.

Currently, MTR is working with Arizona Public Service (APS) to conduct a field test of this membrane CO2 capture process at APS' Cholla, Arizona coal-fired power plant. The demonstration will capture 1 ton CO2/day from real flue gas starting at the beginning of 2010. Lab and field test results to date and future plans will be discussed in this presentation.

See more of this Session: Post-Combustion Techniques for Capture/Separation of Carbon Dioxide II
See more of this Group/Topical: Topical 2: Advanced Fossil Energy Utilization

(142c) Characterization of Coal Flame Impacts Under Oxy-Combustion

Bradley Adams, Reaction Engineering International, Salt Lake City, UT

Andrew Fry, Reaction Engineering International, Salt Lake City, UT

The future use of coal in the United States depends on technologies being made available to capture and store CO2 emissions from power plants. A key candidate CO2 capture technology is oxy-firing of coal. Using oxygen instead of air to combust the coal produces a flue gas stream that it composed primarily of CO2 and water, making subsequent CO2 capture relatively easy. However, application of oxy-firing to existing power plants presents unquantified challenges as the characteristics of oxy-firing compared to air-firing have not been fully determined.

This presentation will draw results from an on-going DOE-sponsored program on oxy-combustion impacts in existing coal-fired boilers. The primary objective of the program is to develop tools to characterize and predict impacts of CO2 flue gas recycle and burner feed design on flame characteristics (burnout, NOx, SOx and fine particle emissions, heat transfer), slagging, fouling and corrosion, inherent in the retrofit of existing coal-fired boilers for oxy-coal combustion. Key elements to be characterized include:

• Appropriate coal feed, oxygen feed, and flue gas recycle (FGR) design for oxy-fired burners or systems;

• Impacts of oxy-firing system design on boiler flame characteristics (burnout, emissions, heat transfer), fouling, slagging and steam tube corrosion relative to the expected behavior in existing air-fired systems.

The proposed research will accomplish this using:

• New experimental data from Sandia National Laboratories' 0.1 kW Entrained Flow Reactor, University of Utah's 100 kW OxyFuel Combustor (OFC), and University of Utah's 1.5 MW L1500 coal-fired furnace;

• Validated mechanisms describing oxy-firing processes;

• Simulation of oxy-combustion in a pilot-scale furnace and a full-scale utility boiler retrofit using oxy-firing design principles.

Information presented here will include background on oxy-combustion flame mechanisms and results from numerical simulations and experiments conducted in a pilot-scale coal-fired furnace that assess flame behavior under air-fired and oxygen-fired conditions.

Three major issues confronted when considering retrofit of existing boilers for oxy-firing are:

• Selection of the flue gas recycle ratio R, here defined as the ratio of the recycle volume to that of the oxygen. Depending on the retrofit objectives, this could be selected to match heat transfer to furnace surfaces rather than the adiabatic flame temperature, of the air-fired furnace.

• Selection of the location from which to recycle the flue gas, that is, selection of the composition of the recycled flue gas based on the location in the flue gas train from which it is recycled back to the furnace.

• Location of oxygen addition and flue gas recycle streams to the boiler.

The selection of R is determined mainly by interest in maintaining the heat flux rate and distribution as close as possible to those achieved with air combustion in order to minimize any design changes to the boiler. Its value depends on the coal, recycle temperature, whether water is condensed or not. It is generally in the range of 2.7 to 3.3 [Kimura et al, 1995; Payne et al., 1989; Wall, 2007; Croiset et al., 2000]. Differences in air and oxy-firing are in large part determined by the composition of the recycled flue gas, which does not depend on R, but does depend on where the recycle stream comes from. Issues related to retrofit of existing boilers to oxy-coal are driven by these differences, which are significant.

Oxy-coal firing provides additional degrees of freedom in the composition of the transport and secondary oxidant streams, namely PO2 in each stream. In order to obtain maximum advantage of the recycle stream on moderating the combustion temperature the recycle stream should be mixed with oxygen as early as possible. Injection early in the furnace is needed to maximize the heat transfer. It is, however, advantage to use some of the CO2 for coal entrainment into the burner and some of the oxygen for flame stabilization. Testing and simulation can quantify the trade-offs for different injection strategies.

The L1500 pilot-scale combustor at the University of Utah is being used to perform experiments detailing firing system design, flame characteristics, together with selected fouling, slagging and corrosion tests. This 5 MBtu/hr (1.5 MW), pulverized coal-fired furnace produces a combustion environment relevant to full-scale utility boilers. In contrast to lab-scale burners, the turbulent length scales in the L1500 are similar to those associated with commercial burners. The furnace is refractory lined, providing realistic boiler temperatures and radiative environment, necessary for heat transfer and corrosion, despite its 10-fold greater surface-to-volume ratio relative to commercial water wall boilers. The furnace has been retrofitted for oxy-combustion with a flue gas recycle system. This modified facility provides maximum flexibility in where and how the FGR, oxygen and coal may be introduced, including staged conditions for reburning of the recycled NOx.

An oxy-research burner has been designed for use in the L1500 pilot-scale furnace. The initial design of the oxy-research burner was based on a patented commercial-scale burner design. This burner design was reduced to a 1.03 MW (3.5 MBtu/hr) firing rate by using a constant velocity scaling technique. The initial design was evaluated and compared to operation of the existing L1500 low-NOx burner through CFD modeling. The modeling results suggested that the constant velocity scaling method did not produce the desired flame behavior for the burner. To correct this behavior, primary gas velocity was reduced, the quarl was lengthened and the burner face setback relative to the quarl was lengthened. These modifications produced the desired flame stabilization location.

The results of this modeling effort suggest that the dominating parameter governing flame shape and stability is not solely velocity. The flame behavior is closely related to particle heating rate, i.e., in order to maintain similar burner behavior when scaling a burner, the particle temperature profile should be maintained. The particle temperature profile is influenced by many factors including: gas velocity, radiation intensity, particle size and hot flue gas recirculation characteristics inside of the quarl. CFD simulations provided an excellent approach to evaluate the combined impacts of all of these factors.

This work will review the burner design and present results from testing of the oxy-research burner in the L1500 furnace under air-fired and oxygen-fired conditions. Results will focus on flame properties (shape, temperature, composition) and heat flux for different recycle gas compositions and oxygen injection strategies.

References

Croiset, E, Thambimuthu KV, Palmer A. Coal Combustion in O2/CO2 Mixtures Compared with Air. The Canadian Journal of Chemical Engineering 2000; 78: 402 - 7.

Kimura, K, Omata K, Kiga T, Takano S, Shikisima S. Characteristics of pulverized coal combustion in O2/CO2 mixtures for CO2 recovery. Energy Conversion and Management 1995; 36: 805-8.

Payne, R, Chen SL, Wolsky AM, Richter WF. CO2 recovery via coal combustion in mixtures of oxygen and recycled flue gas. Combustion Science and Technology 1989; 67: 1 - 16.

Wall, T. F., Combustion Processes for Carbon Capture, Proceedings Combust. Institute, 31 (2007) 31-47.

See more of this Session: Post-Combustion Techniques for Capture/Separation of Carbon Dioxide II
See more of this Group/Topical: Topical 2: Advanced Fossil Energy Utilization

(142d) Equilibrium and Kinetic Studies Carbon Dioxide Capturing On Ca-Based Sorbents

Jorge Pena Lopez, Chemical Engineering Department, University of California at Los Angeles, Los Angeles, CA

Panagiotis (Peter) Smirniotis, Chemical & Materials Engineering Department, University of Cincinnati, Cincinnati, OH

Vasilios I. Manousiouthakis, Chemical Engineering Department, University of California at Los Angeles, Los Angeles, CA

The increase of atmospheric carbon dioxide concentrations in recent years is well documented. Its potential link to an earth wide greenhouse-like effect has led recently to tighter regulations on carbon dioxide emissions from various industrial sources. One of the most common emitters is the fossil fuel based energy production industry which generates high temperature streams containing carbon dioxide at the exit of fossil fuel combustion processes.

We consider the removal of carbon dioxide from high temperature water-air mixtures using calcium-based sorbents. It has been experimentally shown in our prior work that CaO/Zr nanostructured sorbents possess a consistently high capacity to capture carbon dioxide over several high temperature carbonation/ decarbonation cycles. In this work, we provide theoretical support for these experimental results by carrying out both equilibrium and kinetic studies. First, we carry out equilibrium calculations by solving to global optimality a Gibbs free energy minimization problem for the multiphase system: N_2, O_2, CO_2, H_2O, CaO, $CaCO_3$, $Ca(OH)_2$. Then, we formulate a simple semibatch reactor model for a thermogravimetric analysis (TGA) apparatus in which the carbonation process is carried out. The predictions of the resulting ordinary differential equation based kinetic model for the CaO adsorption process are compared to TGA obtained experimental data.

See more of this Session: Post-Combustion Techniques for Capture/Separation of Carbon Dioxide II
See more of this Group/Topical: Topical 2: Advanced Fossil Energy Utilization

(142e) Metal-Adeninate Porous Materials for Selective CO2 Capture

Nathaniel L. Rosi, Chemistry, University of Pittsburgh, Pittsburgh, PA

We describe a new class of porous metal-adeninate crystalline materials. We describe the syntheses of these materials and discuss strategies for optimizing the capacity of these materials as selective sorbents for carbon dioxide. In particular, we describe how we can control the coordination chemistry used to construct the materials in order to expose specific Lewis-basic functional groups to the pore cavities in order to facilitate capture of CO2. We also use these materials to systematically evaluate the impact that pore size and volume has on selective CO2 adsorption.

See more of this Session: Post-Combustion Techniques for Capture/Separation of Carbon Dioxide II
See more of this Group/Topical: Topical 2: Advanced Fossil Energy Utilization

Theoretical Screening Good Sorbents for CO_2 Separation: Application to alkali and alkaline-earth oxides and hydroxides

Yuhua Duan
National Energy Technology Laboratory, United States Department of Energy, Pittsburgh, PA 15236, and URS, South Park, PA 15129, USA

Abstract

By combining density functional theory (DFT) and lattice phonon dynamics, the thermodynamic properties of CO_2 absorption/desorption reactions with oxides (M_2O, NO) and hydroxides (MOH, $N(OH)_2$, where M=Li, Na, K; N=Be, Mg, Ca, Sr, Ba) are analyzed. The energy changes and the chemical potential of these reactions have been evaluated. We have analyzed the thermodynamics of the reactions with and without H_2O present. The predicted heats of reactions and the chemical potentials of these reactions versus CO_2 pressures and temperatures are close to the available experimental data once an empirical correction factor is included. We have compared the thermodynamic conditions for these materials with requirements for both pre- and post-combustion capture of CO_2, and found that MgO, Na_2CO_3/$NaHCO_3$, and K_2CO_3/$KHCO_3$ are promising candidates for both pre- and post-combustion capture, whereas Na_2O, and K_2O can form bicarbonates when exposed to CO_2 and H_2O at conditions consistent with post-combustion capture technology at lower operating temperature comparing with CaO. For alkali metal oxides, their CO_2 capture temperatures could be reduced when steam is presented. The results showed that the alkali hydroxides capturing CO_2 to form bicarbonates are less energy cost comparing to forming carbonates. Our results show that $Mg(OH)_2$ are good candidates for post-combustion CO_2 capture, while LiOH and NaOH can be used as CO_2 sorbents both for pre- and post-combustion technologies. Our computational approach can be used as a general scheme for screening a large number of solids for use as CO_2 sorbents.

I. INTRODUCTION

Carbon dioxide is the major product from combustion and is released into the air and causes global climate warming which we are facing today.[1, 2] In order to solve such environmental problem, there is a need to reduce CO_2 emission to atmosphere by capture and sequestration.[3, 4] Current technologies for capturing CO_2 include solvent-based systems (such as Selexol, Rectisol), and alkanolamine-based materials (such as ethanolamine). It is generally accepted that operation or regeneration of these materials is too energy intensive.[1, 2] Hence, there is critical need for new materials that can capture and release CO_2 reversibly with acceptable energy costs. Accordingly, solid sorbent materials have been proposed for capture of CO_2 through a reversible chemical transformation. Among them, re-generable solid sorbents play an important role for CO_2 capture, especially at above room temperature.[1, 5-11]

The solid sorbents containing alkali and alkaline earth metals have been reported in many literatures.[1, 11-19] Among them, their oxides and hydroxides attract more attention from researchers because of their high CO_2 absorption weight percentage and moderate working temperature.[11, 13, 20, 21] Experimental evidences also showed that mixtures of hydroxides with oxides could improve the CO_2 absorption performance.[11, 13]

Comparing with extensive experimental studies of capturing CO_2 with alkali and alkaline earth oxides and hydroxides,[11, 13, 20, 21] the systematical theoretical investigations on their CO_2

capturing properties are very limited. Based on thermodynamic data, Feng *et al*[22] analyzed 11 simple metal oxides and concluded that CaO is thermodynamically the best candidate among them for CO_2 capture in zero emission power generation systems. Based on density functional theory (DFT), Jensen *et al*[23] investigated the CO_2 adsorption on CaO and MgO. Their results showed that CO_2 adsorbs as monodenate on edge sites and bidentate on corner sites on MgO, while on CaO, CO_2 as monodenate on both edge and corner sites. Through *ab initio* evolutionary simulations and high-pressure experiments, Oganov *et al*[24] investigated the high-pressure structures of $MgCO_3$, $CaCO_3$ and CO_2 and their roles in Earth's lower mantle. Recently, combining DFT with lattice phonon dynamics, we developed a novel methodology to select good solid candidates of CO_2 solid sorbents.[15, 25] For solid sorbent Li_2O, we concluded that although pure Li_2O can absorb CO_2 efficiently, it is not a good solid sorbent for CO_2 capture because the reverse reaction, corresponding to Li_2CO_3 releasing CO_2, can only occur at very low CO_2 pressure and/or at very high temperature when Li_2CO_3 is in liquid phase.[15] This predicted results are in very good agreement with experimental measurements.

In this study, we apply our methodology into alkali and alkaline earth metal oxides and hydroxides to explore their CO_2 capture properties systematically, and compare our results with their available thermo-dynamical data.

II. THEORETICAL METHODS

For the CO_2 capture reaction by oxide or hydroxide solids

$$\text{Solid\_A} + CO_2 + [H_2O] \leftrightarrow \text{Solid\_B} + [\text{Solid\_C}] + [H_2O] \qquad \text{(a)}$$

where [...] are optional, we can ignore the PV contribution and assume that the difference between the chemical potential ($\Delta\mu^o$) of solid phases of A, B(and C) can be approximated by the difference in their electronic energy (ΔE^o) and the vibrational free energy of the phonons.[26] For CO_2, we can treat it as ideal gas. Therefore, the variation of the chemical potential ($\Delta\mu$) for this reaction with temperature and pressure can be defined as[15, 25, 27-29]

$$\Delta\mu(T,P) = \Delta\mu^0(T) - RT\ln\frac{P_{CO_2}}{P_{H_2O}^{\pm1}} \qquad (1)$$

with

$$\Delta\mu^0(T) \approx \Delta E^{DFT} + \Delta E_{ZP} + \Delta F^{PH}(T) - G_{CO_2}(T) \pm G_{H_2O}(T) \qquad (2)$$

where ΔE_{ZP} is the zero point energy difference between the reactants and products and can be obtained directly from phonon calculations using following Eq.(6). If reaction (a) doesn't involve H_2O, the pressure of H_2O in Eq.(1) (P_{H_2O}) should set to 1. If the H_2O is a product in reaction (a), plus "+" sign should be taken in Eqs. (1) and (2), and if it is as a reactant, the minus "-" should be used. The free energy of CO_2 (G_{CO_2}) and H_2O (G_{H_2O}) can be obtained by standard statistical mechanics[30]

$$G_{CO_2}(T) \approx \frac{7}{2}RT + \sum_{i=1}^{4}\frac{N_a h\nu_i}{e^{h\nu_i/kT}-1} - TS_{CO_2}(T) \qquad (3)$$

$$G_{H_2O}(T) \approx 4RT + \sum_{i=1}^{3}\frac{N_a h\nu_i}{e^{h\nu_i/kT}-1} - TS_{H_2O}(T) \qquad (4)$$

where N_a is the Avogadro constant. The entropy of CO_2($S_{CO_2}(T)$) and H_2O($S_{H_2O}(T)$) can be accurately evaluated using the Shomate equation.[31] The vibrational frequencies (ν_i) of CO_2 molecule are $673cm^{-1}(\pi_u)$, $1354cm^{-1}(\sigma_g^+)$, and $2397cm^{-1}(\sigma_u^+)$,[32] and the vibrational frequencies (ν_i) of H_2O molecule are $3657.05cm^{-1}(\nu_1)$, $1594.75cm^{-1}(\nu_2)$, and $3755.93cm^{-1}(\nu_3)$,[33] from which the obtained zero point energies for CO_2 and H_2O molecules are 0.3160eV and 0.5584eV respectively.

In Eq.(2), ΔE^{DFT} is the total energy change of the reactants and products calculated by density functional theory(DFT). In this work, the Vienna *Ab-initio* Simulation Package (VASP)[34, 35] was employed to calculate the electronic structures of solid oxide, hydroxide and carbonate materials. In this study, all calculations have been done using the PAW pseudo-potentials and the PW91 exchange-correlation functional. This computational level was shown to provide an accurate description of oxide systems.[28] Plane wave basis sets were used with a kinetic energy cutoff of 500eV and an augmentation charge cutoff of 605.4eV. The k-point sampling grids of $n_1 \times n_2 \times n_3$, obtained using the Monkhorst-Pack method,[36] were used for these bulk calculations, where n_1, n_2, and n_3 were determined consistent to a spacing of about 0.028Å^{-1} along the axes of the reciprocal unit cells. During calculations, all atoms in the cell were relaxed to the equilibrium configurations. The energies of CO_2 and H_2O molecules are -22.99409eV and -14.27267eV respectively as determined from calculations of an isolated molecule in a cubic box with the length of 20 Å.

Based on phonon calculations under the harmonic approximation, the phonon free energy ΔF^{PH} change between reactant and product solids can be calculated through the free energy of solids described by the Helmholtz form F_{harm} as shown in Eq.(2) and the entropy of the solids (S_{harm})

$$\Delta F^{PH}(T) = \Delta F_{harm}(T) - T\Delta S_{harm}(T) \quad (5)$$

where the F_{harm}, S_{harm} and the internal energy(E_{tot}) of the solids are defined as [26]

$$F_{harm}(T) = rk_B T\int_0^{\infty} g(\omega)\ln[2\sinh(\frac{\hbar\omega}{2k_B T})]d\omega \quad (6)$$

$$S_{harm}(T) = rk_B\int_0^{\infty} g(\omega)\{(\hbar\frac{\hbar\omega}{2k_B T})[\coth(\frac{\hbar\omega}{2k_B T}) - 1] - \ln[1 - \exp(-\frac{\hbar\omega}{k_B T}]\}d\omega \quad (7)$$

$$E_{tot}(T) = \frac{1}{2}r\int_0^{\infty} g(\omega)(\hbar\omega)\coth(\frac{\hbar\omega}{2k_B T})d\omega \quad (8)$$

where r is the number of degree of freedom in the primitive unit cell. It can be seen that the zero-point-energy (E_{ZP}) can be obtained from Eq.(5) by taking $T \rightarrow 0$.

$$E_{ZP} = \lim_{T\rightarrow 0}(E_{tot}(T)) \quad (9)$$

where ω is the phonon dispersion frequency, and g(ω) is the phonon density of state. In this study, we employed the PHONON software package[37] in which the direct method is applied following the formalism derived by Parlinski *et al.*[38] In phonon calculations, a 2×2×2 supercell was created from the optimized unit cell structure that was calculated based on DFT. Structures with a displacement of 0.03Å of non-equivalent atoms were generated from the 2×2×2 supercell and DFT calculations were further performed to obtain the force on each atom due to the displacements. These forces were input into the PHONON package[37] to fit the force matrix and further carry out the phonon dispersions and densities, which can be inputted into Eqs.(6)-(8) to calculate the thermodynamic properties.

In the band structure calculations and phonon dispersion calculations, the symbols and coordinates of the high symmetry points in the first Brillouin zone of the crystals are taken from Bradley and Cracknell's definitions.[39]

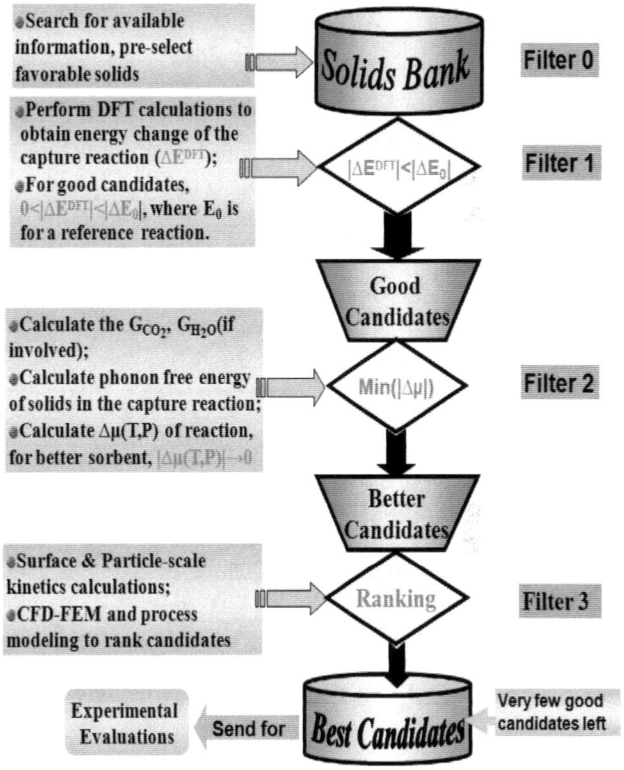

Fig.1 Screening methodology with several filters.

The enthalpy change for the reaction, $\Delta H^{cal}(T)$, can be derived from above equations as

$$\Delta H^{cal}(T) = \Delta\mu^0(T) + T(n_1 S_{CO_2} + n_2 S_{H_2O} + \Delta S_{harm}) \quad (10)$$

We found that the calculated enthalpy and chemical potential are always higher than the experimental or tabulated data by 20~30 kJ/mol. This is likely due to limitations of DFT. We have subtracted a constant value, ΔH_0, from Eq.(2) in order to empirically correct for this apparent bias. We have found that using $\Delta H_0 = 26.2$ kJ/mol, which is the sublimation enthalpy of CO_2, gives reasonable performance for all systems studied here. We initially thought that ΔH_0 was due to a mismatch in the thermodynamic reference states of the Shomate equation and the DFT total energy calculations. However, this is not the case. Nevertheless, the heat of sublimation of CO_2 does provide a practical estimate of the value of the empirical correction factor.

Figure 1 is illustrated our screening methodology. Generally, for a given data bank, we can manipulate our methodology as several filters:

(0) for each candidate, search its available thermodynamic data to make initial judgment: if it is a promised one or there is no available thermodynamic data, continue to next filter. Otherwise, pick another compound;

(i) Performing DFT calculations for both reactants and products of CO_2 capture reaction and obtain the energy difference (ΔE^{DFT}) of the reaction. If 0.5 eV > ΔE^{DFT} > -2 eV, keep this compound and continue to add on free energy of CO_2 and H_2O (G_{CO2}, G_{H2O} as shown in Eqs.(3)-(4)) to obtain the $\Delta E_1(T)$ for the reaction. Otherwise, go back to filter (i);

(ii) Performing phonon calculations for reactants and products to obtain their zero point energy and phonon free energy changes for the CO_2 capture reactions. According to Eqs.(1) and (2) to obtain the chemical potential ($\Delta\mu(T,P)$) of the reaction. For good candidates, their $\Delta\mu(T,P) \rightarrow 0$ at desired working conditions. After this filter, only a short list of compounds left from the given data bank;

Filter (iii) Additional modeling could be performed to rank the remaining short list of better candidates. One is the kinetics of the capture reactions, which could be done by transport and diffusion calculations as well as experimental measurements. Another necessary and doable modeling is the real behavior of solid in the reactor, which can be done by computational fluid dynamics (CFD) based on finite element method (FEM) approach. These simulations are under progressing. Application of these screening filters will ensure that only the most promising candidates will be identified for experimental testing.

This screening methodology provides a path for rapidly evaluating materials for which experimental thermodynamic data are unavailable. One area where this approach could be used to great advantage is in evaluating mixtures and doped materials, where thermodynamic data are not generally available.

III. RESULTS AND DISCUSSIONS

3.1 DFT and Phonon calculated results

The optimized lattice constants and total energies for the 27 alkali and alkaline earth metal oxides, hydroxides, bicarbonate and carbonates considered in this work are presented in Table I, along with experimental structural data. Since there is no crystal structural data for $LiHCO_3$ and $BeCO_3$ available, we take the structures of $NaHCO_3$ and $SrCO_3$ as their initial structures to obtain their optimized structures. From our calculations, the $BeCO_3$ in the structure of $MgCO_3$ resulted in about 0.5eV higher in energy comparing to that in the structure of $SrCO_3$. It should be pointed out that for each compound it may have more than one known phase been investigated, in Table I only the most stable structure with lowest energy is listed. As expected, the agreement between optimized constants and experimental data generally is very good as shown in Table I. The calculated total energy (E^{DFT}) for each compound will be used to estimate the reaction enthalpy of CO_2 capture reaction as described in the next section.

Based on the optimized structures as shown in Table I, a supercell with at least 2x2x2 size will be built to perform phonon calculations. From their phonon dispersion (ω) and phonon density of states ($g(\omega)$), according to Eqs.(6)~(9), the total energy (E_{tot}), phonon free energy (E_{harm}), entropy (S_{harm}) and the zero point energy (E_{zp}) of each compound can be evaluated out with the temperature dependent. The calculated zero-point energy of each compound is given in Table I.

From Table I, one can see that our calculated entropies of these oxides and carbonates are very close to the experimentally measured values. Our calculated electronic properties and phonon dispersion relations (not shown) are also in good agreement with data from the literature for oxides and carbonates.[27,49-52] Therefore, our results can be reliably used to estimate the thermodynamic properties of the reactions of these oxides capturing CO_2 as presented in the following sections.

Table I. Comparison of the experimental and the DFT calculated structural parameters for the compounds in the reactions studied, with all distances in angstroms and angles in degrees. Although calculations are performed on several phases of each compound, only the stable phase, determined by DFT total energy, is listed in the Table. The zero-point energy and entropy calculated from phonon as well as the experimental data are also listed in the Table.

Compound	Space group & Ref.	Structural parameters		Calculated Energy (eV/f.u.)		Entropy (J/mol.K)	
		experimental	calculated	E^{DFT}	E_{ZP}	Phonon (T=300K)	Exp.[a] (T=298.15K)
Li_2O [b]	$Fm\bar{3}m$(No.225)[41]	a=4.62	a=4.631572	-14.43538	0.23628	38.4681	37.610
LiOH	P4/nmmS (No.129)[42]	a=3.549, c=4.334	a=3.55985, c=4.41752	-15.01799	0.41897	37.0148	42.810
Li_2CO_3 [b]	C2/c(No.15)[43]	a=8.35884, b=4.97375, c=6.19377, β=114.789	a=8.405320, b=5.027886, c=6.294445, β=114.1615	-39.55190	0.59912	93.6157	90.169
$LiHCO_3$	P121/c1(No.14)	No experimental data available. Taken same as $NaHCO_3$	a=3.480953. b=8.987688, c=8.099703, β=115.2319	-38.76293	0.80767	60.3772	
Na_2O	$Fm\bar{3}m$(No.225)[44]	a=5.55	a=5.58517	-11.34789	0.12801	76.8426	75.040
NaOH	Cmcm(No.63)[45]	a=3.4013, b=3.3984, c=11.378	a=3.431009, b=3.414088, c=11.39869	-13.72285	0.36924	60.5562	64.430
Na_2CO_3	C12/m1(No.12)[46]	a=9.01029, b=5.23116, c=6.34548, β=96.062	a=8.951803, b=5.335072, c=6.138608, β=102.205	-37.29272	0.49152	122.5343	138.783
$NaHCO_3$	P121/c1(No.14)[47]	a=3.51, b=9.71, c=8.05, β=111.85	a=3.549292, b=9.795588, c=8.125747, β=111.85	-38.01484	0.77735	*74.6295*	101.700
K_2O	$Fm\bar{3}m$(No.225)[48]	a=6.436	a=6.523623	-10.14413	0.08048	112.137	94.100
KOH	P121/m1(No.11)[45]	a=3.957, b=3.995, c=5.742, β=103.93	a=4.010599, b=4.026439, c=5.930733, β=103.4705	-13.38330	0.33355	62.5907	81.250
K_2CO_3	P121/c1(No.14)[43]	a=5.63961, b=9.8312, c=6.83407, β=98.703	a=5.760548, b=9.904778, c=7.181104, β=97.2946	-36.90480	0.45733	160.1208	155.500
$KHCO_3$	P121/a1(No.14)[49]	a=15.1725, b=5.6283, c=3.711, β=104.631	a=15.46219, b=5.733443, c=3.79398, β=105.7123	-37.88604	0.77091	*81.6815*	115.499
BeO	P63mc (No.186)[50]	a=2.6979, c=4.3772, β=120	a=2.710593, c=4.40006, β=120	-14.36657	0.22559	14.7129	13.770
$Be(OH)_2$	P212121(No.19)[51]	a=4.5301, b=4.621, c=7.048	a=4.50825, b=4.650191, c=7.011854	-29.29364	0.90942	39.4651	53.555
$BeCO_3$	Pnma (No.62)	No experimental data available. Taken same as $SrCO_3$	a=6.762928, b=4.851909, c=7.012976	-36.73083	0.57829	50.8681	52.007
MgO	$Fm\bar{3}m$(No.225[52])	a=4.2198	a=4.248879	-12.00759	0.12611	33.2943	26.950

$Mg(OH)_2$	P$\bar{3}$m1(No.164)[53]	a=3.1425, c=4.7665, β=120	a=3.168163, c=4.749008, β=120	-27.14938	0.77531	65.8659	63.137
$MgCO_3$	R$\bar{3}$cH(No.167)[54]	a=4.6338, c=15.0192, β=120	a=4.686491, c=15.13795, β=120	-35.96046	0.53235	69.3531	65.090
CaO	Fm$\bar{3}$m(No.225)[55]	a=4.8152	a=4.819027	-12.98752	0.11088	39.3738	38.10
$Ca(OH)_2$	P$\bar{3}$m1(No.164)[55]	a=3.582, c=4.904, β=120	a=3.613773, c=4.96788, β=120	-28.33593	0.71140	82.8147	83.400
$CaCO_3$	R$\bar{3}$cH(No.167)[56]	a=4.991, c=17.068, β=120	a=5.03979, c=17.12672, β=120	-37.61011	0.4841	95.9951	91.710
SrO	Fm$\bar{3}$m(No.225)[57]	a=5.1326	a=5.1935	-12.19223	0.07903	56.7416	55.580
$Sr(OH)_2$	Pnam(No.62)[58]	a=9.8889, b=6.1202, c=3.9184	a=10.00019, b=6.046296, c=3.947268	-27.72516	0.72379	93.8210	97.069
$SrCO_3$	Pnma(No.62)[59]	a=6.02, b=5.093, c=8.376	a=6.102091, b=5.156528, c=8.490871	-37.33526	0.46027	104.4270	97.200
BaO	Fm$\bar{3}$m(No.225)[60]	a=5.539	a=5.660430	-11.95582	0.06492	74.1841	72.00
$Ba(OH)_2$	Pnma(No.62)[61]	a=11.03, b=16.56, c=7.11	a=11.15059, b=16.7613, c=7.102235	-27.63975			107.280
$BaCO_3$	Pmcn(No.62)[62]	a=5.3293, b=8.9275, c=6.6076	a=5.370614, b=9.007695, c=6.552103	-37.52245	0.44584	116.7325	112.100
CO_2	P1		a=20 (fixed)	-22.99409	0.31598		213.388
H_2O	P1		a=20 (fixed)	-14.27267	0.55841		188.832

[a] Taken from HSC Chemistry Package. [63]
[b] Taken from Ref. [15]

3.2 Thermodynamic properties of CO_2 capture reactions

As an example, the heats of reactions as a function of temperature for eight oxides are plotted in Fig. 2. Data from the HSC Chemistry Package[48] are also plotted in this figure for comparison. The discontinuities in the experimental curves indicate phase transitions.

From Fig.2, it can be seen that the calculated ΔH values for Li_2O, K_2O, CaO and BaO are very close to the experimental data, with differences of less than ±5kJ/mol.[18,55] Na_2O and SrO sorbents have discrepancies of about 12 kJ/mol. Among these 8 oxides, the BeO has the largest difference of 15 kJ/mol. This is probably due to the fact that we do not have experimental data for the crystal structure of $BeCO_3$. We used the $SrCO_3$–based structure for $BeCO_3$, as shown in Table I. Differences between this structure and the real experimental structure may be the reason for the larger errors.

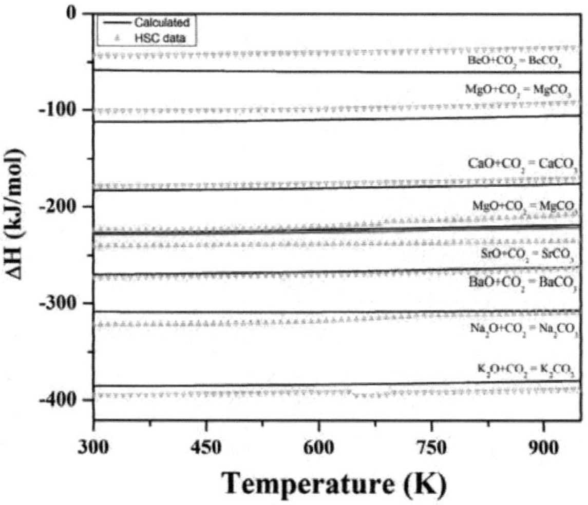

Fig. 2. The calculated heats of reaction as a function of temperature for eight metal oxides. The solid lines were computed from DFT including phonon contributions. The dashed lines were computed from the HSC package. The discontinuities in the HSC curves indicate phase transitions.

As an example, Fig. 3 shows the relationship of calculated chemical potential versus the CO_2

pressure and temperature change. From it, one can see that Overall, by combining DFT with phonon dynamics calculations, we obtained very good predictions comparing with the experimental data. From Fig.3, one can see that for the same reason BeO sorbent has a larger discrepancy at lower temperature range. Our calculated results show that for all of the cases, the DFT-only approach predicted higher CO_2 pressures at each temperature(didn't show in Fig.4) than the DFT+phonon approach.

sorbents are not good candidates for CO_2 capture.

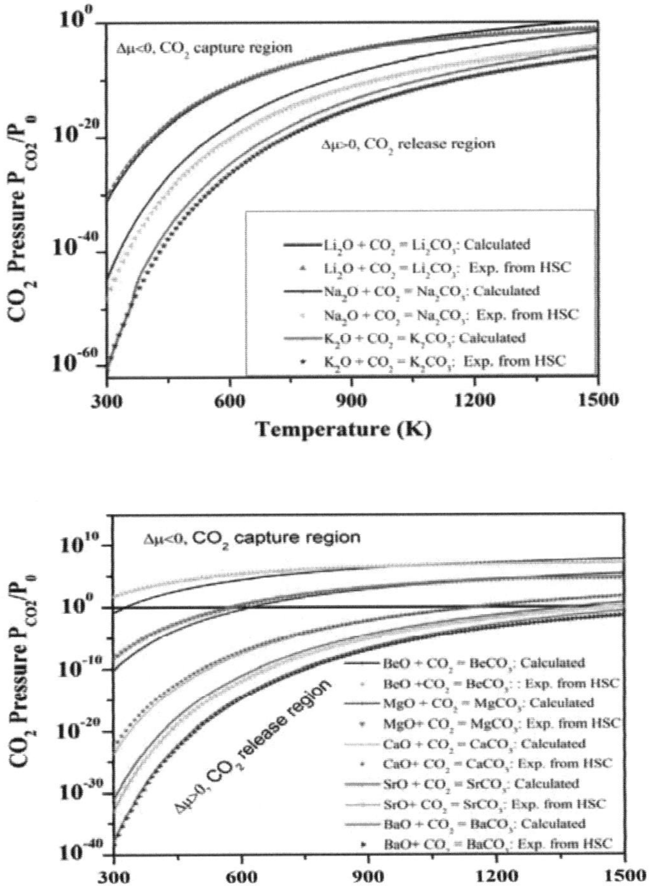

Fig.3. The calculated chemical potentials versus CO_2 pressures (in logarithm scale) and temperatures for the reactions of oxides capturing CO_2 without water involved: (a) alkali metal oxides; (b) alkaline earth metal oxides.

From Fig.3, one can see that up to T=1200K and higher, the Li_2O, Na_2O, K_2O, CaO, SrO and BaO still can absorb CO_2. The CO_2 pressure could be in a big range from very low (~10^{-30} bar) to very high (~10^{10} bar) despite of reaction kinetics, which means during the first half cycle of capturing CO_2, these oxides are very good for CO_2 absorption ($\Delta\mu<0$ region). However, as Fig.3 demonstrated, in the second half cycle of sorbent regeneration to release CO_2, their corresponding carbonates are very stable and only can dissociate ($\Delta\mu>0$ region) at very high temperature. Therefore, these strong

From Fig.3(b), one can see that BeO can absorb CO_2 just above room temperature with lowest pressure of 10^{-2} bar, which is perfect fitting the requirements of the post-combustion capture technology (see section 3.3). Comparing with CaO, MgO is very good for warm temperature CO_2 capture because its turnover temperature is around 600K, which meets the requirements of pre-combustion technology.

Similar results can be obtained for other oxides, hydroxides and carbonates. Table II summarizes our results.

From Table 2, one can see that overall our predicted chemical potentials of the CO_2 capture reactions are quite close to the experimental measurements, which means our theoretical approach can predicted right thermodynamic properties and can be used for investigating unknown systems.

3.3 Applying to pre- and post-combustion CO_2 capture Technology

Under pre-combustion conditions, after water-gas shifting, the gas stream mainly contains CO_2, H_2O and H_2. The partial CO_2 pressure is around 20~25 bar and the temperature is around 300~350ºC. To minimize the energy consumption, the ideal sorbents should work at these pressure and temperature ranges to separate CO_2 from H_2. We have estimated the turnover temperatures for pre-combustion conditions of 12 carbonate and bicarbonate reactions. We have omitted reactions involving Li because of the reported instability of $LiHCO_3$ at ambient conditions. This temperature, denoted T_1, is listed in Table III, and is the temperature above which the sorbent will release CO_2. For post-combustion conditions, the gas stream mainly contains CO_2 and N_2, the partial pressure of CO_2 is around 0.1~0.2 bar, and the temperature range is quite different, however, we prefer to the low-temperature (<423K) capture. We also give the turnover temperature for post-combustion capture in Table III, denoted as T_2.

From Table III we see that in the absence of H_2O, the turnover temperatures (T_1, T_2) for Li_2O, Na_2O, K_2O, SrO and BaO are all over 1000K. This means that these oxides are very good at absorbing CO_2, but that their corresponding carbonates are very stable and only can dissociate to release CO_2 at very high temperatures. Therefore, none of these are good sorbents for CO_2 capture under either pre- or post-combustion conditions. CaO could work at high temperature for CO_2 capture under post-combustion conditions (P_{CO2}=0.1 bar) with temperature around 1030 K, which agrees with the experimental results that under P_{CO2}=1 bar, CaO absorbs CO_2 up to 1123 K and is regenerated above 1173 K.[53-54] For these alkali and alkaline earth metal oxides, as shown in Table III, only BeO and MgO can work at lower temperatures for CO_2 capture and release. Although BeO apparently has favorable thermodynamics it is not a viable candidate for CO_2 capture because the powder or dust could cause chronic beryllium disease. Our results show that MgO could be used for both pre- and post-combustion capture technology due to its low regenerating temperature (T_2=560 K for post-combustion condition and T_1=720 K for pre-combustion conditions) which are close to experimental findings.[60]

Table II. The calculated reaction thermodynamics of CO_2 capture by alkali and alkaline-earth oxides.

Reactions	CO_2 wt%	Calculated thermodynamic properties (kJ/mol) ΔH^{cal} and ΔG^{cal} at T=300 K				HSC data[a] at T=300K	
		ΔE^{DFT}	ΔE^{ZP}	ΔH^{cal}	ΔG^{cal}	ΔH(kJ/mol)	ΔG(kJ/mol)
$Li_2O + CO_2 = Li_2CO_3$	142.52	-204.7859	4.5214	-226.731	-179.261	-224.643	-176.290
$Na_2O + CO_2 = Na_2CO_3$	71.01	-284.7066	4.5860	-308.572	-258.100	-322.153	-277.155
$K_2O + CO_2 = K_2CO_3$	46.72	-363.4242	5.8731	-385.508	-335.698	-394.785	-349.084
$BeO + CO_2 = BeCO_3$	129.35	-33.0514	3.5422	-58.514	-5.155	-43.090	1.665
$MgO + CO_2 = MgCO_3$	109.19	-92.5092	8.7089	-112.254	-58.867	-100.891	-48.206
$CaO + CO_2 = CaCO_3$	78.48	-161.7445	5.5229	-182.950	-135.732	-178.166	-130.127
$SrO + CO_2 = SrCO_3$	42.47	-207.3441	6.2967	-228.490	-178.591	-240.494	-188.850
$BaO + CO_2 = BaCO_3$	28.70	-248.2164	6.2658	-269.536	-218.095	-272.491	-220.394
$Li_2O + 2CO_2 + H_2O = 2LiHCO_3$	285.04	-273.0209	19.4379	-285.961	-142.695	-311.043[b]	-189.930[b]
$Na_2O + 2CO_2 + H_2O = 2NaHCO_3$	142.02	-426.5613	22.8017	-436.196	-290.639	-457.494	-311.281
$K_2O + 2CO_2 + H_2O = 2KHCO_3$	93.44	-517.8530	18.2571	-539.023	-382.926	-537.639	-393.799
$Li_2CO_3 + CO_2 + H_2O = 2LiHCO_3$	59.56	-68.2349	13.6847	-85.429	10.366	-86.400[b]	-13.640[b]
$Na_2CO_3 + CO_2 + H_2O = 2NaHCO_3$	41.52	-141.8547	18.2157	-153.824	-58.739	-135.341	-34.126
$K_2CO_3 + CO_2 + H_2O = 2KHCO_3$	31.84	-154.4288	20.2692	-167.922	-51.733	-142.854	-44.716
$2LiOH + CO_2 = Li_2CO_3 + H_2O(g)$	91.88	-76.6589	0.3483	-98.623	-97.134	-94.567	-88.442
$LiOH + CO_2 = LiHCO_3$	183.77	-72.4469	7.0165	-92.026	-43.384		
$2NaOH + CO_2 = Na_2CO_3 + H_2O(g)$	55.02	-108.6058	-0.4371	-134.416	-127.313	-127.510	-122.984
$NaOH + CO_2 = NaHCO_3$	110.03	-158.3597	9.28817	-140.258	-84.497	-131.426	-78.555
$2KOH + CO_2 = K_2CO_3 + H_2O(g)$	39.22	-136.6990	3.1513	-154.110	-157.036	-150.711	-141.079
$KOH + CO_2 = KHCO_3$	78.44	-145.5639	7.76835	-161.018	-114.758	-146.782	-92.897
$Be(OH)_2 + CO_2 = BeCO_3 + H_2O(g)$	102.28	120.0110	-8.65594	88.238	92.326	8.579	16.532
$Mg(OH)_2 + CO_2 = MgCO_3 + H_2O(g)$	75.46	-8.6502	-0.0512	-32.687	-26.177	-19.665	-12.762
$Ca(OH)_2 + CO_2 = CaCO_3 + H_2O(g)$	59.40	-57.9497	1.4598	-80.212	-76.610	-69.035	-64.033
$Sr(OH)_2 + CO_2 + SrCO_3 + H_2O(g)$	36.18	-85.7463	-2.0349	-111.059	-106.685	-105.424	-97.990
$Ba(OH)_2 + CO_2 = BaCO_3 + H_2O(g)$	25.68	-112.0479	-1.24548	-137.121	-131.611	-121.733	-115.686

[a] Calculated by the HSC Chemistry package[48]; [b] from ref. [15]

When H_2O is present, the alkali-metal oxides could capture CO_2 by forming bicarbonates. From Table III one can see that these oxides can be used for post-combustion capture. Comparing with CaO, these oxide sorbents have much higher CO_2 capture weight percentages and can work at lower temperature. Therefore, they are better candidates for CO_2 sorbents.

From Table III, one can see that the $Na_2CO_3/NaHCO_3$and $K_2CO_3/KHCO_3$ are very good candidates for CO_2 sorbents. These materials have favorable turnover temperatures for both of pre- and post-combustion conditions.

From Table III, one can see that NaOH, LiOH and $KLi_2(OH)_3$ can be used for pre-combustion CO_2 capture technology because their turnover T_1 are 880K, 660K and 690K. The turnover T_1 of KOH is much higher with values of 940K and 1290K which are not good for CO_2 capture. As shown in Table III, the turnover T_1 of other hydroxides are too high to be used for pre-combustion CO_2 separation.

Table III. T The highest temperatures (T_1) for sorbents capturing CO_2 at pre-combustion (P_{CO2}=20 bar) and post-combustion (P_{CO2}=0.1 bar) conditions. For reactions involving H_2O P_{H2O}=1 bar.

Reactions	Pre-combustion **T_1 (K)**	Post-combustion **T_2 (K)**
$Li_2O + CO_2 = Li_2CO_3$	hT[a]	1290
$Na_2O + CO_2 = Na_2CO_3$	hT	hT
$K_2O + CO_2 = K_2CO_3$	hT	hT
$BeO + CO_2 = BeCO_3$	370	300
$MgO + CO_2 = MgCO_3$	720	560
$CaO + CO_2 = CaCO_3$	1390	1030
$SrO + CO_2 = SrCO_3$	hT	1230
$BaO + CO_2 = BaCO_3$	hT	hT
$Na_2O + 2CO_2 + H_2O = 2NaHCO_3$	1000	830
$K_2O + 2CO_2 + H_2O = 2KHCO_3$	1120	940
$Li_2CO_3 + CO_2 + H_2O = 2LiHCO_3$	<300	<300
$Na_2CO_3 + CO_2 + H_2O = 2NaHCO_3$	520	450
$K_2CO_3 + CO_2 + H_2O = 2KHCO_3$	460	410
$2LiOH + CO_2 = Li_2CO_3 + H_2O(g)$	hT[a]	hT
$LiOH + CO_2 = LiHCO_3$	660	500
$2NaOH + CO_2 = Na_2CO_3 + H_2O(g)$	hT	hT
$NaOH + CO_2 = NaHCO_3$	880	680
$2KOH + CO_2 = K_2CO_3 + H_2O(g)$	hT	hT
$KOH + CO_2 = KHCO_3$	1290	930
$Be(OH)_2 + CO_2 = BeCO_3 + H_2O(g)$	<300	<300
$Mg(OH)_2 + CO_2 = MgCO_3 + H_2O(g)$	hT	720
$Ca(OH)_2 + CO_2 = CaCO_3 + H_2O(g)$	hT	1450
$Sr(OH)_2 + CO_2 + SrCO_3 + H_2O(g)$	hT	hT
$Ba(OH)_2 + CO_2 = BaCO_3 + H_2O(g)$	hT	hT

[a] hT means the maximum temperature exceeds our temperature range.

IV CONCLUSIONS

In this study, by combining density functional theory with lattice phonon dynamics, a screening methodology was developed and applied to alkali and alkali earth metal oxides, hydroxides, and carbonates. The preliminary results show that our screening methodology can predict the right thermodynamic properties for CO_2 capture reactions and can be used for screening good CO_2 solid sorbents from vast of solid materials which thermodynamic data are unknown. We have demonstrated that first-principles density functional theory can be combined with phonon density of states calculations to provide a reasonable estimate of the thermodynamics of CO_2 capture reactions involving alkali and alkaline earth metal oxides. We have found that an empirical correction factor is needed to bring calculated data into reasonable agreement with data from the HSC Chemistry software package. A single value of this correction factor of about 26 kJ/mol is adequate for all reactions examined in this work. We found that when steam presented the Na_2O and K_2O are better than CaO and could be used for post-combustion CO_2 capture technology, while MgO can be a good sorbent for both post- and pre-combustion CO_2 capture. We have examined the effect of H_2O on the reaction thermodynamics and have found that our modeling approach can be used to account for partial pressures of CO_2 and H_2O and temperatures. We found that formation of bicarbonates from the alkali metal oxides results in a lower sorbent regeneration temperature and that formation of bicarbonate from the carbonates, by addition of CO_2 and H_2O reduces the CO_2 capturing temperature even further. Indeed, we predict that Na_2CO_3 and K_2CO_3 have turnover temperatures for CO_2 capture through bicarbonate formation that are suitable for operation under both pre- and post-combustion conditions.

The alkali metal hydroxides MOH (M=Li, Na, K) can absorb CO_2 to form either bicarbonates or carbonates, but the former has less energy costs and is favorable to be used for CO_2 separation. Among them, $Mg(OH)_2$, NaOH and LiOH are better candidates for CO_2 sorbents. Our calculated thermodynamic properties of these hydroxides are in good agreement with the available thermodynamic data, which indicates that our methodology can produce reliable results and can be used for predicting CO_2 capture behaviors of other solid materials which thermodynamic data is not available.

We have outlined a screening method that will be useful for assessing potential sorbents for CO_2 capture for which no thermodynamic data exist. The methodology involves applying a series of screening filters of increasing computational complexity in order to identify a small number of systems that can be explored experimentally as promising candidates for reversible CO_2 capture.

ACKNOWLEDGEMENT

This work was performed in support of the National Energy Technology Laboratory's Office of Research and Development under contract number DE-AC26-04NT41817. The author thanks Dr. D. C. Sorescu and J. K. Johnson for fruitful discussions.

REFERENCES

[1] D. Aaron and C. Tsouris, Sep. Sci. Technol. **40**, 321 (2005).
[2] C. M. White, B. R. Strazisar, E. J. Granite, J. S. Hoffman, and H. W. Pennline, J. Air Waste Manag. Assoc. **53**, 645 (2003).
[3] E. Ochoa-Fernandez, H. K. Rusten, H. A. Jakobsen, M. Ronning, A. Holmen, and D. Chen, Catal. Today **106**, 41 (2005).
[4] H. Pfeiffer and P. Bosch, Chem. Mater. **17**, 1704 (2005).
[5] J. C. Abanades, G. Grasa, M. Alonso, N. Rodriguez, E. J. Anthony, and L. M. Romeo, Environ. Sci. Technol. **41**, 5523 (2007).
[6] S. C. Lee, H. J. Chae, S. J. Lee, B. Y. Choi, C. K. Yi, J. B. Lee, C. K. Ryu, and J. C. Kim, Environ. Sci. Technol. **42**, 2736 (2008).
[7] S. C. Lee, B. Y. Choi, S. J. Lee, S. Y. Jung, C. K. Ryu, and J. C. Kim, Stud. Surf. Sci. Catal. **153**, 527 (2004).
[8] S. C. Lee, B. Y. Choi, T. J. Lee, C. K. Ryu, Y. S. Soo, and J. C. Kim, Catal. Today **111**, 385 (2006).
[9] V. Manovic and E. J. Anthony, Environ. Sci. Technol. **41**, 1420 (2007).
[10] M. M. Maroto-Valer, Z. Lu, Y. Zhang, and Z. Tang, Waste Manag. **28**, 2320 (2008).

[11] R. Siriwardane, J. Poston, K. Chaudhari, A. Zinn, T. Simonyi, and C. Robinson, Energy Fuels **21**, 1582 (2007).
[12] V. Nikulshina, N. Ayesa, M. E. Galvez, and A. Steinfeld, Chem Eng J **140**, 62 (2008).
[13] S. C. Lee and J. C. Kim, Catal Surv Asia **11**, 171 (2007).
[14] S. C. Lee, H. J. Chae, S. J. Lee, Y. H. Park, C. K. Ryu, C. K. Yi, and J. C. Kim, J Mol Catal B-Enzym **56**, 179 (2009).
[15] Y. Duan and D. C. Sorescu, Phys. Rev. B **79**, 014301 (2009).
[16] K. Essaki, K. Nakagawa, M. Kato, and H. Uemoto, J. Chem. Eng. Jpn **37**, 772 (2004).
[17] J. Ida and Y. S. Lin, Environ. Sci. Technol. **37**, 1999 (2003).
[18] H. Pfeiffer, C. Vazquez, V. H. Lara, and P. Bosch, Chem Mater **19**, 922 (2007).
[19] N. Togashi, T. Okumura, and K. Oh-ishi, J Ceram Soc Jpn **115**, 324 (2007).
[20] J. C. Abanades, E. J. Anthony, J. Wang, and J. E. Oakey, Environ. Sci. Technol. **39**, 2861 (2005).
[21] G. X. Hu, H. Huang, and Y. H. Li, Int. J. Hydrogen Energy **33**, 5422 (2008).
[22] B. Feng, H. An, and E. Tan, Energy Fuels **21**, 426 (2007).
[23] M. B. Jensen, L. G. M. Pettersson, O. Swang, and U. Olsbye, J. Phys. Chem. B **109**, 16774 (2005).
[24] A. R. Oganov, S. Ono, Y. M. Ma, C. W. Glass, and A. Garcia, Earth Planet. Sci. Lett. **273**, 38 (2008).
[25] Y. Duan, in *Proceedings of 7th Annual Conference on Carbon Capture and Sequestration*, Pittsburgh, 2008.
[26] M. Sternik and K. Parlinski, J. Chem. Phys. **123** (2005).
[27] S. Cristol, J. F. Paul, E. Payen, D. Bougeard, S. Clemendot, and F. Hutschka, J. Phys. Chem. B **106**, 5659 (2002).
[28] Y. Duan, Phys. Rev. B **77**, 045332 (2008).
[29] J. H. Wang and M. Liu, J. Power Sources **176**, 23 (2008).
[30] R. G. Mortimer, *Physical Chemistry* (Academic Press, 2000).
[31] M. W. J. Chase, J. Phys. Chem. Ref. Data, **Monograph 9**, 1 (1998).
[32] F. Gygi and G. Galli, Phys. Rev. B **52**, R2229 (1995).
[33] M. Chaplin, Water Structure and Science, http://www.lsbu.ac.uk/water (2008).
[34] G. Kresse and J. Hafner, Phys. Rev. B **47**, 558 (1993).
[35] G. Kresse and J. Furthmuller, Phys. Rev. B **54**, 11169 (1996).
[36] H. J. Monkhorst and J. D. Pack, Phys. Rev. B **13**, 5188 (1976).
[37] K. Parlinski, Software PHONON (2006).
[38] K. Parlinski, Z. Q. Li, and Y. Kawazoe, Phys. Rev. Lett. **78**, 4063 (1997).
[39] C. J. Bradley and A. P. Cracknell, *The mathematical theory of symmetry in solids* (Clarendon press, Oxford, 1972).
[40] S. V. Alapati, J. K. Johnson, and D. S. Sholl, Phys. Chem. Chem. Phys. **9**, 1438 (2007).
[41] T. W. D. Farley, W. Hayes, S. Hull, M. T. Hutchings, and M. Vrtis, J. Phys.: Condens. Matter **3**, 4761 (1991).
[42] S. L. Mair, Acta Crystallogr A **34**, 542 (1978).
[43] Y. Idemoto, J. W. Richardson, N. Koura, S. Kohara, and C. K. Loong, J. Phys. Chem. Solids **59**, 363 (1998).
[44] E. Zintl, A. Harder, and B. Dauth, Z. Anorg. Allg. Chem. **198**, 88 (1931).
[45] H. Jacobs, J. Kockelkorn, and T. Tacke, Z. Anorg. Allg. Chem. **531**, 119 (1985).
[46] I. P. Swaninson, M. T. Dove, and M. J. Harris, J. Phys.: Condens. Matter **7**, 4395 (1995).
[47] B. D. Sharma, Acta Crystallographica **18**, 818 (1965).
[48] P. Touzain, F. Brisse, and M. Caillet, Can. J. Chem. **48**, 3358 (1970).
[49] J. O. Thomas, R. Tellgren, and I. Olovsson, Acta Crystallogr B **B 30**, 2540 (1974).
[50] D. K. Smith, H. W. Newkirk, and J. S. Kahn, J. Electrochem. Soc. **111**, 78 (1964).
[51] R. Stahl, C. Jung, H. D. Lutz, W. Kockelmann, and H. Jacobs, Z. Anorg. Allg. Chem. **624**, 1130 (1998).
[52] P. Karen, A. Kjekshus, Q. Huang, and V. L. Karen, J. Alloys Compd. **282**, 72 (1999).
[53] T. Nagai, T. Hattori, and T. Yamanaka, Am. Mineral. **85**, 760 (2000).
[54] F. A. Bromiley, T. B. Ballaran, F. Langenhorst, and F. Seifert, Am. Mineral. **92**, 829 (2007).
[55] C. H. Shen, R. S. Liu, J. G. Lin, and C. Y. Huang, Mater. Res. Bull. **36**, 1139 (2001).
[56] H. Sitepu, B. H. O'Connor, and D. Li, J. Appl. Crystallogr. **38**, 158 (2005).
[57] O. J. Rutt, G. R. Williams, and S. J. Clarke, Chem. Commun. **2006**, 2869 (2006).
[58] H. W. Grueninger and H. Baernighausen, Z. Anorg. Allg. Chem. **368**, 53 (1969).
[59] W. Pannhorst and J. Loehn, Z. Kristallogr. **131**, 455 (1970).
[60] L. G. Liu and W. A. Bassett, Journal of Geophysical Research **77**, 4934 (1972).
[61] H. Lueke, Acta Crystallogr. B **24**, 1705 (1968).
[62] S. M. Antao and I. Hassan, Phys. Chem. Miner. **34**, 573 (2007).
[63] HSC Chemistry software 6.1, Pori: Outotec Research Oy, www.outotec.com/hsc (2006).

Utilization of Carbon Dioxide towards Control/Suppression of Underground Coal Fires

Ali S. Rangwala,
Assistant Professor
Fire Protection Engineering, Worcester Polytechnic Institute,
Worcester, MA

Evan J. Granite,
National Energy Technology Laboratory,
Pittsburgh, PA

Abstract
Underground coal fires represent a substantial nuisance in terms of property damage, public safety, greenhouse gas generation, and coal resource destruction. Some of the techniques employed to combat these fires will be illustrated. The magnitude of greenhouse gas generation and coal consumption will be estimated. Future research areas into this important issue such as improved fire fighting, detection, and prevention will be discussed. The possible utilization of carbon dioxide towards control and suppression of underground coal fires will be examined.

Introduction

Uncontrolled underground coal fires have been reported in all major coal-producing countries, including: Australia, China, India, Indonesia, Russia, South Africa, and the United States. The fires result in a direct loss of valuable resources through the unwanted burning of coal reserves. For example Heffern and Coates[1] estimate that somewhere between 33 billion and several hundred billion metric tons of coal has burned naturally in the Powder River Basin (North America) over the last four million years, most of which burned within the last two million years. Such fires have been an important source of greenhouse gas emissions in the geological past. There is also an indirect loss of coal resources in underground coal-mine fires because access to coal seams may be blocked. Underground coal fires in coal mines can also depreciate the quality of groundwater (low pH and high concentration of dissolved solids) seeping from the mine [2].

Coal fires usually start at a surface outcrop of a coal seam with a certain thickness that has been exposed by erosion [3]. Several studies have found that the oxidation rate of coal increases dramatically as the temperature crosses a threshold of ~70 °C to 80 °C [3] This temperature can be easily achieved by strong solar heating[4]. Coal fires can also be initiated in coal mines due to spontaneous combustion, which is the slow oxidation process that can occur due to relatively low exothermic reactions and the consequent build up of heat due to the low conductivity environment which does not allow for heat to be dissipated. This slow gradual build up of heat (which can last for days) ultimately builds up enough temperature to initiate oxidative smoldering combustion of the coal. Spontaneous combustion in underground mines is especially dangerous because the atmosphere may be inherently explosive, and a highly localized incident of self-heating can initiate an explosion involving the entire mine. Underground coal mine fires are hard to locate and extinguish mainly because they so deep seated and difficult to reach.

An illustrative sketch representing the hazards associated with a coal fire is shown in figure 1. The illustration is adapted from recent work on water based suppression of underground coal fires published by Hadden and Rein [5]. Fire fighting efforts are shown on the left hand side. The damage to infrastructure—roads, houses, ground collapse is evident. Such occurrences' are common in most underground fires. Selfheating within the seam can serve as an ignition source for bush fires, forest fires [6] Further problems associated with such fires have been discussed extensively in literature [3, 7, 8], however there has been limited discussion on methodologies used for suppressing or controlling these fires.

Fig 1: Artistic conception of a smoldering fire in an abandoned coal mine and illustration of possible fire damage and suppression attempts (illustration from Hadden and Rein [5])

This paper discusses the option using CO_2 or a combination of CO_2-water vapor as a suppressing agent to extinguish or control these fires. Both are by-products of flue gasses from industrial power plants and thus can be produced abundantly if appropriate technology is implemented. The technique of using CO_2 as an extinguishing agent is definitely not new, with the earliest recorded instance of its use by Walker [9] in 1851. However, there has been limited work in this area from a fundamental viewpoint. Prior work has mostly focused on reporting case studies which are mine specific [9-14]. This paper discusses the relevant parameters that must be taken into account when dealing with suppression of such fires. The current state of knowledge and research areas that must be addressed before this technology can be applied in the field is discussed.

Parameters Involved in the study of Underground Coal Fire Suppression

The earliest use of CO_2 in suppressing an underground coal mine fire was reported by Walker [9] where he discusses a case by the Carbonic Acid Gas Co. of London, coal mine at Clackmannan, about 7 miles from Stirling in Scotland. There had been an underground fire in this mine, in the 9 foot deep coal, extending over an area of 26 acres, for 30 years. The plan adopted was to withdraw the men from the mine, and close all entrances to it, except two, one for the entrance, and the other for the escape of the gas. Carbon dioxide was made by forcing air through a coke furnace, the CO_2, N_2 and SO_2 so formed, in conjunction with a spray of water, was forced into the intake by the aid of a jet of steam. The process was maintained night and day for 3 weeks, the mixed gases and water being forced into the mine, during the whole of that period, and through the burning mass, and allowed to escape by the outlet mentioned. Cold air mixed with water spray was blown through the mass for a further period, and it is stated that at the end of a month, the fire was completely extinguished. Other studies of successful usage of CO_2 in coal mine fire suppression have been discussed by Morris [12] , Ray et al. [14]. and Sinha et al. [15]. The case study by Walker is interesting because it highlights the important parameters that need to be quantified before implementing CO_2 suppression on a larger scale. These parameters are listed in figure 2 and explained further.

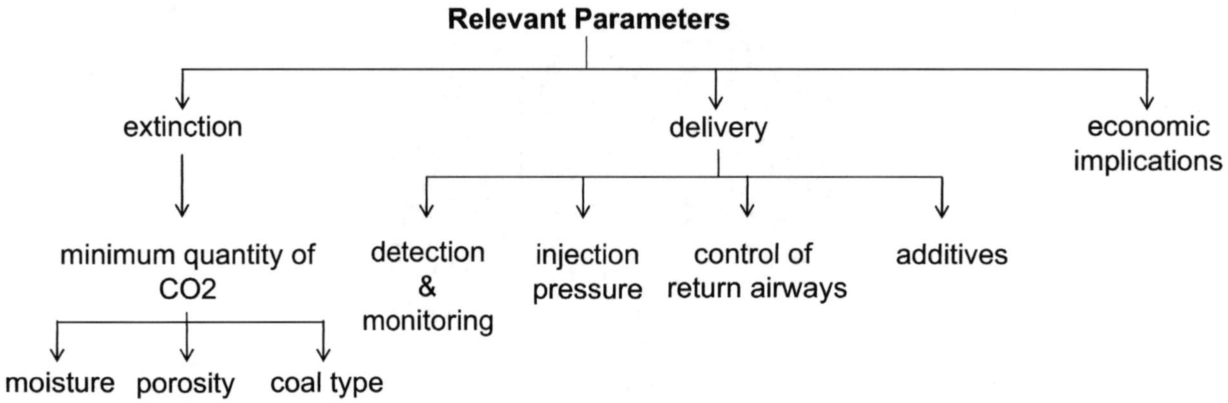

Figure 2: Flow chart representing the relevant parameters that need to be studied to implement CO_2 based control/suppression of underground fires

To simplify the problem, three areas need to be addressed:

1. Extinction and control:
   This refers to the analysis of the minimum quantity of suppressant needed to do the job. Both the quantity and application time need to be analyzed.
2. Delivery mechanism:
   The delivery of the suppressing agent to the seat of the fire is an important problem. Besides determining the optimum injection pressure, there needs to be some study of detecting and monitoring the growth/suppression of the fire. Secondly, additional additives such as water vapor, N2 that would aid in suppression also need to studied. It is also necessary to ensure that the return airways can be modeled.
3. Cost:
   The cost of installing a suppression system will be an important constraint. Ideally, a thorough cost benefit analysis of extinguishment of the fire with respect to the impact on the environment and local community needs to be analyzed. The costs would change based on the availability of CO_2, size of the fire, geography etc.

Each of these areas is further discussed in sections below:

Extinction/Control

Since oxygen has the strongest effect on a smoldering reaction, oxygen displacement is a highly effective method of extinguishing a coal fire. However, the quantity of inert gas required to displace sufficient oxygen needs to be quantified. Further this quantity will depend on the thermophysical properties of the coal, temperature, moisture and porosity. Figure 3 shows a hypothetical plot that shows the influence of CO_2 and autoignition temperature of coal. At a certain CO_2 concentration shown by the dashed line, autoignition is not possible. This zone is shaded and labeled "extinction region."

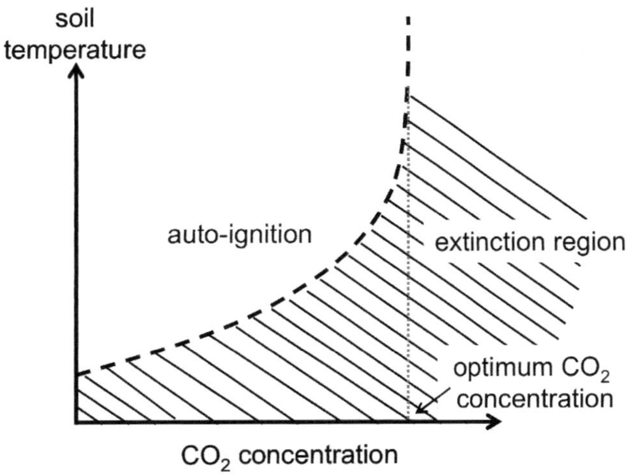

Figure 3: Functional relationship between autoignition temperature of coal and CO_2 concentration (hypothetical)

A functional relationship between minimum quantity of CO_2 and temperature therefore exists and must be determined to set the bounds for the quantity of CO_2 required to extinguish a particular seam of coal. An optimum concentration of CO_2 can then be obtained a plot as shown in figure 3. The authors are not aware of any standard test protocol capable of generating a plot such as figure 3. Further, coal seams extend several km with a thickness in the range of 1–6 m. The vast thickness of strata of low conductivity can maintain temperatures for a long period of time. This causes the inert environment created by CO_2 to merely suspend the smoldering combustion. As soon as CO_2 is stopped autoignition can occur relatively easily due to the trapped heat. Thus, suppression with CO_2 has to be coupled with some kind of monitoring system for temperature distributions in the strata. Typically, smoldering temperatures of coal are in the range of 750 – 950°C [16].with minimum self ignition temperatures varying between 50 – 150°C [17] depending on layer thickness, coal type and thermal conductivity of the strata. To aid cooling, water vapor can be added in addition to CO_2. Water vapor is a good agent to add to CO_2 due to its high specific heat.

Another method is to use liquid CO_2, which would act as a cooling agent as well as an inerting agent. In the US, liquid CO_2 has been used in the Bituner Mine. The quantity used was reported as 18,406 m^3, with an injection rate of 190 m^3/hr [13]. There are some instances where solid CO_2 (dry ice) has been used [18], by placing it directly on the fire scene. Use of dry ice would be useful to suppress smoldering fires in shallow fronts, where the ignition front is open to the atmosphere and having large supplies of oxygen available. In such cases, cooling the coal to a temperature below its critical autoignition temperature would be a more prudent recourse compared to displacement of oxygen.

For deep seated underground fires, oxygen transport is usually the limiting variable and thus CO_2 injection in the form of gas or liquid shows promise in such cases. Since CO_2 has a higher density compared to air, it is a good agent for fire located in depressions compared to other inert agents such ad N_2. If steam is added to CO_2, additional cooling can also be achieved due to the high specific heat of steam. The disadvantage of using CO_2 is the probability of CO production at high temperatures (~1000 – 1200 K) based on the Boudard reaction where CO_2 is converted to CO while consuming carbon $CO_2 + C \rightleftharpoons 2CO$ [17]. The CO produced could then react with steam leading to the water-gas reaction ($CO + H_2O \rightarrow CO_2 + H_2$) which would now promote rather than inhibit the reaction. Thus although CO_2 shows promise of being a cheap and effective extinguishing agent, there is a great need for further research to better understand the extinction mechanisms.

Delivery/Monitoring

Detection is the first step since it is impossible to suppress a coal fire without locating it first. Recent work by Kim [19] from the DOE National Energy Technology Laboratory (NETL) laboratory discusses methodologies that have been used in detecting underground coal fires. Use of satellite imagery has also been implemented to detect and monitor underground coal fires[20]

Cost/Benefit

The costs of containing, controlling, and extinguishing a fire can be so high that the only course of action, particularly for underground fires, is to let the fire burn itself out—this is the case for numerous fires in Pennsylvania (USA). However considering the impact in terms of green house gas emissions alone, this may not be a good strategy of control in today's regulatory environment. Agreed, there will be a cost associated with research on developing an effective suppression methodology, however, it is worthwhile to invest in this area. The need to reduce CO_2 emissions is an obvious motivating factor. A recent study by Keefe et al. [21] shows that CO_2 and Hg emissions from coal-fires in the U.S. are estimated at 1.4×10^{7}–2.9×10^{8} t/yr and 0.58–11.5 t/yr, respectively. These estimated carbon dioxide emissions from underground coal fires represent roughly 1 – 10% of the annual emissions from coal-burning power plants. The world-wide carbon dioxide emissions from coal fires are unknown. The idea of using flue gasses rich in CO_2 from a power plant to extinguish a coal fire therefore deserves some attention in today's times, where environmental concerns are high. It is possible that some of the CO_2 that is utilized will be trapped in the coal bed, thus offering an efficient means of CO_2 sequestration as well. The cost of delivery and additives also need to be factored into the analysis of utilization of carbon dioxide for suppression and control of coal fires.

Conclusions

A brief literature review on the topic of underground coal fire suppression using CO_2 shows that there are several fundamental areas of research that need to be addressed before this idea can be put into practice. However, rapid industrialization, especially in growing economies such as China and India, coupled with the problem of global warming, should be reason enough to invest in this area of research. The ultimate application to a realistic fire requires an analysis of extinction kinetics, several transport issues related to the injection of CO_2 in a coal mine and the migration of the gas to the site of the fire.

References

1. Heffern, E.L. and D.A. Coates, *Geologic history of natural coal-bed fires, Powder River basin, USA, coal fires burning around the world: A global catastrophe.* International Journal of Coal Geology, Stracher, G.B., ed, 2004. **59**(1-2): p. 25-47.
2. Bell, F.G., S.E.T. Bullock, T.E.J. Hälbich, and P. Lindsay, *Environmental impacts associated with an abandoned mine in the Witbank coalfield, South Africa.* International Journal of Coal Geology, 2001. **45**: p. 195-216.
3. Nelson, M.I. and X.D. Chen, *Survey of experimental work on the self-heating and spontaneous combustion of coal.* The Geological Society of America: Reviews in Engineering Geology, 2007. **XVIII**: p. 31-83.
4. Zhang, X., J.L. van Genderen, H. Guan, and S. Kroonenberg, *Spatial analysis of thermal anomalies from airborne multi-spectral data.* International Journal of Remote Sensing, 2003. **24**(19): p. 3727-3742.
5. Hadden, R. and G. Rein, *Small scale extinguishing experiments of coal subsurface fires.* 6th Mediterranean Combustion Symposium, (2009)
6. Bustin, R.M. and W.H. Mathews, *In situ gasification of coal, a natural example: History, petrology, and mechanics of combustion.* Canadian Journal of Earth Sciences, 1982. **19**(3): p. 514-523.
7. Stracher, G.B. and T.P. Taylor, *Coal fires burning out of control around the world: Thermodynamic recipe for environmental catastrophe.* International Journal of Coal Geology, 2004. **59**(1-2): p. 7-17.
8. Krajick, K., *Fire in the hole-paging in mines from Pennsylvania to China, coal fires threaten towns, poison air and water, and add to global warming.* Smithsonian, 2005. **36**(2).
9. Walker, S.F., *The use of carbon dioxide.* Mines and Minerals, June Issue, 1908.
10. Greuer, R.E., *Study of mine fire fighting using inert gasses.* USBM contract report No. 50231075, Bureau of Mines Open File Report, 1974: p. 15-76.
11. Westfield, J., H.S. Brumbaugh, and R.W. Whittaker, *Extinguishing fire with carbon dioxide in the Valier mines.* USBM Information Circular 7563, 1950.

12. Morris, R., *A review of experiences on the use of inert gasses in mine fires.* Mining Sci. Tech., 1987. **6**: p. 37-69.
13. Ray, S.K., A. Zutshi, B.C. Bhowmick, N. Sahay, and R.P. Singh, *Fighting mine fires using gasses with particular reference to nitrogen.* J. South African Inst. Mining and Metallurgy, 2000: p. 265-272.
14. Ray, S.K. and R.P. Singh, *Recent developments and practices to control fire in underground coal mines.* Fire Tech., 2007. **43**: p. 285-300.
15. Sinha, P.R., *Mine fires in Indian coalfields.* Energy, 1986. **11**(11-12): p. 1147-1154.
16. Rein, G., S. Cohen, and A. Simeoni, *Carbon emissions from smoldering peat in shallow and strong fronts.* Proceedings of the Combustion Institute, 2009. **32**(2): p. 2489-2496.
17. Haghighi, M., Z. Sun, J. Wu, J. Bromly, H.L. Wee, E. Ng, Y. Wang, and D. Zhang, *On the reaction mechanism of CO_2 reforming of methane over a bed of coal char.* Proceedings of the Combustion Institute, 2007. **31**(2): p. 1983-1990.
18. Bacharach, J.P.L., A.L. Craven, and D.B. Stewart, *Underground mine fire control with inerting systems.* CIM Bulletin, 1986. **79**: p. 67-72.
19. Kim, A.G., *Locating fires in abandoned underground coal mines.* International Journal of Coal Geology, 2004. **59**(1-2): p. 49-62.
20. Voigt, S., A. Tetzlaff, J. Zhang, C. Künzer, B. Zhukov, G. Strunz, D. Oertel, A. Roth, P. van Dijk, and H. Mehl, *Integrating satellite remote sensing techniques for detection and analysis of uncontrolled coal seam fires in north China.* International Journal of Coal Geology, 2004. **59**(1-2): p. 121-136.
21. O'Keefe, J.M.K., K.R. Henke, J.C. Hower, M.A. Engle, G.B. Stracher, J.D. Stucker, J.W. Drew, W.D. Staggs, T.M. Murray, and M.L. Hammond Iii, *Co2, co, and hg emissions from the Truman Shepherd and Ruth Mullins coal fires, eastern Kentucky, USA.* Science of The Total Environment, 2010.

Possible Use for CO_2 Captured from Coal Derived Flue Gas: Reforming of Methane in Electric Discharge Plasmas

Erik C. Rupp, ORISE Postdoctoral Researcher, National Energy Technology Laboratory, Pittsburgh, PA
Evan J. Granite, National Energy Technology Laboratory, Pittsburgh, PA

Coal-derived energy is perhaps the most important, and accessible, source of energy in the world. According to the United States Energy Information Administration, the US consumed over 1 billion metric tons of coal in 2006, with the entire world consuming over 6.5 billion metric tons. During this same year, the world consumed 3 trillion m^3 of natural gas, with the US contributing 600 billion m^3 to this total. This resulted in the emission in the US of 2 billion metric tons of CO_2 from coal and 1 billion metric tons of CO_2 from natural gas, or just over half the total CO_2 emissions from all sources in the United States during 2006. As the worlds' understanding of anthropogenic climate change increases, it becomes clear that CO_2 emissions must be abated, with CO_2 formed from stationary sources such as power plants potential the easiest to capture or convert to useful products.

The current direction for the use of CO_2 involves sequestration in geological formations after capture from flue gas. The US Department of Energy's Fossil Fuel Program's carbon sequestration core and the FutureGen Initiative are major spearheads for this technology. Other proposals for the use of the captured CO_2 involve advanced catalysts, engineered photosynthesis systems, biomimetic systems and biological systems. The use of CO_2, natural gas and electrical plasma discharges could provide another option for the use of captured CO_2, through the production of syngas, which is a useful product for either energy generation or in the Fischer-Tropsch process to produce synthetic liquid hydrocarbons.

The interaction of plasma discharges with CO_2, and the CO_2 – CO equilibrium in the plasma has been studied since the late 19th century with research picking up in the early 20th century (Wendt, Hunt). This work typically focused on corona discharges and equilibrium conditions. This work was so early that it tended to be a series of observations without attempts at explanation. In the middle of the 20th century, it was recognized that energetic electrons were the likely cause of CO_2 conversion and equilibrium with CO (Mizushima). It was eventually understood that CO_2 – CO equilibrium in a plasma discharge involved complex chemistry, involving the surfaces of the electrodes (Barton) and the input power (Brown). Although there is theoretical importance to this work, the equilibrium of CO_2 and CO has little practical value as the conversion of CO_2 is typically low. More recent work has focused on the use of CO_2 with CH_4 in a variety of different plasma discharges.

In general, there are two categories of plasma discharges, thermal and non-thermal. The major difference between the two is whether the gas stream is heated or if only the electrons are at a higher temperature, with the gas stream unheated. In non-thermal plasmas, the majority of the electrical energy provided to the gas stream is used to produce energetic electrons, with little heating of the gas stream. There is also a difference in operating pressures, as thermal plasma discharges tend to operate at pressures, while non-isothermal plasma discharges can operate at atmospheric or near-atmospheric pressures.

The differences in the two operation categories allow for different uses and create specific advantages and disadvantages. Thermal plasma reactors, typically created by arcs or radiofrequency (RF) inductively coupled discharge, provide high power output at high pressures. However, they are limited by low energy efficiencies due to low selectivity, high gas temperatures and electrode design issues. There is a variety of non-thermal plasma discharges, including low pressure glow, corona, microwave, dielectric barrier, gliding arc and spark discharge. Fridman et al. reviewed several of these non-thermal discharges, discussing the physics and primary features of the discharges. A number of other sources provide key background information for plasma discharges(van Veldhuizen, Lieberman, Razier).

This review is intended to discuss the state of technology for CO_2 reforming of CH_4 in plasma discharges. A number of discharge technologies (corona discharge, dielectric barrier or silent discharge, radio frequency discharge, thermal arc discharge) will be presented, with particular emphasis placed on CO and H_2 selectivity, removal of CO_2 and CH_4 and energy efficiency. A discussion about the viability of the range of technologies for the use of CO_2 and CH_4 to produce syngas, as well as key technological hurdles, concludes the review.

References

H. Hunt and W.C. Schumb, *J. Am. Chem. Soc.*, 1930, **52**, p. 3152

G. Wendt and M. Farnsworth, *J. Am. Chem. Soc.*, 1925, **47**, p. 2494

S. Mizushima, Y. Morino, H.H. Hirabayashi, K. Kakihara and Y. Mizushima, *J. Am. Chem. Soc.*, 1950, **72**, 51, p. 5176

S.S. Barton, S.T. Balke, R.J. Beardmore, R.T. Perry, M. Daniel and W. Newstead, *J. Phys. Chem.*, 1967, **71**, p. 4573

L. Brown and T. Bell, *Ind. Chem. Eng. Fund.*, 1974, **13**, p. 203 and p. 210

E.M. van Veldhuizen, *Electrical Discharges for Environmental Purposes: Fundamentals and Applications*, 2000, Nova Science Publishers; Huntington, NY

M.A. Lieberman and A.J. Lichtenberg, *Principles of Plasma Discharges and Materials Processing,* 1994, Wiley; New York

Y.P. Razier, *Gas Discharge Physics*, 1991, Springer-Verlag; New York

A. Fridman, A. Chirkov and A. Gutsol, *J. Phys. D: Appl. Phys.*, 2005, **38**, p. R1

(157c) Survey of Potential Uses for Carbon Dioxide Captured From Coal-Burning Power Plants

Evan J. Granite, National Energy Technology Laboratory, Pittsburgh, PA

Erik C. Rupp, National Energy Technology Laboratory, Pittsburgh, PA

Bryan D. Morreale, National Energy Technology Laboratory, Pittsburgh, PA

Henry W. Pennline, National Energy Technology Laboratory, Pittsburgh, PA

Dirk Link, National Energy Technology Laboratory, Pittsburgh, PA

David Luebke, National Energy Technology Laboratory, Pittsburgh, PA

Todd H. Gardner, National Energy Technology Laboratory, Morgantown, WV

Approximately 50% of the electricity in the United States is generated through the combustion of coal. One billion tons of coal is burned annually for electricity production, resulting in over two billion tons of carbon dioxide. Carbon dioxide is a potent greenhouse gas that can contribute to global warming. The United States Department of Energy is sponsoring research on the separation and sequestration of carbon dioxide from coal-derived flue gas. A major issue is determining good uses for the massive quantities of carbon dioxide that could potentially be separated from flue gas.

Some of the current uses of carbon dioxide that will be examined include fire suppression, food preservation, refrigeration and cooling, beverage carbonation, inert gas needs such as for welding, enhanced oil and coal bed methane recovery, supercritical cleaning, biomass production, and polymer-plastic manufacture. Future research areas for carbon dioxide utilization such as fuel production, pharmaceutical chemicals, coal gasification, oxy-combustion of coal, and passivation of alkaline wastes will also be discussed.

Literature Cited

1. Edwards, J.H. Potential sources of CO2 and the options for its large-scale utilization now and in the future, Catalysis Today, 23, 59-66, 1995.

2. White, C.M.; Strazisar, B.R.; Granite, E.J.; Hoffman, J.S.; Pennline, H.W. Separation and Capture of CO2 from large Stationary Sources and Sequestration in Geological Formations – Coalbeds and Deep Saline Aquifers, Journal of the Air & Waste Management Association, 53, 645-715, 2003.

See more of this Session: Beneficial Utilization of Carbon Dioxide Separated From Flue Gas
See more of this Group/Topical: Topical 2: Advanced Fossil Energy Utilization

AUTHOR INDEX

AUTHOR INDEX